NO LONGER PROPERTY OF
DUPONT LAVOISIER LIBRARY

CONVERSION AND UTILIZATION OF WASTE MATERIALS

Applied Energy Technology Series

James G. Speight, Ph.D., *Editor*

Khan **Conversion and Utilization of Waste Materials**
Mushrush and Speight **Petroleum Products: Instability and Incompatibility**
Speight **Environmental Technology Handbook**

IN PREPARATION

Lee **Alternative Fuels**
Wyman **Handbook on Bioethanol: Production and Utilization**

CONVERSION AND UTILIZATION OF WASTE MATERIALS

Edited by

M. Rashid Khan
*Texaco Research and Development
Beacon, New York*

USA	Publishing Office:	Taylor & Francis 1101 Vermont Avenue, N.W., Suite 200 Washington, D.C. 20005-3521 Tel: (202) 289-2174 Fax: (202) 289-3665
	Distribution Center:	Taylor & Francis 1900 Frost Road, Suite 101 Bristol, PA 19007-1598 Tel: (215) 785-5800 Fax: (215) 785-5515
UK		Taylor & Francis Ltd. 1 Gunpowder Square London EC4A 3DE, UK Tel: 0171 583 0490 Fax: 0171 583 0581

CONVERSION AND UTILIZATION OF WASTE MATERIALS

Copyright © 1996 Taylor & Francis. All rights reserved. Printed in the United States of America. Except as permitted under the United States Copyright Act of 1976, no part of this publication may be reproduced or distributed in any form or by any means, or stored in a database or retrieval system, without the prior written permission of the publisher.

1 2 3 4 5 6 7 8 9 0   BRBR   9 8 7 6

This book was set in Times Roman by Pro Image. The editors were Mary Prescott and Holly Seltzer. Cover design by Michelle Fleitz. Printing and binding by Braun-Brumfield, Inc.

A CIP catalog record for this book is available from the British Library.
∞ The paper in this publication meets the requirements of the ANSI Standard Z39.48-1984 (Permanence of Paper)

Library of Congress Cataloging-in-Publication Data

Conversion and utilization of waste materials/edited by M. Rashid Khan
   p. cm.
   Includes index.

   1. Waste products.   I. Khan, M. Rashid.
TP995.C66   1996
629.4′458—dc20   95-40392
                                                                                  CIP

ISBN 1-56032-382-5

Dedicated to Nilu.

CONTENTS

Contributors   xiii
Preface   xvii

Part One   General   1

1   UTILIZATION OF WASTE MATERIALS: EXAMPLES OF
BUSINESS SUCCESSES   3
M. Rashid Khan and Christine A. Gorsuch

Background	3
Plastic and Rubber Recovery	4
Wood, Paper, Biomass—Reuse and Composting	7
Construction—Recycled Products and Supplies	10
Creative Recycling of Other Wastes	10
Conclusion	12
Acknowledgments	13
References	13

2   DESIGNING RECYCLING TO PRESSURE
PACKAGING INNOVATION   15
Wayne E. Pearson

	Introduction	15
	Package Technology and Resource Recovery Technology	17
	Complexity of Municipal Solid Waste	20
	Enter the Garbagemen	21
	Economic Realities	22
	Economic Driving Forces	23
	Current Recycling Technology Is Too Expensive	27
	Economic Goals for Recycling	28
	Material Recovery Facility May Hold the Key	29
	Markets	33
	Summary	36
	How Everyone Can Win	37
3	COLIQUEFACTION OF WASTE MATERIAL WITH COAL: A RESEARCH PROGRAM OF THE CONSORTIUM FOR FOSSIL FUEL LIQUEFACTION SCIENCE G. P. Huffman	39
	Background	39
	Coliquefaction of Waste Plastics with Coal	41
	Results and Discussion	43
	Conclusions	48
	Summary	48
	References	49

Part Two Plastics, Polymers, Tires, and Automotive Wastes   51

4	ADVANCED RECYCLING TECHNOLOGIES FOR PLASTICS Mark W. Meszaros	53
	Background	53
	Advanced Recycling Overview	55
	Conrad Advanced Recycling Process	64
	Conclusion	74
	References	75
5	RECYCLING OF POLYMERS FROM AUTOMOBILE SHREDDER RESIDUE B. J. Jody, E. J. Daniels, and A. P. S. Teotia	77
	Introduction	77
	Disposal/Treatment Methods	83
	The Argonne National Laboratory Process	93

	Conclusions	100
	Acknowledgments	101
	References	102

6	PYROLYTIC REPROCESSING OF SCRAP TIRES INTO VALUE-ADDED PRODUCTS Michael A. Serio, Marek A. Wójtowicz, Hsisheng Teng, and Peter R. Solomon	105
	Introduction	105
	Process Description and Discussion	109
	Conclusions	120
	Acknowledgments	120
	References	120

7	COPROCESSING OF COAL WITH WASTE TIRES AND POLYMERS: IN SITU ELECTRON SPIN RESONANCE INVESTIGATIONS M. M. Ibrahim and M. S. Seehra	123
	Introduction	123
	Apparatus for In Situ ESR Measurements	124
	Experimental Results and Discussion	126
	Acknowledgments	132
	References	132

Part Three Biosolids, Biomass, and Municipal Solid Wastes   135

8	FUNDAMENTALS OF HYDROTHERMAL PRETREATMENT OF BIOSOLIDS (SLUDGE): FEEDSTOCKS FOR CLEAN ENERGY M. Rashid Khan, C. Gorsuch, R. Zang, and S. DeCanio	137
	Introduction	137
	Experimental	138
	Conclusions	142
	References	145

9	LAND APPLICATIONS OF BIOSOLIDS M. Rashid Khan and K. Mitall	147
	Introduction and Background	147
	Nutrient Contribution	148

	Soil and Sludge Parameters	149
	EPA Regulations	152
	Opinions and Support for Land Applications	153
	Practical Land Applications	153
	Concluding Remarks	155
	References	155
10	**ETHANOL FROM BIOMASS BY GASIFICATION/ FERMENTATION** E. C. Clausen and J. L. Gaddy	157
	Introduction	157
	Microbiology of Ethanol Production	161
	Conclusions	165
	References	167
11	**GENETIC ENGINEERING FOR PRODUCTIVITY IN THE FERMENTATION OF XYLOSE TO ETHANOL** Jie Xu, Anu Das, Jarrod Erbe, L. M. Hall, and Kenneth B. Taylor	169
	Introduction	169
	Materials and Methods	173
	Results	176
	References	178
12	**GAS RECOVERY AND UTILIZATION FROM MUNICIPAL SOLID WASTE LANDFILLS** Todd A. Potas	181
	Introduction	181
	Discussion	184
	References	188
13	**EMISSIONS FROM BURNING FOREST FUELS: DEVELOPMENT OF A MODEL** Wei Min Hao, Lisa M. McKenzie, and Geoffrey N. Richards	189
	Introduction	189
	Experimental	190
	Results	192
	Discussion	196
	Acknowledgments	196
	References	196

14	COPROCESSING OF LIGNOCELLULOSIC WASTES AND COAL TO LIQUID FUELS Palaniraja Sivakumar, Heon Jung, John W. Tierney, and Irving Wender	199
	Introduction	199
	Coal Liquefaction	201
	Components of Lignocellulosic Wastes	204
	Liquefaction of Lignocellulosic Wastes Containing Lignin In Situ	207
	Liquefaction of Chemically Modified Lignin	209
	Coliquefaction of Lignocellulosic Wastes and Coal	210
	Experimental	212
	Results and Discussion	213
	Acknowledgments	216
	References	216
15	MATHEMATICAL ANALYSIS OF A MUNICIPAL SOLID WASTE ROTARY INCINERATOR James T. Cobb, Jr., and Kunal Banerjee	221
	Introduction	221
	Modeling of Rotary Kiln Treaters	222
	Chemical Rate Expressions	224
	Kiln Reactor Model	225
	Results from Using the Model	226
	Conclusions	229
	References	230

Part Four   Coal Conversion By-Products   233

16	SELECTIVE LEACHING OF COAL AND COAL COMBUSTION SOLID RESIDUES Catherine A. O'Keefe	235
	Introduction	235
	Background	238
	Chemical Fractionation Procedures	239
	ASH Leaching Procedures	246
	Summary	250
	References	251

17 CHARACTERIZATION OF CONCRETES FORMULATED WITH BLENDS OF PORTLAND CEMENT AND OIL SHALE COMBUSTION ASH 253
James T. Cobb, Jr., and C. P. Mangelsdorf

Introduction	253
Background	254
ASH Properties	255
Concrete Formulation and Compressive Strengths	256
Mechanical Properties of Concrete	258
Zero-Slump Materials	259
Conclusions	261
Acknowledgments	261
References	261

18 UTILIZATION OF COAL-TAR PITCH IN INSULATING-SEAL MATERIALS 263
Janusz Zieliński and George Górecki

Background	263
Experimental Studies	266
Results and Applications	269
References	270

19 BRIQUETTING ANTHRACITE FINES FOR RECYCLE 271
Salustio Guzman and John T. Price

Introduction	271
Fines Generation	272
Briquetting of Anthracite Fines	273
Results	274
Proposed Process Flowsheet	280
References	281

INDEX 283

CONTRIBUTORS

KUMAL BANERJEE, Ph.D.
Department of Chemical and Petroleum
  Engineering
Energy Resources Program
School of Engineering
University of Pittsburgh
1137 Benedum Hall
Pittsburgh, PA 15261

EDGAR C. CLAUSEN, Ph.D.
Department of Chemical Engineering
University of Arkansas
Fayetteville, AR 72701

JAMES T. COBB, JR., Ph.D.
Department of Chemical and Petroleum
  Engineering
Energy Resources Program
School of Engineering
University of Pittsburgh
1137 Benedum Hall
Pittsburgh, PA 15261

EDWARD J. DANIELS, M.B.A.
Process Evaluation
Energy Systems Division
Argonne National Laboratory
ESD/362
9700 S. Cass Avenue
Argonne, IL 60439

ANU DAS, B.S.
Department of Biochemistry
University of Alabama at Birmingham
University Station
Birmingham, AL 35294

STEVEN DECANIO, Ph.D.
Texaco Research and Development
P.O. Box 509
Beacon, NY 12508

JARROD ERBE, B.S.
Department of Biochemistry
University of Alabama at Birmingham
University Station
Birmingham, AL 35294

JAMES L. GADDY, Ph.D.
Bioengineering Resources, Inc.
Industrial Park
1650 Emmaus Road
Fayetteville, AR 72701

GEORGE GORECKI, M.S.
Industrial Systems Group
Brent America, Inc.
921 Sherwood Drive
Lake Bluff, IL 60044-2203

CHRISTINE A. GORSUCH, M.S.
Texaco Research and Development
P.O. Box 509
Beacon, NY 12508

SALUSTIO GUZMAN, Ph.D.
Research and Development Laboratories
QIT-Fer et Titane Inc.
P.O. Box 560
Sorel, Quebec J3P 5P6
CANADA

LEO M. HALL, Ph.D.
Department of Biochemistry
University of Alabama at Birmingham
University Station
Birmingham, AL 35294

WEI MIN HAO, Ph.D.
The Shafizadeh Center for Wood and
  Carbohydrate Chemistry
University of Montana
Missoula, MT 59812

GERALD P. HUFFMAN, Ph.D.
Consortium for Fossil Fuel Liquefaction
  Science
University of Kentucky
341 Bowman Hall
Lexington, KY 40506-0059

MANJULA M. IBRAHIM, Ph.D.
Physics Department
West Virginia University
P.O. Box 6315
Morgantown, WV 26506-6315

BASSAM J. JODY, Ph.D.
Energy Systems Division
Argonne National Laboratory
9700 South Cass Avenue
Argonne, IL 60439

HEON JUNG, Ph.D.
Korea Institute of Energy Research
Yusung, Taejon 305600
KOREA

M. RASHID KHAN, Ph.D.
Texaco Research and Development
P.O. Box 509
Beacon, NY 12508

CLARK P. MANGELSDORF, Ph.D.
School of Engineering
University of Pittsburgh
Pittsburgh, PA 15261

LISA M. MCKENZIE, B.A.
Shafizadeh Center for Wood and
  Carbohydrate Chemistry
University of Montana
Missoula, MT 59812

MARK W. MESZAROS, Ph.D.
Flinn Scientific, Inc.
131 Flinn St.
P.O. Box 219
Batavia, IL 60510

KAREN MITALL, B.S.
Vassar College
Poughkeepsie, NY 12602

CATHERINE A. O'KEEFE, M.S.
Energy and Environmental Research
  Center
University of North Dakota
15 North 23rd Street
P.O. Box 9018
Grand Forks, ND 58202-9019

WAYNE E. PEARSON, M.S.
Packaging Research Foundation
P.O. Box 189
Kennet Square, PA 19348

TODD POTAS, M.S., P.E.
Rust Environment and Infrastructure
3033 Campus Drive, Suite 175
Minneapolis, MN 55441

JOHN T. PRICE
Energy Mines and Resources Canada
Energy Research Laboratories
CANMET
555 Booth Street
Ottawa, Ontario K1A 0G1
CANADA

GEOFFREY N. RICHARDS, Ph.D.
Shafizadeh Center for Wood and
   Carbohydrate Chemistry
University of Montana
Missoula, MT 59812

MOHINDAR S. SEEHRA, Ph.D.
Physics Department
West Virginia University
P.O. Box 6315
Morgantown, WV 26506-6315

MICHAEL A. SERIO, D.Sc.
Hydrocarbons Group
Advanced Fuel Research, Inc.
87 Church St.
East Hartford, CT 06108

PALANIRAJA SIVAKUMAR, M.S.
Department of Chemical and Petroleum
   Engineering
University of Pittsburgh
Pittsburgh, PA 15261

PETER R. SOLOMON, Ph.D.
Advanced Fuel Research, Inc.
87 Church St.
East Hartford, CT 06108

KENNETH B. TAYLOR, M.D., Ph.D.
507A-CHSB
933 South 19th Street
Birmingham, AL 35294-2041

HSISHENG TENG, Ph.D.
Department of Chemical Engineering
Chung Yuan Christian University
Chung-Li, Taiwan 32023

ARVIND P. S. TEOTIA, Ph.D.
Energy Systems Division
Argonne National Laboratory
9700 South Cass Avenue
Argonne, IL 60439

JOHN W. TIERNEY, Ph.D.
Department of Chemical and Petroleum
   Engineering
University of Pittsburgh
Pittsburgh, PA 15261

IRVING WENDER, Ph.D.
Department of Chemical and Petroleum
   Engineering
University of Pittsburgh
Pittsburgh, PA 15261

MAREK A. WOJTOWICZ, Ph.D.
Advanced Fuel Research, Inc.
87 Church St.
East Hartford, CT 06108

JIE XU, Ph.D.
Department of Biochemistry
University of Alabama at Birmingham
University Station
Birmingham, AL 35294

RICHARD ZANG, M.S., P.E.
Texaco Alternate Energy and
   Technology Department
White Plains, NY 10650

JANUSZ ZIELINSKI, Ph.D.
Institute of Chemistry
Technical University of Warsaw
Branch at Plock
Plock, POLAND

PREFACE

Waste is a terrible thing to waste in a sustainable environment. The public has recognized this sentiment, and, as a result, there are more people in the United States who recycle than who vote. During the past several decades, there has been a growing recognition of the impact of the environment on our lives. The United Nations Conference in Brazil in 1994 endorsed the need to conserve and manage resources and acknowledged the potential impact of waste on public health and the environment. However, the media have hyped recycling to be the sole cure-all to all environmental problems, although the problems associated with the waste business are not that simple. Riding high on a wave of public enthusiasm, many communities adopted materials recovery and recycling programs, at any expense, without evaluating the alternatives and assessing the total impact of these programs on the environment. Deriving a sensible recycling policy is complicated. Uninformed recycling programs can waste a large amount of resources. As a result, sorted, cleaned, and transported wastes have accumulated, often with a low or sometimes negative value. The processes of collecting, recovering, separating, and reprocessing can be expensive, and the products can be difficult to market. As an example, in 1993, the city of Philadelphia recognized that its municipal recycling program was too expensive and discontinued it. The environmental and energy costs associated with collecting and transporting small amounts of materials to be recycled can exceed any environmental benefits. Communities need to recognize this problem and weigh the costs and benefits of recycling.

To effectively utilize waste materials, environmentally acceptable and economical waste-based fuel forms or materials need to be developed by appropriate pre- and post-treatments. The objective of this book is to identify problems and opportunities in using various wastes, with an emphasis on the conversion of wastes into use materials or fuels. This book presents chapters on the following types of waste: municipal solid waste (MSW), biosolids (sludge), biomass, plastics, tires, automotive shredder residue, and coal wastes. Some chapters incorporate data on selected conversion processes. A few chapters of the book are aimed at providing information on the characterization, mitigation, and use of by-product streams (ash or minerals) formed during conversion of wastes. The book also incorporates data on the coprocessing of coal and wastes. *Conversion & Utilization of Waste Materials* covers a field that is growing rapidly throughout the world and includes chapters written by distinguished contributors to this emerging industry. This book is intended to serve as a forum for scientific collaboration, and it will be useful for researchers, students, academics, and other individuals who are interested in better understanding how this field is developing. The topics discussed will be of interest to those who are either trying to dispose of wastes (municipalities) or the engineers who are trying to incorporate wastes in their conversion processes (e.g., power plants). The cutting-edge research covered in this book will help guide future development and innovations.

The book is divided into four parts. Part I presents a discussion of general issues involving waste utilization and recycling. Chapter 1 provided a number of examples of businesses that deal successfully with waste materials. In spite of inherent difficulties involved in recycling and utilizing waste materials, a large number of small businesses have carved their niches and are managing to find markets and make a profit. Along with the financial successes, the technical successes achieved in various industries around the United States are briefly outlined in the first chapter. The recycling industry faces the problem of contaminated and mixed wastes, which poses significant challenges in recycling. Impurities can diffuse into products that are packaged in recycled containers. These issues and the need for new technology development are discussed in Chapter 2.

Costs associated with processing equipment to use reclaimed plastics are high and may not be economical for the small plastic molding industry. This issue is also complicated by the fact that the average price of virgin high-density polyethylene in the United States is about 30 cents per pound, and that the price of recycled high-density polyethylene is between 32 and 33 cents per pound. The current rate of disposal of plastics in the united States would represent about 80 million barrels of oil per year, if these plastics were converted to liquid fuels. Coliquefaction of coal and plastics offers the interesting possibility of synergism. Chapter 3 shows that relatively high quality oil consisting of mainly straight chain alkanes and minor alkenes and some light hydrocarbon gases can be produced from the liquefaction of plastics and of plastic-coal mixtures. Such reactions are further enhanced by catalysts such as highly dispersed metal oxides.

Part II addresses issues related to plastics, polymers, tires, and automotive wastes. Chapter 4 provides an excellent review of plastic recycling and outlines some advanced technologies for plastic conversion. Today, obsolete automobiles are the single largest source of scrap iron and contribute to over 25% of the 36 million metric tons of scrap ferrous metal used in the production of finished steel products. Over one million tons of nonferrous metals are contributed to recycling. For every ton of metals recovered, about 500 pounds of nonmetallic waste, commonly termed automotive shredder residue (ASR), is coproduced. ASR is a heterogeneous mixture of materials. Chapter 5 offers a description of a three-step process that is used to separate thermoplastics and other potentially recyclable products from ASR. The process involves treatment of a concentrated plastic stream with organic solvents at ambient temperatures to dissolve the desired plastics. In Chapter 6, the authors evaluate the option of processing tires into valuable products such as activated carbon and other carbon forms (e.g., carbon black) and into liquid and gaseous fuels. The results show that producing activated carbon and carbon black from scrap tires is viable. The activated carbon derived from tires can be used in a number of applications.

Determining the free radicals present in hydrocarbons, as monitored by electron spin resonance (ESR) spectroscopy, is useful for evaluating the fundamentals involved in a conversion process. Chapter 7 provides a description of a state-of-the-art high-temperature and high-pressure technique for evaluating the free radical chemistry of coprocessing and for better understanding the role of free radicals in wastes such as tires and various plastics during thermal treatment. Such a study is highly relevant because disposal of over 200 million tires by landfilling is a serious problem in the United States.

Biosolids, biomass, and municipal solid waste issues are addressed in Part III. Ocean dumping of sludge is banned, landfills are closing, and the public is demanding that sewage sludge be put to beneficial reuse. Chapter 8 discusses a technique that is used to prepare dewatered sewage or industrial sludge as a feedstock for clean energy via gasification. One of the most promising pretreatments of sludge is hydrothermal pretreatment which improves the sludge rheology and slurry solids loadings. These are important for the overall heating value of the mixture and for pumping and transportation. Chapter 8 describes the fundamentals of hydrothermal pretreatment of sewage or industrial sludge. Chapter 9 outlines the fundamental issues related to land application of biosolids.

The United States has about 1.5 billion tons of biomass residue and wastes that could serve as a source of energy. This residue could furnish 10 quads, or about 15% of our energy requirements, if converted at a 50% efficiency. If an energy crop is grown on idle arable rangeland and forestland (about 200 million acres), another 25 quads could be produced. Therefore, the United States could supply half of its energy from renewable biomass and wastes.

The major components of cellulosic biomass are hemicellulose, cellulose, and lignin. Synthesis gas, mostly CO, H_2, and some CO_2, could be formed by biomass gasification. The components of synthesis gas may be converted into ethanol by certain anaerobic bacteria. Because nearly all the biomass can be converted anaerobically into a gaseous fuel, ethanol yields of about 50% of the total biomass are possible. This compares to only about 30% for enzymatic or acid hydrolysis/fermentation processes. Chapter 10 explores the fundamental aspects of converting biomass into ethanol by gasification and fermentation.

An attractive option for the production of ethanol is its derivation fro corn syrup, sugar, or crop residue (lignocellulosic) materials. A cost effective way of processing lignocellulosic material consists of using acid hydrolysis to release a xylose fraction followed by strong acid treatment to release glucose. The sugars are then fermented into ethanol; however, hydrolysis and fermentation are relatively slow. The hydrolysis step can be catalyzed by enzymes, as described in Chapter 11.

Chapter 12 discusses the potential gas recovery and utilization from municipal solid waste landfills. Landfill gas, which is mostly methane, is generated by anaerobic treatment of municipal solid waste (MSW) in place. The MSW gas will eventually migrate out of the landfills to be emitted to the atmosphere, unless the gas is effectively recovered. Gas recovery systems installed in landfills have proven to capture high quality gas (with greater than 500 BTU/ft^3). Chapter 12 discusses the standards proposed by the Environmental Protection Agency and outlines the minimum control requirements for gas recovery and subsequent combustion or utilization of non-methane organic compounds generated in the landfill. Comparisons of various technologies that derive clean energy from landfill gas is provided in this chapter.

Chapter 13 addresses the nature of emissions that are of concern for combustion of biomass. In Chapter 14, the potential coprocessing of cellulosic waste and coal is described. The process involves the treatment of coal and cellulosic waste with carbon monoxide or synthesis gas ($CO + H_2$) in the presence of an alkali metal compound as a catalyst. The authors report that conversion of coal is increased by addition of paper in the H_2/tetralin/Mo system; however, the quality of the product is significantly improved when coprocessing is performed in the CO/H_2O/alkali system at 400C. Chapter 15 provides a mathematical analysis of a municipal solid waste incinerator.

Part IV of the book addresses the potential for the use of coal (or shale) conversion by-products. Characterization of waste by-products is similar to the conventional techniques utilized in coal and solid fuels industries. Chapter 16 describes many of the techniques that can be applied in evaluating the leaching characteristics of solid residues. Coal fines represent a significant reserve of waste resource in North America. Chapter 17 explores the potential for the use of concrete formulated with blends of Portland cement and oil shale combustion ash. This study is important because the United States has a huge reserve of oil shale, a source of petroleum that is not economical to extract at the present time. One major problem is the disposal of the spent shale.

The use of coal-tar pitch as a potential insulating-seal material (for building, road construction, and machinery industries) is investigated in Chapter 18. By combining coal-tar with polymers, the authors attempt to produce materials with improved physical and chemical properties. The small size of anthracite fines make them unsuitable for direct usage. Chapter 19 provides an interesting way of briquetting these fines with a binder pitch.

Conversion and Utilization of Waste Materials provides an interdisciplinary treatment of waste related topics, although no one book can single-handedly address all of the issues related to such a vast field. This book would not have been possible without the dedicated effort of the individual authors and reviewers and without support from Taylor & Francis, particularly Holly Seltzer, and Texaco. I thank you all.

M. Rashid Khan

PART ONE

GENERAL

CHAPTER
ONE

UTILIZATION OF WASTE MATERIALS: EXAMPLES OF BUSINESS SUCCESSES

M. Rashid Khan and Christine A. Gorsuch

BACKGROUND

Over the past decade there has been a growing movement toward environmental awareness and recycling. Recycling has dramatically affected the general population's perception of waste handling and disposal. This is partly evidenced by the fact that today in the United States there are more people who recycle than who vote. The media often consider recycling to be the solution for all wastes that are generated. However, recycling technology is still developing. There is a large discrepancy between the amount of waste collected for recycling and the amount that is actually recycled or reused. Recycling is not a simple task. There is a need for volume collection and separation. Some mixed products cannot be recycled, so sorting of different types of wastes is desirable. Perhaps a more critical reason is the lack of markets for recovered items. Factors such as transportation, waste recovery, cleaning, compacting, storage, and the integrity of recycled goods complicate recycling attempts. Keeping these factors in mind, it is often concluded that recycling is not necessarily a profitable undertaking. However, some businesses have found success in the use of recovered items.

The objective of this chapter is to outline a number of waste utilization strategies with selected examples of technical or business successes. This review is not comprehensive; small businesses are emphasized. The technical successes are exemplified because, given the right circumstances, the technical successes may ultimately lead to business successes. The research and development work reported in various chapters of this book is not discussed in this chapter. The following classes of materials and their uses are considered here: plastic and rubber, wood, paper composting, construction materials, and some other miscellaneous wastes.

PLASTIC AND RUBBER RECOVERY

About 20% of municipal solid waste comes from plastic waste [1]. With only 2% of our plastics being recycled, there is a great need to find alternative uses for the 16 million tons of waste plastics generated annually in the United States [2]. Because of the demand for their recycled products, polyethylene terephthalate (PET), the plastic used for soft drink containers, and high-density polyethylene (HDPE), used for milk jugs, are the most commonly recycled plastics. PET is being recycled into residential and commercial fiberglass, bathtubs, shower stalls, sinks, and portable spas by companies such as Aqua Glass. Lumber made from waste plastic is maintenance free and strong; it is used to make piers, fences, and decks.

Advanced Recycling Technology, Inc. (Plastics)

Some early strategies in mixed plastic waste utilization were extrusion methods. Advanced Recycling Technology, a company from Belgium, created a process called Extruder Technology I to use mixed plastics as received from the municipal waste stream. First the plastics are shredded; then they are put through the extruder. The material is headed to between 200° and 300°C by the friction in the extruder. This temperature range is appropriate to melt the major components of the mixed plastic stream, the polyolefins, which make up about 60% of the feedstock. Plastics that melt at higher temperatures and foreign materials like metal and dirt are trapped in the molten phase and help strengthen the final product. The product is called Syntal and is molded into fence posts, marine pilings, planking for boardwalks, docks, and animal troughs. Syntal is superior to wood, concrete, and metal in its applications because of its durability. It is resistant to bacteria, fungi, extreme temperatures, salt water, and animals (they find it inedible). Syntal also absorbs shocks and sounds and is good for playground projects because it does not splinter. Similar mixed plastic extrusion products are marketed by Plastic Recycling, Inc. and Recycloplast Inc.; all three technologies can recycle their end products if they fail quality specifications or have served their purpose [3].

Mobil Chemical Co., Ford, Chrysler (Plastics)

Mobil Chemical Company mixes used polyethylene grocery bags equally with waste wood sawdust to make a rot-resistant lumber called Timbrex [4]. Ford Motor Company used recycled PET in molded plastic car parts made by a company called Allied Signal. Allied Signal mixes the waste polyethylene terephthalate with glass to produce PETRA. PETRA is used by Ford to make molded plastic grills, luggage racks, and other pieces for cars. Chrysler also includes molded recovered PET products in its car fenders, upholstery fiber, and seat cushions [5].

Texaco, Inc. (Plastics, Tires, Sewage)

Another large corporation involved in waste plastic and rubber utilization is Texaco, Inc. Texaco's Alternate Energy Research Department has patented technologies to produce energy from waste plastics and tires. By liquefying or slurrying these wastes, fuels are produced that can be used in the environmentally clean Texaco Gasification Process. In gasification, a technology more commonly associated with coal, the fuel feedstock is converted to synthesis gas, which can be burned for electricity or sold for manufacture of chemicals. The solid waste from gasification is a glassy material called slag, which traps metals in a nonleachable matrix. This nonhazardous slag can be used in such applications as road bed aggregate. Texaco researchers have also developed strategies to convert sewage sludge (mixed with coal) to energy via gasification [6]. Texaco has also developed a process for converting used lubricating oil into marine fuel.

Rising Star Futon Company (PET)

An excellent example of a successful recycled-product business is seen in the Rising Star Futon Company. Their futons are made with fibers from recycled (PET) soda bottles. The bulk of their waste plastics comes from plastic manufacturing rejects with about 25–30% postconsumer wastes making up the difference. Working with a plastic recycling company, Rising Star chose three fibers that together make a comfortable futon mattress. Two of the fibers feel like cotton; they are called Well-Spring and Cloverfill. The third is Ecocore fiber, which is used as a substitute for the rubber in the futon core. The recycled fibers have additional benefits in that they are more durable than cotton and foam rubber and they weigh less than cotton, so transportation costs are reduced. Each of Rising Star's futons uses about 480 PET bottles [7].

Advanced Recycling, Inc. (Computers)

In Belleville, New Jersey, Advanced Recycling, Inc., has developed a program in which computers are dismantled and their components collected, sorted, and resold to various industries. Finding a market for all the components was not easy; in the beginning, a great deal of money was spent on disposal costs for items such as glass monitors and engineering-grade plastics. Once markets for these components were found, disposal costs were drastically reduced. Advanced Recycling landfills only 800 to 1000 pounds of waste per month from the 400,000 pounds of equipment they process. Their feedstock is obtained mostly from commercial companies that are upgrading their computers and are looking for a way to dispose of their hardware. Depending on the quality and value of the machinery, Advanced Recycling either charges or pays for the feedstock. Carnegie-Mellon University recently published a study which estimates that "if the current U.S. pace of discarding 10 million computers per year continues, around 150 million old machines will have found a landfill as their final resting place by the year 2005. At that rate, disposal costs for the computers could reach $1 billion and the space required would be equivalent to an acre of land dug $3\frac{1}{2}$ miles deep" [8]. Looking at these numbers, it is obvious that there is a large potential for growth in the computer recycling business [8].

Deja Shoes (Plastic and Rubber Wastes)

A recycled shoe line called Deja is made entirely with recycled materials, including tires and plastics. The inventor of Deja Shoes worked with top shoe manufacturers Nike and Avia to produce high-quality footwear from 20 waste feedstocks. Part of the rubber they use has been engineered by a rubber recycling process called surface modification, developed by Composite Particles, Inc. Surface-modified rubber can be used to replace new rubber and plastic; it can be bonded as if it were virgin polymer. In their advertising Deja Shoes state that one can return the old shoes to them; they will recycle them. Other used-tire apparel includes rubber neckties from designer Denny LaShier and handbags that were inner tubes in their former lives from Used Rubber USA [9]. People will soon be able to purchase recycled jeans to complement their recycled shoes. Joint research between Burlington Industries Denim Division and North Carolina State University has resulted in a process that produces new denim fabric from old jeans [10].

EnviroTech, Inc. (Tires)

An example of land application of scrap tires is in irrigation systems, such as HydroCulture, from a Texas company called EnviroTech, Inc. EnviroTech was founded in 1991 to make products from waste tires, which was made economically feasible by environmental legislation in Texas. The HydroCulture layer system conserves water and finds a value in waste tires. The company modifies fields by using a 6-foot layer of shredded tires on top of a linear and a base of sand. Between the topsoil and tire layer is another liner that allows water through, but not soil. A drainage system is built in with culverts made of large whole tires that have been lined up and connected. The layer of tire pieces allows water to percolate through the soil, allowing rain and irrigation water to be filtered and collected. At present, the HydroCulture system is in the demonstration stage on land near the EnviroTech plant in Hutchins, Texas. EnviroTech has other applications for tire rubber; they use shredded tires for playground fill. A product called EnviroFlex consists of powdered rubber from scrap tires sprayed over polyester fabric along with latex and asphalt and is used to repair roofs [11].

WOOD, PAPER, BIOMASS—REUSE AND COMPOSTING

The municipal solid waste (MSW) stream contains 37.5% paper waste by weight [2]. Roughly 10% of the total stream is newspaper. For the average American home this is equivalent to 28 pounds per month, 35% of which is now recycled [1]. Foot wastes and yard clippings account for nearly 25% of the MSW stream; composting technologies could help reduce the amount landfilled. Wood wastes, such as pallets and construction items, can be composted, used for fuel, or mixed with other wastes to produce recycled items like lumber and road signs.

Hon Industries (Wood)

Wood wastes can be used to make fuel pellets, like those made by Hon Industries of Iowa. With help from the Iowa Department of Natural Resources, Hon Industries received a grant that will help them use nearly 10 tons of wood wastes per day. Hon produces commercial fuel pellets for use in wood stoves, and they have found that their waste pellets burn better than regular chopped wood. They plan to increase their process capacity in the future [12].

Wood Fibers International (Wood)

Wood Fibers International has been able to develop high-added-value products from urban wood waste. Raw material is obtained from old pallets and shipping dunnage that often consists of fine woods, such as birch, cherry, and mahogany. These woods are used to produce office and home furnishings, jewelry boxes, and gift items [13].

Community Environmental Services, Inc. (Wood)

On the other end of the spectrum is a pilot program of Community Environmental Sevices (CES) in Austin, Texas for the treatment of raw septage with waste wood chips. The sewage from a nearby treatment facility is allowed to pass over a tank containing wood chips and the water is drained to the bottom. The solid portion of the septage is thereby separated from the water; the used chips are composted, and the water can be taken for municipal water treatment. This project does not appear to be making a profit, simply because virtually no money is involved. However, if this is found to be an effective way to treat sewage, similar projects could be quite lucrative [14].

Forest Products Laboratory (Wood, paper)

Along similar lines, the Forest Products Laboratory in Madison, Wisconsin, a branch of the U.S. Department of Agriculture's Forest Service, is researching other aspects of wood and paper recycling. One project that is showing promising results uses microbial enzymes to improve the recycling of paper fibers. These enzymes effectively digest the small fibers that are typically found in recycled pulp. Recent studies have also shown that these enzymes reduce the amount of mechanical action needed for deinking. The laboratory also mixes waste wood with newspaper and cement to make fire- and fungus-resistant cement board [15].

Champion Recycling Corporation (Paper)

Champion Recycling Corporation is a private company that has found profit in paper recycling by entering a partnership with the city of Houston, Texas. Champion built a $3.5 million paper sorting and processing plant adjacent to the city's Solid Waste Recycling Processing Facility. The location of the plant has enabled the company to minimize transportation costs. The city supplies Champion with curbside-collected magazines and newspapers, which are sorted, processed, and prepared for sale to a deinking plant. Champion receives the recovered material free of charge and sells it to the deinkers for a profit. The partners in the undertaking have signed a 20-year contract, ensuring that Champion will have a monopoly on the recovered paper for a long time [16].

Gardenville (Compost)

Whereas the success of the Champion Corporation was in part due to the carefully planned nature of the undertaking, the story of Malcolm Beck and his business differs greatly. Beck is the owner of Gardenville, originally a small store selling natural gardening supplies and organically grown produce. As more buyers wanted to purchase the compost he used in his fields, he bought more land and increased production. Beck has agreements with HEB Grocery Company and Pace Picante, who transport their waste produce to his land to be used to make compost. Gardenville is now a multimillion dollar business with seven centers in Texas [17].

Green Mountain Technologies (Compost bins)

As this example illustrates, composting is a practical way to handle the produce waste that normally enters landfills. Composting can be done on a small scale in the privacy of your own home—in fact that is just what is being done in Wilmington, Vermont. Green Mountain Technologies (GMT) helped provide the community with demonstrations on composting and low-cost bins for individual household use [18].

Ceres Environmental, Inc. (Compost)

Production of compost bins may be a profitable venture, but Ceres Environmental, Inc. has shown that there is also a big business in the sale of compost. During the summer season, over 200 tons of wood waste per day pass through their facility in Maple Grove, Minnesota. When the plant opened in 1990, there were few markets for the end-product composts and the company operated under government subsidies. By 1993 most government subsidies were removed because Ceres had found suitable markets for its products and was able to turn a profit. The company collects a tipping fee ranging from around $9 to $18 a ton for clean wood and brush to $39 a ton for stumps. Most of their compost is sold to yard and garden stores. The Minnesota Department of Transportation also purchases the chips for use on its metropolitan road projects. The composting process is organized so that special types of mulches can be made for specific uses. For example, wood chips cut against the grain are used in playgrounds because they have more bounce. Ceres, which now grosses over $3 million in annual sales, has grown into one of the largest processors of wood waste in the United States [19].

State University of New York at Syracuse (Compost)

Along with wood waste, other new materials are being tested as raw material for composting. Researchers at the State University of New York (SUNY) in Syracuse have been testing the use of pulverized construction dry wall as compost material. The initial results seem positive; they have found that the use of these materials tends to increase soil fertility and nutritional content. Samples of the soil were tested for trace metal pollutants such as arsenic, barium, cadmium, chromium, lead, mercury, selenium, and silver, because there is a risk of contamination from construction sites. However, soil treated with drywall compost had amounts of these metals that were very low or below detection limits. If further studies on the efficiency and environmentally benign nature of drywall continue to find positive results, it may become a major feedstock for composting, as enormous amounts of waste drywall are generated by the construction industry [20].

CONSTRUCTION—RECYCLED PRODUCTS AND SUPPLIES

In the field of new home construction, one can find a way to make recycling a lucrative business. An interesting example is the Resource Conservation House. This house is part of a larger development consisting of four experimental houses: one has high-efficiency gas appliances, the second has masonry construction, the third is handicapped accessible, and the fourth, the Resource Conservation House, consists of about 80% recycled materials. The materials and labor were donated by manufacturers and industry associations. Steel beams were used in the frame instead of lumber; the beams contain two thirds recycled material from old automobiles, cans, and other scrap. Lite-Form, Inc., of Sioux City, Iowa, donated foam sheets that hold the poured concrete for the foundation. The sheets are bound together by recycled plastic and provide layers of insulation for the basement. Cellulose Insulation Manufacturers Association supplied the insulation for the walls; their insulation contains 100% recycled newsprint. FiberBond Wallboard, which contains 30% newsprint, was donated to take the place of sheetrock. Eiger Building Products, Inc. made the roofing material for the project, panels containing 50% recycled computer casings that fit together like siding material and are made to resemble shingles. Timbrex lumber, mentioned previously, was used for the deck. All the pipes were made from recycled copper, and some of the siding was made from recycled aluminum. Of course, no Resource Conservation House would be complete without a recycling system built into the kitchen. Builders designed a cabinet that pops out with the tap of a foot to reveal bins for source separation [21].

Design Impact (Building materials)

Stores like Design Impact in California sell recycled and environmentally friendly construction products to help you create a Resource Conservation home of your own. The products they market include carpet made from recycled PET fibers, recycled glass bricks, and tiles from recycled plastics. Many of these items have more advantages than just being environmentally friendly. The recycled PET carpeting, called Enviro-Teck, resists stains better than conventional carpet, and the Poly-Mar HD recycled plastic tiles are scratch resistant [22]. You could build your energy-efficient resource conservation home from whole scrap tires. The tires are stacked up and filled with dirt, then coated with adobe. There are about 50 of these tire homes in New Mexico [9].

CREATIVE RECYCLING OF OTHER WASTES
Chicagoland Processing/Enviromint (Film)

A striking example of a business success in the nonconventional recycling industry is the Chicagoland Processing/Enviromint founded by John Obie. In 1975

Obie was a college student with a one-man recycling business. By 1991 Obie's business had grown into a $14 million business that is one of the world's largest recyclers of photographic waste. Having amassed a large amount of scrap film while in the recycling business, Obie became interested in the potential of photographic waste. However, at the time, the most common method of extracting silver from photographic film involved the use of cyanide. Obie eventually came up with a nontoxic enzymatic process that separates silver from photographic film. The technology was patented in 1983, approved by the Environmental Protection Agency (EPA), and subsequently implemented on a large scale. An added benefit of this reclamation method is that it does not leave cyanide residue on the plastic, so the PET can be reused by the photo industry [17]. Chicagoland Processing/Enviromint was also able to use the recycled silver from the film to generate a completely new product: silver minted coins commemorating sports teams, players, and hit movies.

Amour Hydro Press (Fiberglass)

Fiberglass waste is made into recycled products by Amour Hydro Press in Washington. Most of the state's fiberglass waste is currently landfilled. Because of tipping fees associated with disposal of fiberglass, manufacturers of fiberglass products, like boats and bathtubs, find Amour Hydro Press's fiberglass reuse process quite attractive. Currently this company accepts fiberglass preconsumer and postconsumer wastes; Amour does charge a fee for disposal and charges shipping costs and cutting costs only if required. The chopped waste fiberglass is mixed with dyes and resins, then molded into the desired product. Pure fiberglass is used for applications in which strength is important, such as in railroad ties and car bumpers. Alternatively, the fiberglass can be mixed with wood and paper to create a light but less strong material that is used to make durable fence posts and road signs. At present, the first steps of the Amour process limit its application to noninsulation fiberglass, but at least one large insulation manufacturer has started research into insulation recycling with Amour Hydro Press. Experimentation has also begun to use this type of process to recycle acrylics and scrap tires.

Digital Equipment Corp., Envirocycle (Leaded Glass)

EPA efforts to limit the amount of lead allowed into the environment have taken the form of new land disposal restrictions. These regulations mandate disposal of leaded glass in hazardous-waste landfills. Cathode ray tubes used for computer monitors and television screens are the largest source of leaded glass. The lead is added to the tubes to shield the viewers from x-rays generated during normal operation of the machines. Manufacturers, who use different formulas and vary the content of lead in the glass present a real challenge to the recycling of such products. Digital Equipment Corporation and Envirocycle have undertaken a joint program to solve the problem of leaded glass recycling. In the process developed, monitors are pulled apart and the plastics and components are recycled. Then the monitors are sorted by manufacturer's imprint and the glass is crushed and cleaned. The crushed glass is sold to manufacturers like Asashi Corning (a partnership between Corning and Asashi Glass in Japan) and remelted for a new batch of computer screens.

CONCLUSION

The markets for recycled trash products are growing with consumer awareness of environmental concerns, shrinking landfill space, landfill bans on various materials, and government legislation and incentives. As of 1994, 27 states gave tax incentives to recycling businesses, and numerous states have agencies and task forces to find and improve recycling markets (see Figure 1). Because transportation of waste for recycling and reuse can be a large expense, centralizing businesses involved in waste utilization can save money. With this in mind some states, like California, are creating "recycling market development zones." Companies located in these zones are eligible for state grants and loans and other local incentives. West Virginia encourages recycling by waiving its solid waste fees for industries recycling 70% of their wastes.

With the potential for fiscal incentives and a cheap supply of raw materials from trash, the time is right for waste utilization businesses. We are seeing them pop up all over the country. Many are using raw materials from tires, plastics,

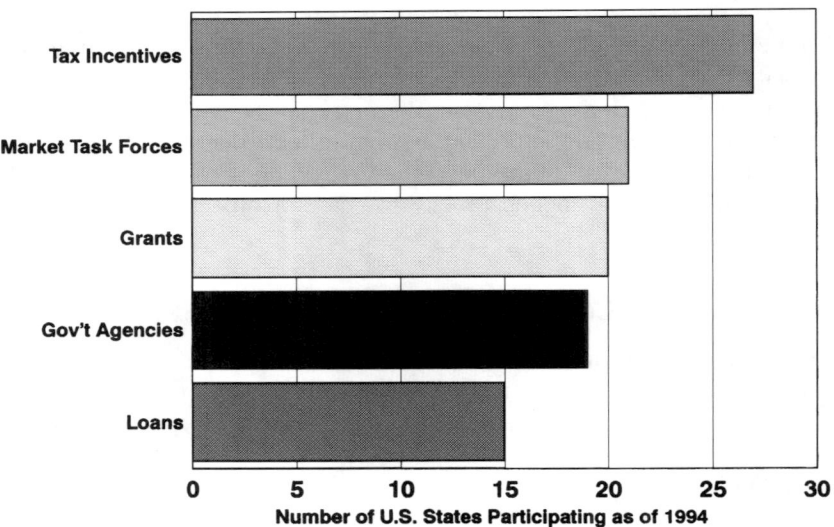

Figure 1 State participation in incentives and programs to develop the recycling industry. Data from Steuteville, 1994.

and various other wastes, such as fiberglass and wood, in simple processes that hardly change the raw material and in complex applications in which new raw materials are being created from scrap. Although waste minimization efforts are being initiated in nearly every state, no shortage of garbage is expected in the near future, so one can expect these businesses to continue to arise and to expand.

ACKNOWLEDGMENTS

This review is not exhaustive. We are certain that a large number of successful businesses were not included here. The authors did not verify the accuracy of the facts stated herein, and no warranty is provided regarding the information presented. The authors wish to thank Vandana Purohit and Saskia VanBergen, who reviewed some of the literature cited in this chapter as part of their term papers in Advanced Environmental Chemistry at Vassar College.

REFERENCES

1. U.S. Environmental Protection Agency, Office of Solid Waste, Environmental Fact Sheet, EPA/ 530-SW-91-024, July 1992.
2. Steuteville, R. The state of garbage in America: Part II. *BioCycle,* 30–36, May 1994.
3. Riggle, D. Finding markets from scrap tires. *BioCycle,* 41–55, March 1994.
4. Martin, A. The bumpy road to tire recycling. *Garbage,* 28–34, May/June 1991.
5. Tesla, M. R. Scrap tire process turns waste into fuel. *Power Engineering Barrington III.* Vol. 98, No. 5. Tulsa, OK: PennWell.
6. Riggle, D. High end markets for recycled rubber. *BioCycle,* 50–51, January/February 1994.
7. Barrier walls from used tires. *Biocycle,* 87, February 1994.
8. Sound barriers from litter. *Biocycle,* 10, June 1993.
9. 10 trend setting businesses for the 90s. *In Business* 16(3):18–24, 1993.
10. Tire recycling firms come up with new products. *Biocycle,* 26, October 1993.
11. O'Sullivan, D. A. New waste recycling processes show promise. *Chem. Eng. News,* 60(15):58–59, April 12, 1982.
12. Swanekamp, R. Ridge station eases Florida's waste disposal problems. *Power,* 84–86, October 1994.
13. Scrap tyres to fuel oil without shredding. *Process Engineering,* 17, August 1984.
14. Farcasiu, M., and C. M. Smith. Coprocessing of coal and waste. Report No. HD226KA, Merton Allen Associates, March 1993.
15. Brewer, G. Mixed plastics recycling. *Waste Age,* 153–160, November 1987.
16. Utilizing mixed wastes. *BioCycle,* 21, May 1994.
17. Pieper, P. Wood waste alchemy. *BioCycle,* 41–42, August 1994.
18. Comello, V. Smoother sailing. *R&D Magazine,* 20–22, October 25, 1993.
19. Raftery, M. Toxic-free products in growing demand. *BioCycle,* 27–29, July/August 1993.
20. Wood scraps to fuel pellets. *BioCycle,* 22, May 1994.
21. Recycled house reveals emerging industry. *BioCycle,* 28–30, May/June 1993.
22. Freebourne, J. The second time around. *BioCycle,* 44–45, March/April 1994.

CHAPTER
TWO
DESIGNING RECYCLING TO PRESERVE PACKAGING INNOVATION

Wayne E. Pearson

INTRODUCTION

Many municipalities in the United States are recognizing rapid increases in the cost to dispose of trash. This is caused by the rising costs of building and maintaining environmentally appropriate landfills and/or waste-to-energy plants. A secondary factor is that many communities do not wish to have a landfill or waste-to-energy plant in their vicinity, the so-called NIMBY (not in my backyard) syndrome. As a consequence, it is becoming increasingly difficult to find appropriately priced waste disposal facilities, if, in fact, they can be found at all.

Recycling is one alternative for managing solid waste. The public feels comfortable about recycling because they have been doing it as both adults and children in one way or another in Boy Scouts, Girl Scouts, community, and church groups. Principally, they have been collecting newspapers and beverage containers, such as aluminum beer and soft drink cans and glass bottles. Often these materials are collected to redeem the deposit value or for their value as scrap. An ancillary benefit of recycling is that it permits some people to feel better about what they consider excessive consumption. So, recycling is met with enthusiasm across the nation. This is reflected in municipal, state, and federal laws mandating recycling rates with near-term dates. We can expect recycling legislation to gain momentum.

Recycling affords the opportunity to divert from the waste stream useful materials that can become feedstocks for a wide variety of products that can theoretically compete with virgin materials. What recycling really means, however, is that we are going to try to sell our trash to someone. The problem is that trash has little value. No one wants to buy garbage. Consequently, a great deal of work must be done to trash to turn it into viable products. This work includes processing the garbage into raw materials that can be manufactured into products that can be sold to a buyer who will buy them again and again. We are learning that there is a great deal more to recycling than just collecting potentially valuable materials.

According to a study by Franklin Associates, Ltd., commissioned by Keep America Beautiful (KAB), the United States recycled and/or composted 21% of the 203 million tons of municipal solid waste (MSW) generated in 1992 (Figure 1). Of the 21% "recycled," roughly half (11%) came from commercial operations and half from residences. The 10% that came from residences breaks down as follows:

Buy-back/drop-off programs	4.5%
Composting yard debris	3.0%
Curbside collection programs	2.5%

According to the study, the practical overall limits of recycling and composting range from 25 to 35% of total MSW generated. The study further shows that curbside programs are expensive relative to other waste management alternatives and divert only a small amount of material from waste to energy and landfill options. Nevertheless, many states are legislating 25-50% rates of recycling by the year 2000. The task of processing more than 100 million tons of trash per year in less than 10 years is an enormous challenge.

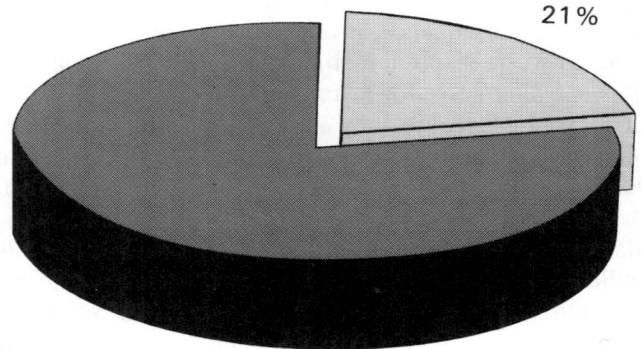

Figure 1 Management of MSW in the United States, 1990 (203 million tons). (Source: KAB 1994.)

However, physical recovery of materials for reuse, which is the traditional definition of recycling, need not be the only method for managing resources. Recovery of materials for refinery feedstocks or for energy, as in coliquefaction with coal, may be just as environmentally sound. Moreover, recovery of resources will develop with time and will improve as a result of research aimed at developing new technology and with changes in the economics of the manufacture of virgin materials. What is critical in the interim, however, is that the methodology for recovery of resources from our discards be designed so as to preserve product innovation. The strategic goal of society should be to design products for the environmentally sound management of the resources used, from their origin as raw materials to how the resources are managed after the product has served its useful purpose. Consideration should be given first to what the product is supposed to do and to the minimum amount of material and energy required to make the initial product.

PACKAGE TECHNOLOGY AND RESOURCE RECOVERY TECHNOLOGY

Packaging materials represent about one third of the MWS (Figure 2). Packaging presents the largest challenge for recycling because of the complexity of the mix, and packaging has attracted the bulk of criticism and negative legislation, including legislation that mandates huge recycling rates within the next 2 to 7 years.

The technology for the design of packaging has grown exponentially (Figure 3) from the glass jar and welded steel can to very sophisticated multilayer packages and composite packages that permit a large spectrum of products to be

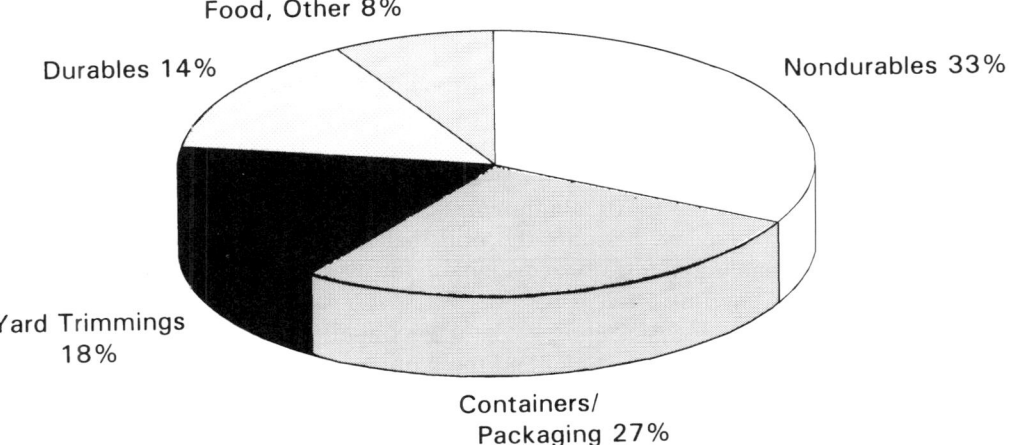

Figure 2 Products generated in MSW by percent, 1990.

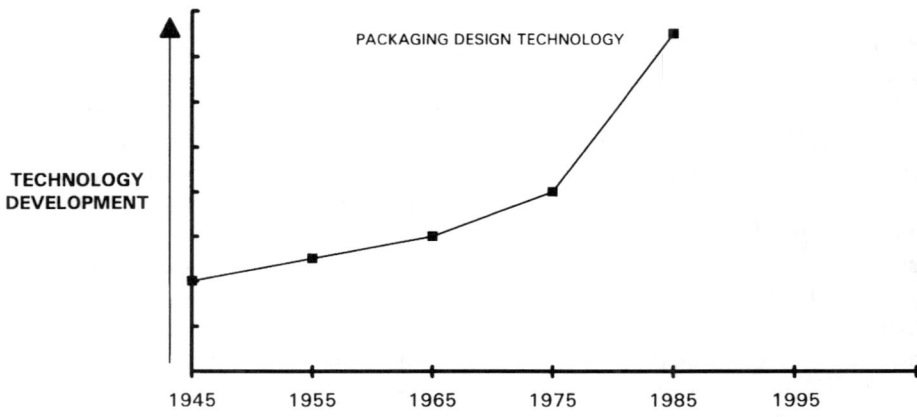
Figure 3 Packaging design technology. (Source: Plastics Recycling Foundation.)

delivered to the consumer, such as liquid detergents and soap, pharmaceuticals, and aseptic packaged foods. The products and the required packaging have been designed to meet the consumer's needs for products that are safe and healthy, have a long shelf life, and are convenient to enhance our expanding standard of living.

On the horizon are gamma-irradiated foodstuffs that have the potential to require little or no energy-consuming refrigeration during manufacture, distribution and storage in the consumer's home. Furthermore, form, fill, and seal packaging and the development of lightweight multilayer and linear low-density polyethylene films have permitted great amounts of source reduction in the packaging of consumer products.

By contrast, resource recovery means recycling and, more narrowly, closed-loop recycling. Recovery of the energy quotient in plastics and paper or conversion of these materials to sophisticated fuels is currently unacceptable to those in our society who are convinced that recycling is the only way to recover resources from our discards.

Yet, recycling technology and the infrastructure for recycling have changed little from the 1940s, when we were recycling newspapers, steel, and glass containers (Figure 4). By the 1970s, aluminum, steel, and plastic beverage containers were included, and today these are the principal items in municipal trash that are technically and economically recyclable. However, the state of the art has changed little during this 50-year period.

Another problem for packaging is that the economics of recycling are driven by weight. The heavier the material and the more expensive to manufacture the original product, the more likely it is to be selected by the public and the waste hauler. This approach favors high-cost and heavy materials. Plastics and modern packaging, which have contributed much to weight and cost reduction in our

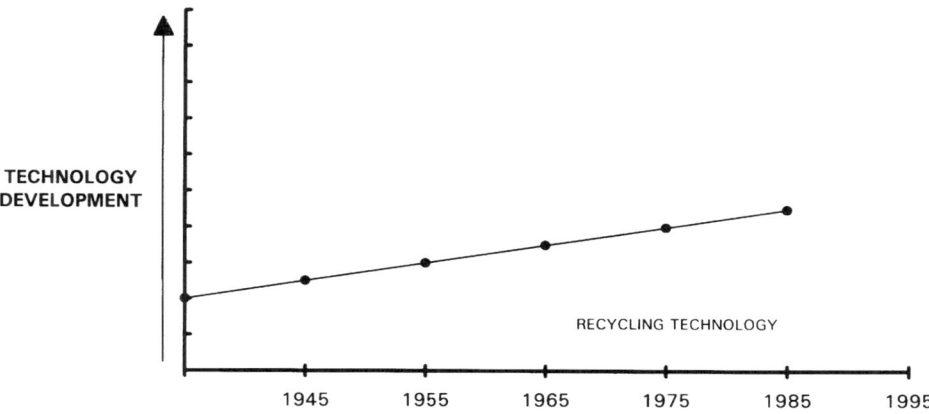

Figure 4 Recycling technology. (Source: Plastics Recycling Foundation.)

modern products, suffer in this misplaced comparison. Two easy examples help to focus the point:

One pound of plastic replaces about 15 pounds of glass for packaging liquids in bottles. Yet glass is nearly always selected for recycling; plastic rarely is. A polystyrene cup is 93% air when it is in use but it can hold coffee at 180°F without burning the consumer's hand, something paper, glass, metal, and even ceramic cannot claim.

It is precisely the kinds of innovations that have contributed to the reduction in weight, energy, and cost that would be banned and reversed if we were to stay with the current recycling technology. Packaging design would have to regress to match the current primitive and embryonic technology (Figure 5).

The current recycling infrastructure is economically suitable only for beverage containers, which creates a problem for the rest of the packaging materials. This in turn creates a problem for consumer products companies, which in turn creates a problem for society at large. There is a tendency to "force fit" products into current crude state-of-the art recycling. Exacerbating the issue is the perception of much of the public that the term "recyclability" applies only to what is recycled here and now. This tends to negate the concept of technical recyclability and the future development of recycling. Because political correctness mandates that recyclability means recycled, pressure is being placed on consumer products companies to shift their packaging to materials that are perceived to be recyclable.

Moreover, public pressure is forcing a shift from colored to clear packaging materials. New Jersey, for example, has considered banning green glass containers for imported beer, which have become a glut on the recycle market, and

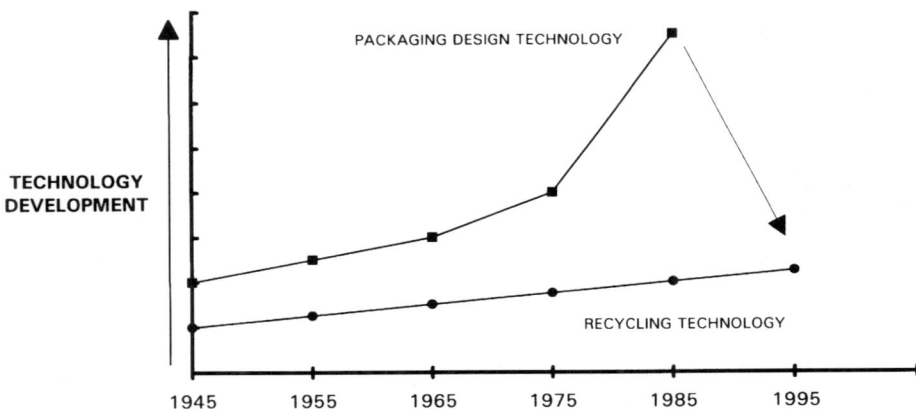

Figure 5 Packaging design versus recycling technology. (Source: Plastics Recycling Foundation.)

several polyethylene terephthalate (PET) reclaimers are arguing for elimination of the green PET soda bottle.

COMPLEXITY OF MUNICIPAL SOLID WASTE

Our municipal solid waste is very complex. It contains literally thousands of materials. The materials we have traditionally recycled include such familiar items as the aluminum soft drink can, the various colors of glass bottles, the tin can, and newspapers. Recently, plastic containers, such as the PET soda bottle and the high-density polyethylene (HDPE) milk, water, and juice bottle, have been introduced to expand this elite group to 10 items that are being recycled.

To go beyond these top 10 items, the problem increases by orders of magnitude. Beneath the top 10, for example, are at least 100 items, including Pyrex glass, various and sundry kinds of paper reading materials, a wide variety of plastic containers and colors of plastic containers, to say nothing of bimetal and trimetal containers (Figure 6).

Immediately under the hundred items are at least 1000 items consisting of multilayer materials including combinations of paper and plastic (as found on the cardboard milk container); aluminum, plastic, and paper (as found on the drink box); and multilayer plastics (as found on the wide variety of form, fill, and seal pouches for such uses as packaging potato chips).

When we start to think about disassembling automobiles, appliances, and houses to recover recyclable components, we suddenly discover that the number has increased to thousands and thousands of subitems. Each of these subitems would have some value if it could be separated and segregated completely from the rest of the thousands and thousands of items.

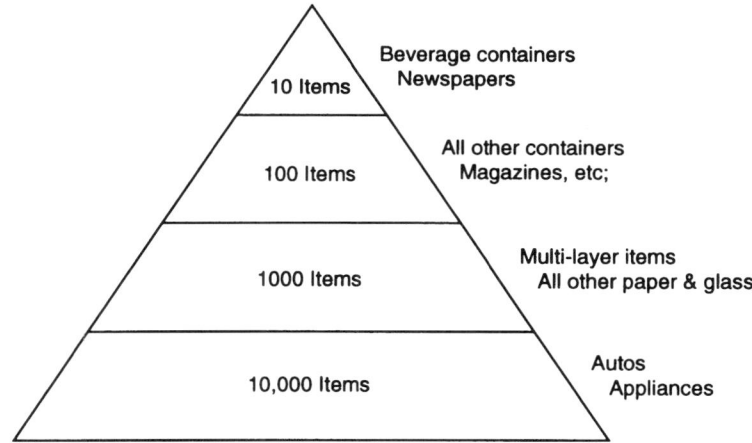

Figure 6 The trash pile. (Source: Plastics Recycling Foundation.)

Based on the foregoing discussion, it might occur to the reader that one way to improve recycling as an economical business would be to reduce the number of items in the trash pile to as few as possible and to raise the intrinsic value of the remaining items as high as possible. For example, if we were to use gold foil to wrap hamburgers versus polystyrene foam, paperboard boxes, or paper, there is no doubt that when the gold foil was discarded, it would be recycled, whereas there is little chance that plastic foam, cardboard boxes, or paper would be.

ENTER THE GARBAGEMEN

The garbageman, the person or organization who has conveniently picked up our trash and disposed of it efficiently in a landfill or in a waste-to-energy plant, has been paid a fee for this service. The fee has been nominal and the service has been excellent. However, we have placed the "sale of trash" in his hands. Traditionally, he has never sold anything and has little understanding of how markets work. However, he feels that if he collects the material, it has value and people should be beating a path to his door in order to get this valuable material, despite the fact that contaminents have not been separated, which lowers the value.

The current methodology for processing waste into valuable things is primitive, labor intensive, and expensive, which makes it difficult to produce a high-quality component from trash at a reasonable cost. This forces industry to refine the material further, which often duplicates handling and sorting efforts and adds additional expense.

We are in a situation in which the garbageman may well be describing how products our society uses should be designed so that the waste would have an extremely high value and cost little to process. However, such an approach sacrifices low-cost, low-energy packaging methodology for higher cost materials so as to make more money for the garbageman.

This simple approach seems to be logical to the public at large, who are often uninformed about the intricate interrelationship between the packaging material and the product it is designed to protect. The public probably does not recognize what they give up when such simple approaches are implemented. For example, when Maine banned the milk and juice carton because it is a multi-component package (aluminum foil, paper, and several different types of plastics) and therefore not recyclable with the current infrastructure, they probably did not recognize that they were banning the aseptic package that permits perishable foodstuffs, including milk, to be packaged without a need for refrigeration from the time the material is placed in the package through the distribution of the product, arrival on the shelf of the supermarket, and arrival at the home. Refrigeration requires enormous amounts of energy and furthermore requires chlorofluorocarbons (CFCs), the chemicals that are perceived to be causing the destruction of the ozone layer. Therefore, it would seem advisable for the public to want to retain the drink box and force the development of technology for recycling it rather than banning it. This logic was finally understood by the citizens of Maine, and they rescinded the ban.

ECONOMIC REALITIES

For recycling to be economically viable, including the recovery of refinery feedstocks or fuels, the full costs of collecting, sorting, and reclaiming must be recovered. The costs of collecting and sorting are generally borne by the community, or public sector. The costs of reclaiming are borne by the private sector. The costs to the community include the costs to collect, sort, package, and ship the recyclables. These costs can be offset in part or in total by the costs avoided for collection of the trash and the costs avoided for alternative disposal, such as by incineration or landfilling. The sum of these costs would be the price the community should receive for material they sell to a reclaimer. If the reclaimer paid that price as a raw material cost, the reclaimer would have to add the operating costs for refining the material, the cost for capital, and the cost of paying dividends to stockholders.

With the technology that is being used today, the sum of these costs is generally higher than the costs of comparable virgin materials. The small scale and high labor intensity of curbside collection and subsequent separation and packaging of recyclables at material recovery facilities (MRFs) are the reason for the high cost of current state-of-the-art recycling. As a rule, curbside collection of recyclables usually costs twice as much as the collection of trash, and the high cost of separation at the MRF usually exceeds the value of the recovered material in competition with similar virgin commodities. The problem does not end at the MRF. In the case of plastics, the materials from the MRF are sorted at low levels of quality; that is, many materials are not separated completely before they are baled for shipment to the reclaimer. Consequently, when the bale arrives at the reclaimer, it must be debaled and resorted before the reclaimer can process the plastics to meet quality standards required by manufacturers who are accustomed to the high quality of virgin plastics.

On the other hand, if the virgin material is expensive to manufacture in the first place, the high cost of recovering such material from the waste stream may be competitive with the high cost of manufacturing the virgin material. An example of a high-cost virgin material is aluminum. Aluminum must be manufactured from bauxite ore to an ingot. An enormous amount of energy is needed to convert this material to a metal. Moreover, the bauxite is often imported, and a large amount of ingot is now imported as well. As the dollar has weakened substantially, the cost of importing ingot has risen. As a result, aluminum metal derived from waste is substantially more economical. This is the driving force for separation of aluminum beer and soft drink cans from the waste stream.

Some plastics have value as well. The PET soda bottle is one. The polycarbonate water bottle is another, and some other materials used in automobiles, airplanes, and houses would have similar high value. However, most of the plastic materials and a great amount of paper and glass, as well, do not have this threshold value (Figure 7).

ECONOMIC DRIVING FORCES

Several scenarios would make the economics of recycling more favorable. Some can be controlled, and some cannot. Some are attractive and some are not.

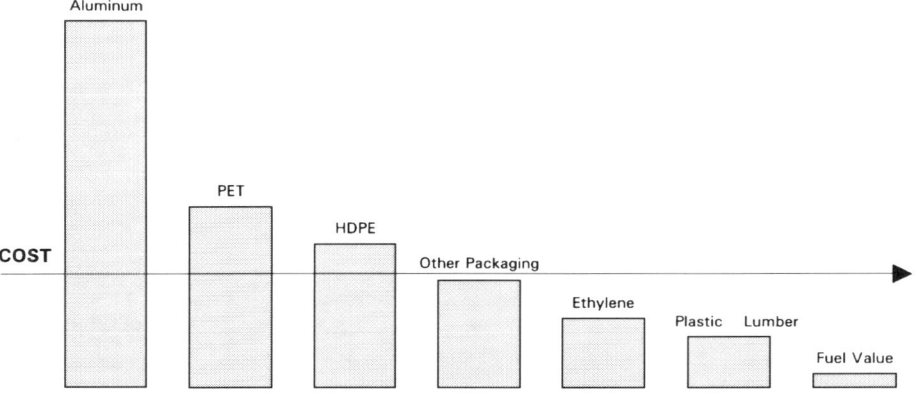

Figure 7 Economics of recycling (oil at $20 per barrel). (Source: Plastics Recycling Foundation.)

Higher Priced Oil

An economic driving force that would make the current recycling technology become viable would be a significant increase in the price of oil, say to $35 per barrel. The price of virgin materials would rise correspondingly, whereas the cost of recycled materials would not rise as abruptly (Figure 8). Consequently, virtually all polymer recycling could become economically sound.

There is a slim likelihood that this will happen during the next several decades simply because the United States does not control the supply or the price of oil. Saudi Arabia does today, and just around the corner are additional supplies from Iraq and Russia.

Higher Alternative Disposal Fees

Another economic driving force that could make the current recycling technology viable would be an increase in alternative disposal fees. Because the avoided cost for alternative disposal can be legitimately applied as though it were revenue, it can be determined how high these fees need to be to make the lower price polymers become viable.

It can be seen that high-value items such as aluminum and PET require no offsetting subsidy in the form of an avoided tipping fee. However, a tipping fee of about $200 per ton would be required to make the low-price polymers such as low-density polyethylene, polypropylene, and polyvinyl chloride become viable, and a tipping fee of about $400 per ton would be required to make it attractive to convert the polymers to the monomer ethylene (Figure 9).

Realistically, the actual costs to build and safely operate a modern landfill or waste-to-energy plant are not nearly high enough to provide a driving force.

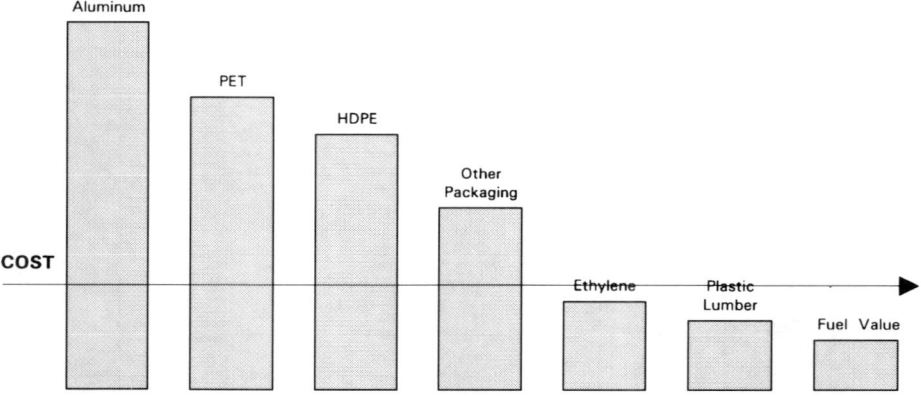

Figure 8 Economics of recycling (oil at $35 per barrel). (Source: Plastics Recycling Foundation.)

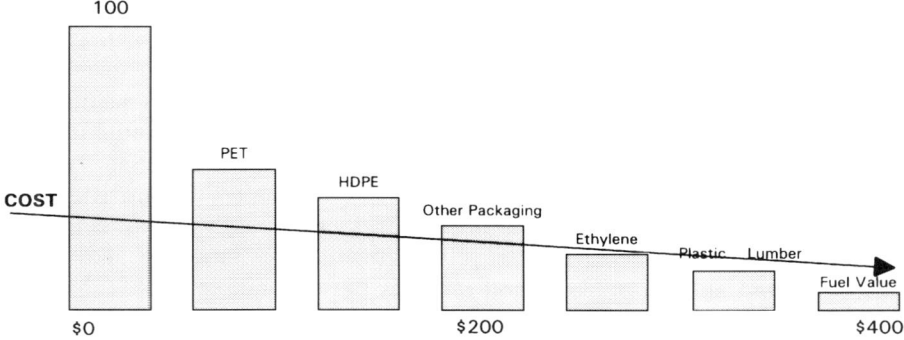

Figure 9 Economics of recycling (oil at $20 per barrel). (Source: Plastics Recycling Foundation.)

Design for Recycling

A driving force that could make the current recycling technology become viable would be conversion of all present materials that are not economically viable to those that are. This simplistic logic is what is involved in the idea of "designing for recycling." It can be accomplished by product bans or by encouraging that products be switched from those that are not recycled to those that are. Because consumers often translate the term recyclable to mean what is recycled here and now in their neighborhoods, it can only make sense to a consumer products company, who wishes to respond to the wants of the consumer, to change to a material that is recycled now despite the fact that the material to be replaced is technically recyclable. For example, companies that have switched from polyvinyl chloride or polypropylene to PET did so to design their products for recyclability. The biggest negative factor is that this approach causes virtual abandonment of innovation. New products and product concepts would not fit the current recycling technology and infrastructure. Examples of products that are already being threatened by this philosophy are the drink box, which is a composite of paper, aluminum, and plastic, multilayer plastics for form-fill-and-seal packaging, colored products, foamed plastics, polyvinyl chloride, and green glass.

The problem with this redesign approach is that innovative products in the 1990s would be switched to those that match the current recycling technology, which is not significantly more sophisticated than that in use in the 1940s (Figure 5).

Legislation Mandating Recycle Content

A third economic driving force that could make the current recycling technology become viable would be legislation that mandates a recycled content of products.

If companies are prohibited by law from marketing a product that does not contain a prescribed amount of recycled material, their choices are to abandon the product or to pay whatever is required to procure the recycled material. In this case, the supply and price of virgin material have no bearing on the decision. The price can be higher than that of virgin material, and the substantial increase in price permitted by this legislation would allow materials that ordinarily could not compete with virgin materials to become viable instantly (Figure 10). Many states and municipalities are passing or are considering passing such legislation.

Although this will accelerate recycling, it will be inflationary and, worse, this "command and control" method removes the free market forces that have a long history of effectiveness and efficiency.

Reduce the Cost of Recycling

The preferred economic driving force is one that reduces the cost of recycling to make recycled materials competitive with virgin materials. This will require a new approach to collection and sorting and will affect reclamation technology as well. To accomplish this, the goals shown in Table 1 must be met.

The result of such an approach would be to permit virtually all materials to become economically viable versus virgin materials without the public having to subsidize the process or pay premium prices in the marketplace (Figure 11). Moreover, it would permit the public to retain the wide range of choices that

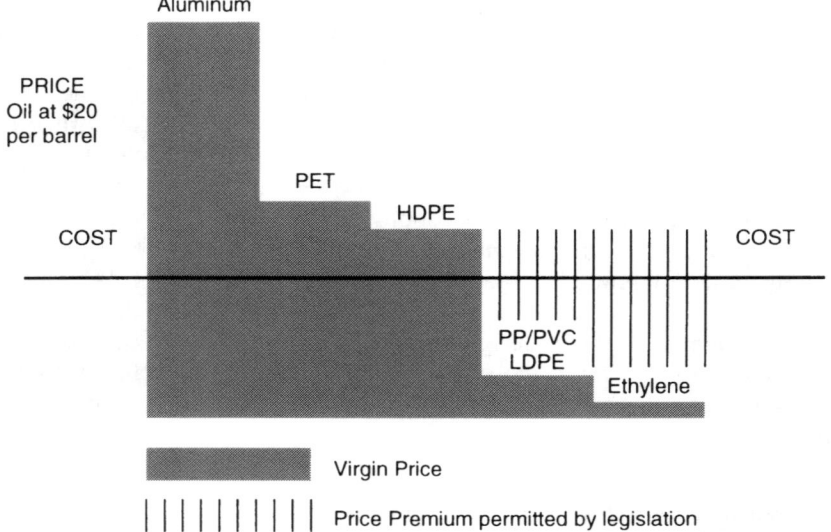

Figure 10 Economics of recycling, legislative impact. (Source: Plastics Recycling Foundation.)

Table 1 Goal costs

Collection	Equal trash collection costs
Sorting	Equal landfill/incineration costs after subtracting revenue
Reclamation	Equal virgin costs as maximum

Source: Plastics Recycling Foundation.

they currently enjoy as a result of the continued supply of innovative new products.

CURRENT RECYCLING TECHNOLOGY IS TOO EXPENSIVE

At base, the issue that looms largest in restricting the amount of material that can be diverted from the trash through recycling is that the state-of-the-art of collection, sorting, and reclamation is too costly because it is small in scale and extremely labor intensive. This system must compete with the well-established, more automated, large-scale, capital-intensive, minimum-labor-content technology for gathering virgin raw materials, for producing product from the raw materials, for distribution of the products, and for disposal of the products. It is capital intensive to refine the oil at major refineries. The basic chemicals manufactured at the refinery are converted to polymers in capital-intensive polymer plants at volumes ranging from 100 million pounds per year up to 1 billion pounds per year. The products are manufactured in large volumes and the distribution systems are highly capital intensive.

The disposal system developed in the United States is equally capital intensive. Modern waste management has permitted the last human hand to be re-

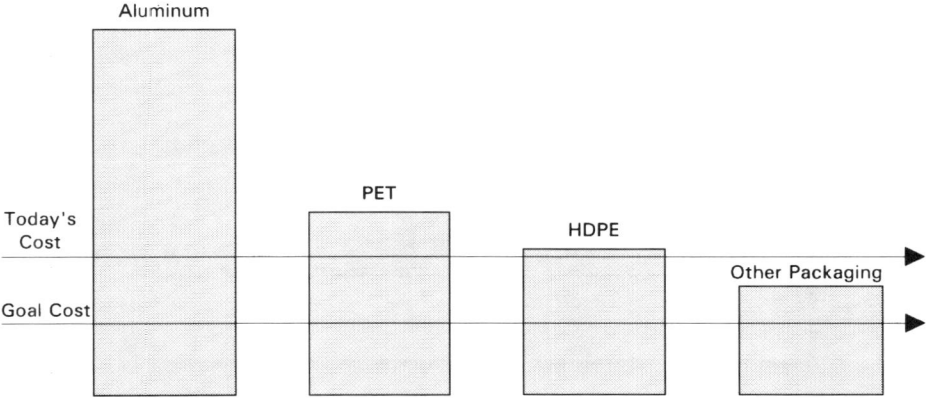

Figure 11 Economics of recycling (oil at $20 per barrel): today's cost versus goal cost. (Source: Plastics Recycling Foundation.)

moved from touching garbage. The latest concepts use a single-operator truck. A robot arm picks up as much as 100 gallons of trash from a resident and deposits the material into a compactor truck. The compactor truck drives to a landfill, where the material is handled with large front-end loaders and bulldozers, or it is deposited in an incinerator, where a crane with a huge claw gathers up an enormous quantity and deposits it in the incinerator.

By contrast, the current technology for recycling the top 10 items in municipal solid waste is inefficient and costly. It involves curbside pickup in expensive trucks that require much labor to operate. This results in a cost of collection for recycling that can be two to five times higher than the cost to collect trash. Moreover, current curbside pickup captures only about one third of the recyclables (Figure 12).

The recyclables are then transported to a material recovery facility that is about one tenth the scale of a landfill or an incinerator, where much hand labor is employed to sort the top 10 items as well as to remove the unwanted trash and debris that are indigenous to the collection system. A plastics reclamation plant is a labor-intensive and inefficient operation on about one tenth the scale of the capital-intensive polymer plant with which it must compete.

It is small wonder that materials derived from the garbage pile have difficulty competing with virgin materials. In the case of plastics, this is obviously exacerbated by the deliberate and strategic positioning of the price of the fundamental raw material for virgin polymer, namely oil, by the Middle East suppliers.

ECONOMIC GOALS FOR RECYCLING

If we are to achieve our goal of recycling 25% of our trash by the late 1990s, we must develop and implement low-cost collection and sorting technology to process desirable components from the trash. As a goal, the cost to collect

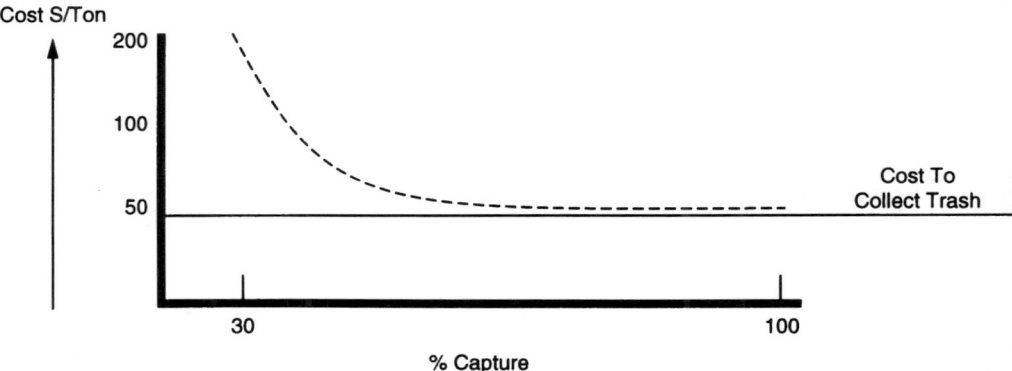

Figure 12 Collection cost versus capture rate. (Source: Center for Plastics Recycling Research.)

recyclables should not exceed the cost to collect trash. The cost to sort should be offset by the revenue derived from the sale of high-quality sorted materials. Under these ideal conditions, recycling costs would be lower than alternative disposal costs, regardless of the cost of the alternative disposal.

Expressed another way, the cost of collecting, sorting, and refining should be below the cost for manufacturing a virgin material as illustrated for plastics. This can be expressed in cents per pound as shown in Table 2.

The key to minimizing costs of collecting and sorting is to maximize the amount of recyclables processed per hour. This requires that the amount of recyclables captured from each household be maximized to reduce the capital (truck) and labor employed per unit of recyclables captured. It requires scaling the collection equipment and routing to maximize the amount of recyclables that can be collected in a shift.

MATERIAL RECOVERY FACILITY MAY HOLD THE KEY

A large-scale MRF is required to make recycling become commercially viable. The present small-scale, labor-intensive MRFs are acting more like transfer stations for trash. They have neither the appropriate technology nor the processing scale to meet the needs of the future. Moreover, little marketing effort is being funded to cultivate customers and to determine their needs and interests. Rather, the public believes that if the material is collected, no matter how poor the quality of the separation, people must buy it or laws will be passed to force them to buy it.

Just Collect Trash

If the consumer delivers only a third of the recyclables, at a cost two to three times what it costs to collect trash, and if 100% of the recyclables are

Table 2 Recycling costs in cents per pound

Item	Today	Goal
Collection	10	2.5
Separation	9	5
Reclamation	20	10
Total	39	17.5
Revenue	−2	−2
Landfill avoidance	−4	−4
Total	−6	−6
Grand total	33	11.5
Price for virgin	28	

in the trash, why not simply collect and ship trash to the material recovery facility?

The answer is that we do not know how to separate the recyclables from the trash! However, this is not an acceptable answer for a nation that:

Developed and harnessed the atom for energy, medical, and other uses.
Placed a human being on the moon.
Invented and exploited the transistor.

If we are ever to recover major portions of the resources in our trash, we must approach it as a societal problem and combine the forces of government, waste haulers, and industry while drawing on academia for the development and dissemination of the needed technical knowledge, in a partnership.

There is no alternative. To expect individual material suppliers to develop a system to recover their own materials is unrealistic given the complexity of the trash. If each of the eight categories shown in Figure 13 were composed of a single material, it might be reasonable for the consumer to separate materials and direct them to the appropriate industry. The problem is that each subset is of no value unless it is separated into a multitude of subsets. Metal is not just aluminum beer cans, glass is not just glass containers, paper is not just newspaper, and plastic is not just the PET soda bottle.

The Future MRF

The future MRF must be designed to process "all" of the trash to obtain the maximum amount of suitable commodities for sale. This means that garbage

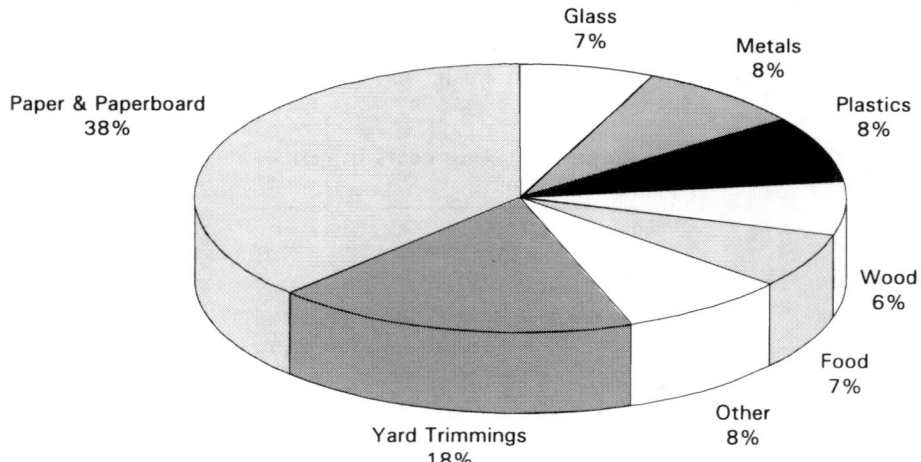

Figure 13 Percent of materials generated in MSW by weight, 1990. (Source: Environmental Protection Agency.)

will be the incoming feedstock. Foodstuff and products left in containers as residue or attached to films and paper will have to be acceptable to the MRF and to the neighbors of the MRF. In addition, the residue from the MRF processing will be classed as garbage and, as such, must be disposed of as garbage. This strongly suggests that the MRF should be located adjacent to either a landfill or a waste-to-energy plant.

Dealing with the question of how to process raw trash requires a fresh and bold approach. The goal should be to gather raw trash, transport it to a processing facility, and process the raw trash into economically viable commodities at the lowest cost and in an environmentally sound manner. A research lead suggests that ore processing technology may be applicable for the separation of the commodities in the raw trash. This technology is used on a massive scale to clean and separate the minerals in ores and is used to clean coal. The feed for such a system is material ground into fine particles that can be separated in froth flotation aqueous systems. It is conceivable that raw trash could be comminuted and slurried in water, whereupon the "grindables" (read recyclable resources) could be separated from the nongrindable material. The grindables slurried in water could be transported by pipeline literally hundreds of miles from the source of the trash to appropriately scaled processing facilities, which could have the ore separation technology as a centerpiece of the material recovery process.

With this fresh approach in mind, one could imagine combining the collecting and transporting of municipal solid waste in aqueous form with the transporting of municipal sewage. The resources portion of the waste would be transported by pipeline to the processing facility (future MRF), and the nonresource portion of the municipal solid waste could be combined with the sewage using the same large-scale facilities and technologies that are in place for converting the organic portion of our waste to sludge, which could, in turn, be converted to fertilizer and mulch.

It is essential that the scale of the future MRF be large enough to justify automating as much of the sorting as possible to minimize costs, including transportation. Present-day MRFs range in size from 100 tons per day to about 250 tons per day. By contrast, the modern landfill or waste-to-energy plant has a capacity of at least 2000 tons per day, roughly a 10-fold scale difference. The future MRF should be scaled at a minimum of 2000 tons per day. It is also essential that the MRF be designed and operated to produce uniform and high-quality products. This will require the development of a vision of the purpose of the MRF, and it will require a commitment on the part of the public and the MRF management to take a long-term approach to the venture.

The future MRF and its management will insist on separating high-quality commodities from the trash, and the materials will be competitive in price. In addition, the MRF management will recognize that they must create the markets; the markets will not create themselves. To create markets, the MRF mush produce new products and price them to attract prospective buyers. All of the skills that the private sector employs to develop and sustain markets and customers must be used. Consequently, one of the first organizational tasks of MRF management should be to hire marketing personnel who have a proven track record in the field of marketing. In addition, MRF management should be prepared to fund marketing development programs. This investment in the "business" to sell trash is as important as, if not more important than, the investment in the hardware for collecting and sorting the trash.

In scoping a future MRF, it would seem wise to attempt to define the products that the MRF would ultimately sell. Therefore, the first task would be to assay the trash in terms of the types of materials that could ultimately be separated. From this breakdown, it would be possible to define some types of markets and potential customers for these materials. It would also be possible to assign some tentative specifications and market prices. Furthermore, it would be possible to determine the form in which the potential customers would like to receive the materials. For example, would they like it in bales or chipped? Would they want it transported by truck, by rail, or by ship? Chipped products of plastics could be shipped by rail hopper car or by hopper truck, as virgin resins are handled today.

It would be highly desirable for the future MRF to include material chipping and then sorting of chips rather than whole items. This lends the system to automation, and it has the added advantage of producing a product that does not have to be baled to be shipped. Elimination of the bale will save money at the MRF and also at the reclamation plant. The great challenge, or course, is that the MRF must have good enough technology and control to produce a high-quality end product.

The future MRF could also implement some simple cleaning of dirty materials to remove foodstuffs or contaminants that would create effluent problems for the end user. By locating the MRF at a landfill or waste-to-energy plant, the effluent could be readily treated and disposed. On the other hand, disposing of the effluent from a reclamation plant that is located in an urban area can be very difficult. The chemical oxygen demand (COD), biological oxygen demand (BOD), and possible other contaminants may require the reclaimer to put in a secondary treatment facility at huge capital and operating expense. Often a reclaimer is forced to shut down because the plant, as initially approved by the public, was not authorized to handle garbage.

An ancillary benefit of the MRF washing the materials would be elimination of inappropriate use of potable water by the home owner, who is instructed to wash recyclables before placing them at the curb. Under current curbside practice, the homeowner washes food, such as peanut butter, from containers by using hot potable water from the tap. Detergent is often employed, usually at 100 to 1000 times the amount needed. The wastewater is then sent through the sewer to the water treatment plant. This process uses expensive potable water, energy to heat the water, and excessive amounts of detergent. The heat and detergent are lost.

By contrast, the MRF would use process water, essentially raw from, for example, a river, and precisely the amount of detergent needed to do the job. The water, the heat from the water, and the detergent would be recovered for reuse. This is much more economically sensible and, more important, is more environmentally sensible.

Technologies that must be developed and demonstrated for the future MRF will include methods to:

Bring compactor truckloads of trash to the tipping area of the 2000-ton-per-day MRF.
Separate metals from the rest of the trash.
Separate different metals from each other.
Separate glass from the rest of the trash.
Separate different glasses from each other.
Separate paper from plastics.
Separate different papers from each other.
Separate different plastics from each other.

New technology that can be imagined for MRFs or reclamation plants will be required. Such technology includes:

Substitution of automated macrosorting technology for manual sorting.
Inclusion of microsorting technology.
Use of characteristics and properties such as melt point, dielectric constant, and x-rays for the detection of different plastics and for their separation at both the macro and micro levels.
Washing of recyclables and their separation by hydrocyclone and froth flotation technologies.

MARKETS

Roughly three types of products can be derived from trash:

1. *Pure molecules* such as aluminum, iron, glass, and monomers from plastics.
2. *Mechanically separated polymers* such as paper fiber and clean polymers such as PET, HDPE, and polyvinyl chloride (PVC) from beverage containers.
3. *Commingled or mixed products* such as mixed glass colors, mixed paper fibers, and mixed plastics.

The time it takes to create a market is usually proportional to the time it takes to demonstrate to a potential buyer that the product can replace the virgin material the buyer is currently using. Usually, this takes precedence over the issues of price and availability.

A *pure molecule* can be refined to a very high level of purity. Usually, any impurities are measured as 50 ppm or less. This translates to 99.9995% pure. The markets for materials at the molecular level are huge and established. Therefore, there is little problem identifying markets and buyers for ferrous, aluminum, copper, gold, or silver metals derived from trash. Likewise, chemical monomers, such as ethylene and propylene derived from trash, would find multibillion pound annual markets. Performance can be judged by matching specifications. The only questions that must be answered for the buyer before there can be a commercial transaction, or sale, are related to price and availability. These can usually be answered quickly, and the sale can be consummated in a relative short period of time.

A *mechanically separated polymer,* on the other hand, will probably have a substantially higher level of contamination than the virgin competitive material. Therefore, specifications have to be changed before a sale or substitution for a generic virgin material can take place. This can be time consuming and usually requires a performance test before a buyer will agree to begin negotiations on price and availability. Therefore, it will take more time to persuade a buyer to purchase the generic polymer, and the selling expense associated with making this type of product will be higher than that associated with a monomer.

An example of a quality problem is the contamination of PET with PVC and, vice versa, the contamination of PVC with PET. Either polymer is difficult to fit into existing uses if contamination with the other polymer exceeds 50 ppm. The difficulty in selling contaminated polymers is a reflection of this.

Commingled or mixed polymers are orders of magnitude more difficult to describe. The problem is that it is difficult to define a mixed material in any terms except those of what it does. For example, "plastic lumber" could replace wood or concrete, but it does not have the specifications of wood or concrete. Therefore, it takes a long time to prove to a potential buyer that the new product, plastic lumber, will be satisfactory as a replacement for wood or concrete. Of importance also is the fact that it can be expensive to run an experiment to prove the worth of the new product. The final consequence is that there is much elapsed time between getting in touch with the potential buyer and consummating a sale. This time is reflected in a much higher selling expense associated with the commingled product than with the generic polymer or monomer.

The volume of markets for various products derived from plastics and the time to achieve the needed communication to penetrate them can be illustrated graphically. The markets for monomeric and clean generic materials are huge because the existing markets for the virgin generic materials are huge (Figure 14). For this reason, it can be predicted that it will be possible to penetrate those markets in a reasonable period of time. On the other hand, the market for mixed plastics or other mixed materials is small today. Nevertheless, the potential is believed to be substantial and thus worthy of pursuit (Table 3).

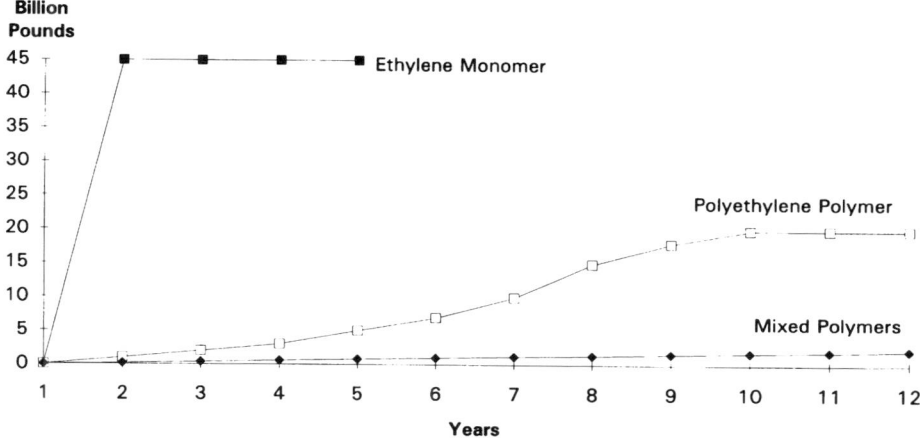

Figure 14 Ease of market penetration. (Source: Plastics Recycling Foundation.)

As the costs to recover valuable resources from the waste stream are reduced by lowering the cost of collection and lowering the cost to separate and refine the resources, the markets for the refined products will expand rapidly. For example, recovery of feedstocks suitable for making building-block monomers for plastics is largely dependent on competitive costs for the raw material. Similarly, the use of polyolefins for coliquefaction with coal depends on the low cost of the raw material. Current data suggest that if collection costs were equal to those for collecting trash, this new developing use could compete with the cost of disposing of the material in a landfill.

Products that the MRF might market would include but not be limited to:

Glass for cullet (as today)
Glass for fiberglass or asphalt modification
High-quality aluminum (as today)
High-quality tin cans (as today)
Bimetal products separated for specific markets
Paper fiber of high quality such as bleached kraft

Table 3 Potential markets—1995 (millions of pounds)

PET	631
HDPE	528
PVC	494
PP	767
PS	477
Commingled	500

Source: Center For Plastics Recycling Research.

Other plastics in addition to PET and HDPE from beverage containers
Various commingled plastic feedstocks
    To replace virgin materials
    As feedstocks for solvent separation processes
    As raw materials for plastic lumber
Various mixtures of plastics (free of PVC and oxygenated compounds)
    As refinery feedstocks
    As hydrogen donors in coliquefaction with coal
Combinations of paper and plastics for
    Refuse-derived fuel
    Raw material for composite board for walls and furniture

Looking beyond mere retrofitting of the existing MRFs are "new departure" technologies that include separation of materials at the molecular level by depolymerizing and/or dissolving polymers in organic solvents to separate different plastics from one another. Such commingled plastics could be used as feedstock for a refinery, which would reduce them to original building block materials such as naphtha or gas oil from which ethylene, butylene, propylene, and styrene monomers can be derived or use them in the coal liquefaction process to produce high-grade fuels.

New plastics products the MRF might market will include but not be limited to:

Other resins in addition to PET and HDPE from beverage containers.
Various feedstocks for commingled plastics processors (e.g., plastic lumber and refined commingled uses).
A mixture of plastics suitable for separation by solvent dissolution methods.
A mixture of plastics suitable for a refinery feedstock (e.g., free of chlorinated and, possibly, oxygenated compounds).
A mixture of plastics as a hydrogen donor in coliquefaction with coal.

Another potential for the future MRF is to mine materials from landfills. As old landfills are recycled to make room for new wastes, the mined materials could be salvaged for feedstock or fuel. Once this is realized, it is a small step for the MRF, which would be located on a landfill, to begin to store materials for future sale when the markets are more fully developed or when the price is more propitious.

SUMMARY

The MRF is the key to the ultimate large volume and large percentage of recycling. Today, most MRFs are being designed and run by waste haulers. The business strategy of the waste hauler is to charge a fee to cover the cost of picking up trash and hauling it to its final destination, namely a landfill or waste-to-energy plant. The MRF is not a landfill or a waste-to-energy facility.

The MRF of the future will be a manufacturer of raw material commodities derived from discards. The price the MRF receives will be in large measure related to the efficiency of the separation and the quality of its products. It will also depend on whether the commodity is packaged and marketed in a way to attract and hold prospective buyers.

The successful MRF will be one that recognizes that its business is to satisfy customers by offering material derived from a new "ore," called municipal solid waste, and it will seek out new customers and markets in the same way that successful companies market commodities derived from virgin sources.

Technology is available or can be developed to separate trash into viable commodities that can be sold to industry as feedstocks for existing businesses. The MRF will make obvious separations of metals, glasses, papers, and plastics with technologies that exist. It is to be hoped that the MRF will add technologies that exist or can be developed to separate into their subtypes, glasses, metals, papers, and plastics. Then the reclamation plants, be they metal smelters, paper or pulp makers, or plastics processors, will add technology for upgrading the quality of generic materials received from the MRF or will add technology to utilize commingled materials. Commingled plastics, cleaned of debris and non-plastic materials, could be sold by the MRF to plastics processors who, by mixing and making them compatible, could process the material with existing plastics processing equipment to compete with virgin plastics. Some organic materials, notably plastics, will be suitable as feedstocks for refineries and as fuel components.

HOW EVERYONE CAN WIN

Society will best be served when recycling technology, including recovery of energy and chemical feedstocks, is expanded to keep pace with the modern innovative packaging design needed for the 1990s rather than having packaging and product design reversed to match the recycling technology of the 1940s (Figure 15). Preliminary evidence from research suggests that development of economically viable technology is feasible.

Society will be the loser if packages and products are designed to match what the recycler or trash hauler wants us to do based on the current state of the art. It does not make sense to throw out the innovative engineering concepts that are possible in order that garbage processing be more profitable. Do we want a society designed by the garbageman? That may be where we are heading. But that need not be the case!

A society that can develop the technology to place a man on the moon need not have its future structure based on a primitive method for managing its waste. We certainly can employ our technological expertise to figure out how to use

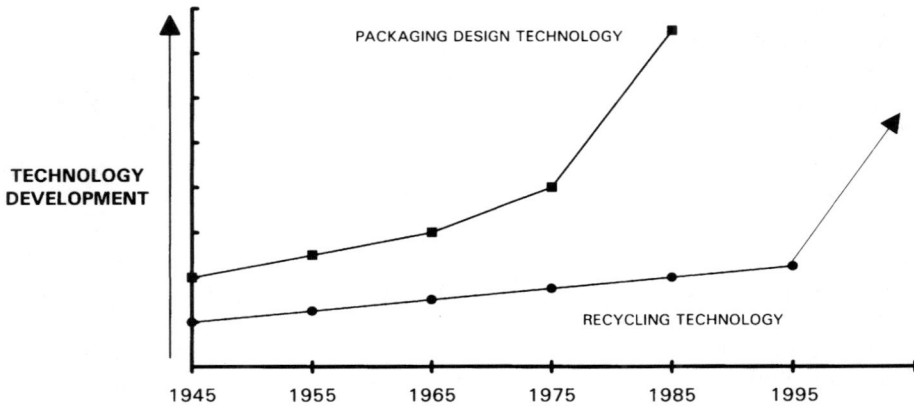

Figure 15 Packaging design versus recycling technology. (Source: Plastics Recycling Foundation.)

our trash in an appropriate way so as to protect the innovations of the past, present, and future, while utilizing in the most effective way the resources that are contained in the trash.

This is the challenge to society. It will require a partnership of government, industry, and the public. It must make use of the best technological brainpower the nation has in all these areas.

Because it will take time to create the new technology, we need credible voices to describe the developing technology in recycling. Society will best be served by finding such voices. The radical activist who wants to take us back to the technology of the 1940s is not the appropriate voice. The appropriate voice will most likely be found in academia, with support from industry, government, and the public, and with everyone pulling together for a feasible and sustainable future.

CHAPTER
THREE

COLIQUEFACTION OF WASTE MATERIAL WITH COAL: A RESEARCH PROGRAM OF THE CONSORTIUM FOR FOSSIL FUEL LIQUEFACTION SCIENCE

G. P. Huffman

BACKGROUND

The Consortium for Fossil Fuel Liquefaction Science (CFFLS) is a research organization supported by the U.S. Department of Energy that includes approximately 100 participants from five universities—the University of Kentucky, the University of Pittsburgh, the University of Utah, West Virginia University, and Auburn University. Since 1985 the CFFLS has been engaged in a broad research program aimed at developing innovative, economical methods for converting coal into oil by direct liquefaction. During the past 2 years, the CFFLS has been investigating methods for converting waste materials into oil using direct coal liquefaction technology. Although the program is still young, CFFLS scientists have obtained promising results in the liquefaction of plastics, rubber tires, paper, and other wastes and the coliquefaction of wastes with coal. In the current chapter we present a brief summary of the principal results for all of the waste materials investigated and a somewhat more detailed summary of the research on waste plastics conducted by CFFLS investigators at the University of Kentucky.

Waste Plastics

Currently, over 44 billion pounds of plastic waste material are disposed of in the United States each year [1]. The dominant components of plastic waste—polyethylene, polyethylene terephthalate (PET), polypropylene, and polystyrene—have a high hydrogen-to-carbon ratio and a molecular chain structure that are well suited for liquefaction. Experiments conducted by Consortium scientists have yielded a number of promising results [2–7]. Under typical direct liquefaction conditions (~400–450°C, 1500–2000 psi of hydrogen, 30–60 minutes reaction time), in the presence of an HZSM-5 zeolite catalyst, plastics have oil yields of 80–98% and total liquefaction conversions of 90–100%. When they were coliquefied with coal of several different ranks, oil yields as high as 60–80% were observed and the total liquefaction conversions reached levels of over 90%. Simulated distillation analysis of the oil products [7, 8] from such experiments indicates that they are high in quality and could readily be upgraded to transportation fuel (gasoline, jet fuel, diesel fuel) or used as the feedstock to produce new plastics or rubber.

Rubber Tires

Over 280 million automotive tires are discarded in the United States each year [9]. Major problems have been encountered with fires involving old tires, which are very difficult to extinguish. Retreading, combustion, and conversion to rubberized asphalt have all been considered as disposal solutions. Liquefaction is an attractive alternative option. Liquefaction of tires alone yields approximately 65% high-quality oil and 35% a solid residue that is primarily carbon black [10–12]. As tire manufacturers typically add 30–35% carbon black to aromatic oils to make up the feedstock from which tire rubber is prepared, this indicates that essentially all of the oil in tires can be recovered by direct liquefaction. Furthermore, high hydrogen pressures do not appear to be required to convert tires to oil by hydrotreatment [10, 11]. Coliquefaction of tires with coal gives higher liquid product yields than expected on the basis of the results obtained from the liquefaction of rubber and coal separately, possibly because of catalytic activity by the carbon black [10].

Waste Oils

Almost 1.2 billion barrels of waste oil are generated in the United States each year, posing an environmental hazard from metal-bearing compounds, possible halogenated hydrocarbons, and other toxic materials. Investigation of the coliquefaction of coal with waste oil by CFFLS scientists [13] indicates that this process provides many benefits including increased conversion of coal to oil, removal of metals from the oil [14], markedly reduced oil viscosity, and reduction of the amount of recycle solvent required.

Waste Cellulosic Materials

Cellulosic wastes (paper, yard wastes, food wastes, agricultural wastes) constitute more than half of the material in landfills, both by volume and by weight. Earlier research [15] demonstrated that such materials could be converted to oils, and a small pilot plant was operated for some time at Albany, Oregon. Recently, CFFLS scientists have investigated two different methods for coliquefying paper with coal [16]. In one approach, standard direct liquefaction technology was used and a molybdenum catalyst was employed. In the second approach, an alkali catalyst was used, and the reaction took place under a CO-H_2O atmosphere. Hydrogen is produced under these conditions by a catalytic reaction. In both approaches, nearly 100% of the paper alone was converted to hydrocarbon products, with oil yields of 20–25% and gas yields of 50–70%. When paper was mixed with coal, the total conversion of the coal was 80–90%, with oil yields of 20–30% and gas yields of 20–30%. Although the gas yield is higher than desired, the gas is rich in hydrogen that could be used in upgrading coal-waste coliquefaction products.

COLIQUEFACTION OF WASTE PLASTICS WITH COAL

Polyethylene (PE), PET, polypropylene (PPE), and plastic wastes from such items as milk jugs, soft drink bottles, plastic wraps, and plastic flatware have been successfully converted to oil in direct liquefaction experiments with coal [3, 4]. Comparative experiments were performed with and without the presence of coal under typical direct liquefaction conditions (420–450°C, 60 minutes reaction time, 800 psig H_2 cold). Two types of catalysts were used: highly dispersed iron-based catalysts and an HZSM-5 zeolite catalyst. Using PE, PPE, PET, and a mixed waste plastic with the zeolite catalyst, oil yields of 80 to 98% and total conversions of 90 to 100% were obtained at liquefaction temperatures of 420–430°C. A nanoscale ferrihydrite catalyst in a sulfided state was less active but also gave similar results at somewhat higher temperatures. Coliquefaction experiments were performed on coal-plastic mixtures (usually 50:50 mixtures) using a bituminous coal, a subbituminous coal, and a lignite. The HZSM-5 zeolite catalyst and nanoscale iron catalysts were used, separately and together. The oil yields for these coliquefaction experiments were as high as 60 to 80%, and the total conversions reached levels of over 90%. Oil yields for coal-plastic mixtures were higher, typically by ~10%, than the average of the oil yields for the coal and plastic alone, implying synergistic effects.

Experimental Procedure

Medium-density polyethylene (MDPE), high-density polyethylene (HDPE), PET, PPE, actual plastic wastes such as milk jugs and soft drink bottles, and a mixed waste plastic (MWP) prepared in our laboratory from a variety of items (milk jugs, soft drink bottles, yogurt containers, motor oil bottles, disposable plastic flatware, plastic sacks and wraps, etc.) were liquefied. Coliquefaction experiments were performed on mixtures of MDPE and MWP with Blind Canyon bituminous coal, iron ion-exchanged Beulah lignite [17], and iron ion-exchanged Black Thunder subbituminous coal [18]. Proximate and ultimate analyses for these coals and the MWP prepared in our laboratory are shown in Table 1. The experiments used several types of catalysts: ion-exchanged iron [13, 14], ultrafine ferrihydrites [19, 20], and an HZSM-5 zeolite catalyst. The structure of the ion-exchanged iron and ferrihydrite catalysts and their activity in coal liquefaction have been discussed in detail elsewhere [17–20].

Most of the liquefaction experiments were conducted in 50-mL tubing bombs charged with 5 g of feedstock (plastic or coal plus plastic). Tetralin was added with feedstock in most experiments in a 3:2 tetralin/feedstock ratio. Several experiments were run using a waste motor oil as a solvent (waste oil/plastic ratio = 3:2). Unless otherwise noted, the zeolite and ferrihydrite catalysts were added at a concentration level of 1 wt % of the feedstock. In order to sulfidize the iron when iron catalysts were present, dimethyl disulfide (DMDS) was added with a sulfur-to-iron ratio of 2. The reactors were then pressurized to 800 psig hydrogen (cold) and heated to the desired temperature in a fluidized sand bath while being agitated at 400 cycles per minute.

At the end of the reaction period, the microreactor was quenched in a room-temperature sand bath. The liquefaction products were determined by Soxhlet extraction and the gaseous products collected and analyzed by gas chromatography. The liquefaction products were separated into oil [tetrohydrofuran (THF) and pentane soluble], asphaltene and preasphaltene (THF soluble but pentane insoluble), and insoluble residues or IOM (THF insoluble). Total conversion was defined as [100 − %(THF insolubles)], and oil yield was defined as [100 − %(pentane insolubles) − %(THF insolubles) − %(gas)]. Although the gas yield

Table 1 Proximate and ultimate analyses of coal samples and waste plastics used in this research

Proximate[a]	Beulah lignite	Black Thunder coal	Blind Canyon coal	Mixed waste plastic
% Ash	9.56	6.34	6.57	0.45
% Volatile	56.08	43.62	46.75	98.80
% Fixed carbon	34.36	49.83	46.68	0.74
Ultimate[b]				
% Carbon	73.14	73.49	81.61	84.65
% Hydrogen	4.47	3.09	6.21	13.71
% Nitrogen	1.01	1.29	1.38	0.65
% Total sulfur	0.77	0.60	0.47	0.01
% Oxygen	20.61	19.33	10.33	0.98

[a] Dry basis.
[b] Dry mineral matter–free basis.

was not determined for some experiments, it was generally quite low (<1–2%). PET was the only feedstock having a significant gas yield (13–14%). The reproducibility of experiments run by a single operator on a single system was normally ±1%, and measurements by different operators on different systems were repeatable to within ±5%.

Soxhlet extraction results for various plastic resins and the MWP prior to reaction are summarized in Table 2. It is seen that the MDPE exhibited significant solubility in THF prior to reaction, which increased with extraction time. The pentane solubility of MDPE is small but not negligible. The extraction percentages for all of the other resins and the MWP are quite low (<1–2%).

RESULTS AND DISCUSSION

Direct liquefaction results for individual plastic resins are presented in Table 3. The experiments were carried out with a 3:2 solvent/feedstock ratio at temperatures of 420° to 450°C, 800 psig hydrogen (cold), for a reaction time of 60 minutes. Two types of catalyst were used: an HZSM-5 zeolite catalyst and a nanoscale ferrihydrite treated with citric acid (FHYD/CA) [20] added at a concentration of 1 wt % of the feedstock. DMDS was added to convert the FHYD/CA to pyrrhotite during the reaction. The noncatalytic oil yields for MDPE and HDPE at 430°C with a tetralin solvent are 26 and 11%, respectively. It is seen that the HZSM-5 zeolite catalyst was highly active, increasing the oil yields for HDPE and MDPE to over 90% under the same conditions. The FHYD/CA catalyst was significantly less effective at 420° and 430°C. PPE had very high thermal oil yields, which increased to nearly 100% at 420°C in tetralin with the addition of either catalyst. The oil yield of PET at 430°C in tetralin was increased only slightly by addition of either catalyst, from 62 to 64%. The gas yield of PET was also significant (13–14%); this is not surprising in view of the ester linkages (carbonyl groups) present in its structure.

Experiments with a waste motor oil solvent also indicated that zeolite was the most effective catalyst. However the FHYD/CA catalyst was nearly as active,

Table 2 THF and pentane solubility of plastics (Soxhlet extraction time 20 hours)

Material	THF soluble (%)	Pentane soluble (%)
PPE	1	1
PET	2	2
MDPE	24	5
MDPE	38[a]	
HDPE	1	1
MWP	0.1	0

[a] Extraction time 32 hours.

Table 3 Liquefaction yields for pure plastic resins: medium-density and high-density polyethylene, polyethylene terephthalate, and polypropylene

Sample	Catalyst	Solvent	T (°C)	Oil (%)	Gas (%)	Total (%)
MDPE	None	Tetralin	430	26	<1	65
MDPE	None	Tetralin	450	33[a]		100
MDPE	FHYD/CA	Tetralin	420	14	<1	74
MDPE	FHYD/CA	Tetralin	430	34	<1	68
MDPE	HZSM-5	Tetralin	420	79	<1	92
MDPE	HZSM-5	Tetralin	430	96	1	99
HDPE	None	Tetralin	430	11	<1	33
HDPE	HZSM-5	Tetralin	430	96	1	98
PET	None	Tetralin	430	62	13	91
PET	HZSM-5	Tetralin	430	64	14	96
PET	FHYD/CA	Tetralin	430	64	13	92
PPE	None	Tetralin	420	83	<1	89
PPE	HZSM-5	Tetralin	420	98	2	100
PPE	FHYD/CA	Tetralin	420	98	2	100
PPE	HZSM-5	Waste oil	420	93	2	95
MDPE	HZSM-5	Waste oil	420	44	1	81
MDPE	HZSM-5	Waste oil	430	55	1	93
MDPE	HZSM-5	Waste oil	450	93	4	97
MDPE	FHYD/CA	Waste oil	420	50	1	82
MDPE	FHYD/CA	Waste oil	450	91+		96

[a] Oil + gas; gas not determined.

particularly at 450°C, where both catalysts produced over 90% oil yield and nearly 100% total conversion.

The oil yields for MDPE and HDPE are shown graphically in Figure 1. Figure 2 shows the oil yields for the MWP (liquefaction conditions 430°C, tetralin/feedstock ratio 3:2, reaction time 60 minutes, 800 psig H_2 cold) and for milk jug and coke bottle samples. These results demonstrate that an acid cracking catalyst such as HZSM-5 zeolite is highly effective in depolymerization of plastic under direct liquefaction conditions.

Coliquefaction results for 1:1 mixtures of a bituminous coal (Blind Canyon, DECS-6) and the MWP (430°C, 60 minutes, 800 psig H_2 cold, tetralin/feedstock ratio 3:2) are shown in Figure 3. The largest increase in oil yield (~20%) was achieved by addition of 1 wt % of the HZSM-5 zeolite catalyst to the coal and plastics mixture. Addition of 1 wt % of a binary Zn-ferrihydrite [$(Fe_{0.95}Zn_{0.05})$OOH with added DMDS] to the Blind Canyon coal increases the total conversion of the coal by 12% and the oil yield by 8%. Mixtures of Blind Canyon coal and the MWP, both with and without the FHYD/Zn catalyst (1 wt %), gave oil yields that were approximately 10% higher than those for either the MWP or Blind Canyon coal alone under the same liquefaction conditions, implying synergistic effects.

Significant enhancement of the liquefaction conversion of low-rank coals was achieved in our previous studies [17, 18] by incorporating iron in the coals

COLIQUEFACTION OF WASTE MATERIAL WITH COAL 45

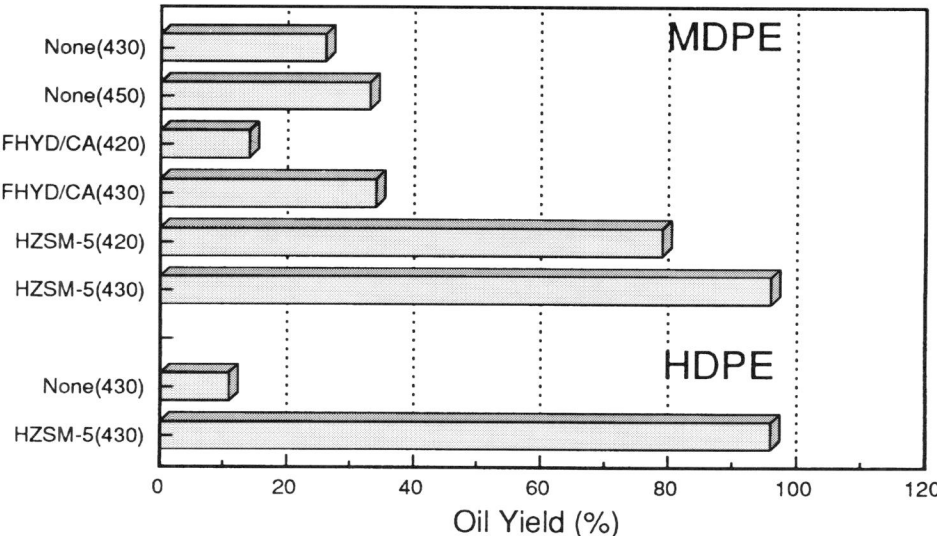

Figure 1 Oil yields for medium-density and high-density polyethylene. Catalysts and reaction temperatures (°C) are indicated.

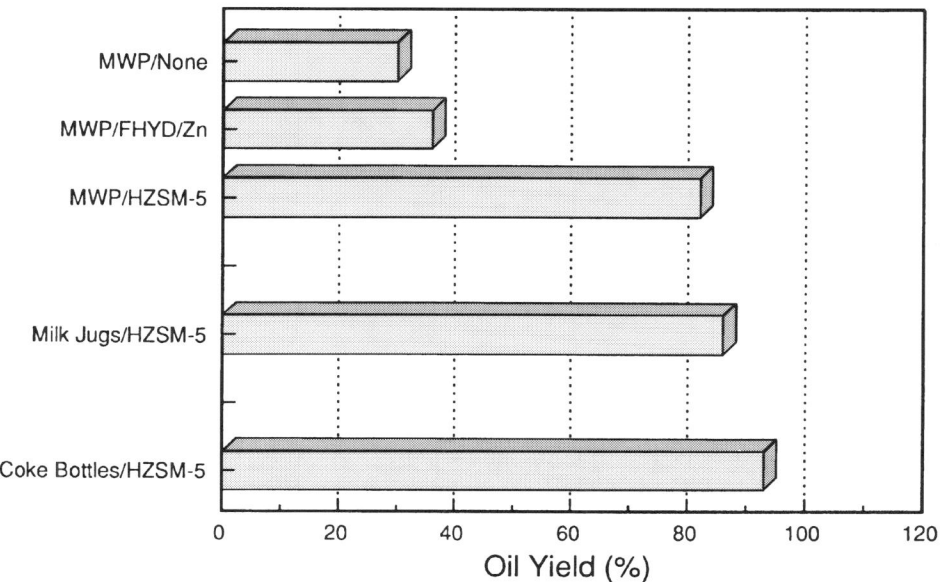

Figure 2 Oil yields for a mixed waste plastic, milk jugs, and coke bottles.

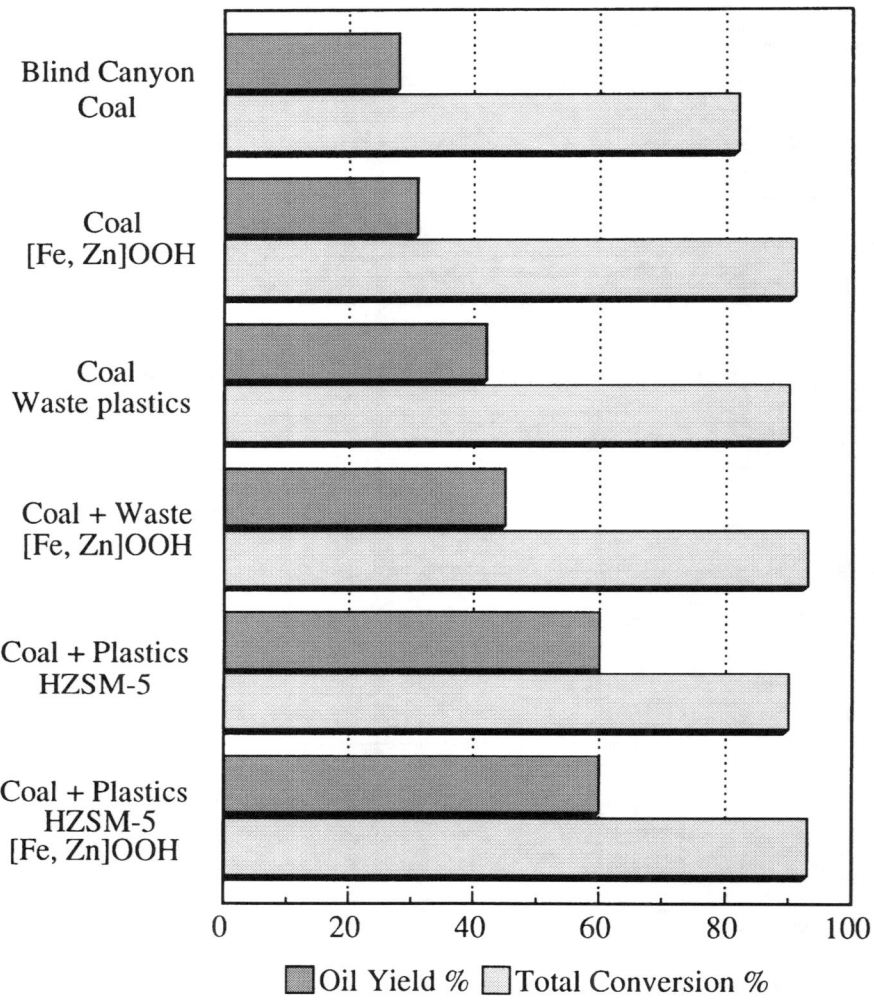

Figure 3 Coliquefaction results for a bituminous coal and a mixed waste plastic.

by an ion-exchange process. The liquefaction behavior of low-rank coals containing ion-exchanged iron in terms of both total yield and desirable products is superior to that obtained by mixing an ultrafine iron oxide with the coal. Figure 4 illustrates the results obtained for coliquefaction of Black Thunder coal containing 0.5 wt % ion-exchanged iron and mixed waste plastics with added tetralin (coal/plastic ratio 1:1, tetralin/feedstock ratio 3:2, 430°C, 800 psig H_2 cold, 60 minutes). The liquefaction results for the MWP and iron ion-exchanged coal in

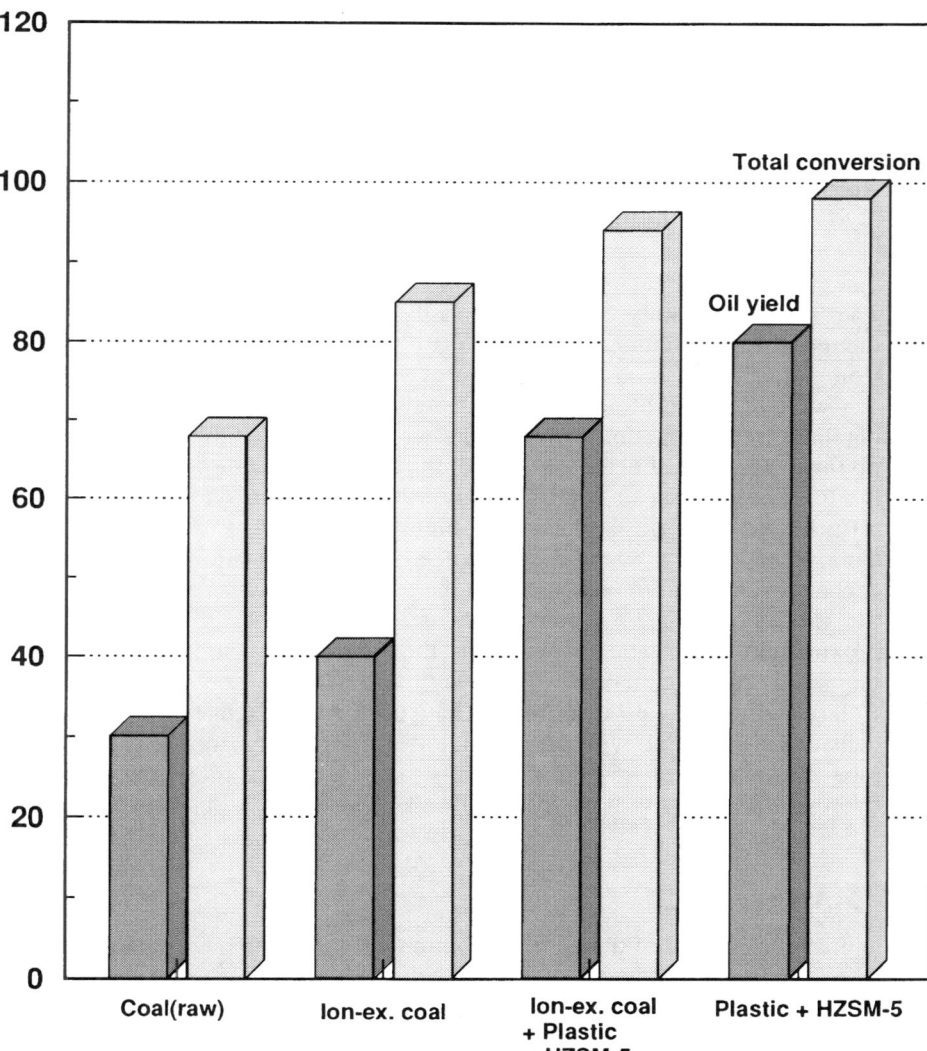

Figure 4 Coliquefaction results for an iron ion-exchanged Black Thunder coal and a mixed waste plastic.

the presence of HZSM-5 (1.0 wt %) show significant enhancements of both total conversion and oil yield. Synergistic effects also occur here. If the oil yields for the ion-exchanged coal alone and the waste plastic plus HZSM-5 alone are averaged, the anticipated oil yield from the 50:50 mixture would be 59%; however, the oil yield obtained was about 68%.

CONCLUSIONS

The following conclusions can be drawn from this investigation:

1. Catalytic direct liquefaction appears to be an attractive recycling alternative for waste plastics, giving very high oil yields and total conversions.
2. An HZSM-5 catalyst is much more active at moderate temperatures (420–430°C) for the liquefaction of plastic than nanoscale, sulfided, iron-based catalysts that are effective for the liquefaction of coal. This indicates that acid cracking catalysts will be useful for direct liquefaction of plastics. The iron-based catalysts are significantly more effective for plastics liquefaction at higher temperatures (450°C).
3. Coliquefaction of mixed waste plastics with a bituminous coal, both with and without the addition of an iron catalyst, exhibits higher oil yields than obtained by liquefaction of either the waste plastics or the coal alone, implying synergistic effects. Addition of the HZSM-5 zeolite catalyst increased the total conversion to approximately 90% and the oil yield to 60% for a 50:50 mixture of waste plastic and bituminous coal.
4. Coliquefaction of a 50:50 mixture of mixed waste plastic and an iron ion-exchanged subbituminous coal with the addition of 1.0 wt % HZSM-5 zeolite catalyst gave a total conversion of over 90% and an oil yield of approximately 70%. The oil yield of the mixture appeared to have been increased approximately 10% by synergistic effects.
5. Coliquefaction of an iron ion-exchanged Beulah lignite with medium-density polyethylene with no added solvent gave very good oil yields and total conversions at 450°C, indicating that the plastic plays the role of a hydrogen donor solvent for the coal under those conditions [4].

SUMMARY

To summarize, the goal of the CFFLS research program is to develop economical processes for liquefying waste materials with coal and transform a major environmental disposal problem into a valuable transportation fuel resource. The size of this potential resource is substantial. By coliquefaction of petroleum-based wastes (plastics, rubber, and waste oil) with coal, 280 million barrels of oil per year could be produced. This is about the amount of oil currently imported into the United States each month. If we were to utilize just 25% of this currently wasted resource, it would be sufficient to provide 2% of our transportation fuel needs. Preliminary economic analysis [21–23] of the coliquefaction of plastic wastes with coal indicates that oil could be produced from this process at about $20 per barrel with tipping fees for waste disposal of $25 per ton. These encouraging economics, coupled with the promising experimental results summarized in this chapter, suggest that this program could eventually alleviate two of the country's greatest problems: oil imports and waste disposal.

Future studies will attempt to establish optimum conditions for coliquefaction of waste plastics and coal, including the effect of temperature, pressure, reaction time, and coal/plastic ratio. A variety of potential acid catalysts, such as different types of zeolite, clays, and highly dispersed metal oxides, will be investigated. More representative commingled plastics will be emphasized. Experiments with mixed wastes (plastics, rubber, waste oil, paper) are also planned.

Note added in press: New research results on this topic by the CFFLS and others will appear in a special issue of *Fuel Processing Technology* in early 1996.

REFERENCES

1. Franklin Associates, Ltd., Characterization of Plastic Products in Municipal Solid Waste, final report to the Council for Solid Waste Solutions, February 1990.
2. Anderson, L. L., and W. Tuntawiroon. Coliquefaction of coal and polymers to liquid fuels. *Am. Chem. Soc. Div. Fuel Chem. Preprints* 38(3):816–822, 1993.
3. Mehdi Taghiei, M., F. E. Huggins, and G. P. Huffman. Coliquefaction of waste plastics with coal. *Am. Chem. Soc. Div. Fuel Chem. Reprints* 38(3):810–815, 1993.
4. Mehdi Taghiei, M., Zhen Feng, F. E. Huggins, and G. P. Huffman. Coliquefaction of waste plastics with coal. *Energy & Fuels* 8:1228–1232, 1994.
5. Xiao, X., W. Zmierczak, and J. S. Shabtai. Depolymerization-liquefaction of plastics and rubber. 1. Polyethylene, polypropylene, and polybutadiene. *Am. Chem. Soc. Div. Fuel Chem. Preprints* 40(1):4–8, 1995.
6. Huffman, G. P., Z. Feng, K. V. Mahajan, P. Sivakumar, H. Jung, J. W. Tierney, and I. Wender. Direct liquefaction of plastics and coliquefaction of coal-plastic mixtures. *Am. Chem. Soc. Div. Fuel Chem. Preprints* 40(1):34–37, 1995.
7. Orr, E. C., W. Tuntawiroon, W. B. Ding, E. Bolat, S. Rumpel, E. M. Eyring, and L. L. Anderson. Thermal and catalytic coprocessing of coal and waste materials. *Am. Chem. Soc., Div. of Fuel Chem. Preprints* 40(1):44–50, 1995.
8. Liu, K., W. H. McClenne, and H. L. C. Meuzelaar. Catalytic reactions in waste plastics and coal studied by high pressure thermogravimetry with on-line GC/MS. *Am. Chem. Soc. Div. Fuel Chem. Preprints* 40(1):9–14, 1995.
9. Scrap Tire Management and Recycling Opportunities, Hearing before the Subcommittee on Environment and Labor and Subcommittee on Regulation, Business Opportunities and Energy of the Committee on Small Business: House of Representatives, Serial No. 101-52, April 18, 1990.
10. Farcasiu, M., and C. M. Smith. Coprocessing of coal and waste rubber. *Preprints Div. Fuel Chem.* 37(1):472–479, 1992.
11. Liu, Z., J. W. Zondlo, and D. B. Dadyburjor. Tire liquefaction and its effect on coal liquefaction. *Energy & Fuels* 8:607–612, 1994.
12. Sharma, R. K., D. B. Dadyburjor, J. W. Zondlo, Z. Liu, and A. H. Stiller. Coal-tire co-liquefaction. *Am. Chem. Soc. Div. Fuel Chem. Preprints* 40(1):56–60, 1995.
13. Sanjay, H. G., A. R. Tarrer, and C. Marks. Iron-based catalysts for coal/waste oil coprocessing. *Energy & Fuels* 8:99–104, 1994.
14. Huggins, F. E., Zhao, J., Hoffman, G. P., and Tarrer, A. R. Investigation of zinc additives in colique faction of waste lubricating oil and a bituminous coal. *J. Environ. Sci. & Health*, in press.
15. Appell, H. R., I. Wender, and R. D. Miller. Conversion of urban refuse to oil. *Chem.Ind.(London)* 47:1703, 1963.

Acknowledgement: This research was supported by the U.S. Department of Energy as part of a cooperative research agreement with the Consortium for Fossil Fuel Liquefaction Science.

16. Jung, H., J. W. Tierney and I. Wender. Coprocessing of cellulosic waste and coal. *Am. Chem. Soc. Div. Fuel Chem. Preprints* 38(3):880–885, 1993.
17. Taghiei, M. M., F. E., Huggins, B. Ganguly, and G. P. Huffman. Liquefaction of lignite containing cation-exchanged iron. *Energy & Fuels* 7:399–405, 1993.
18. Taghiei, M. M., F. E. Huggins, V. Mahajan, and G. P. Huffman. Evaluation of cation-exchanged iron for catalytic liquefaction of a subbituminous coal. *Energy & Fuels* 1:31–37, 1994.
19. Zhao, J., Z. Feng, F. E. Huggins, and G. P. Huffman. Binary iron oxide catalysts for direct coal liquefaction. *Energy & Fuels* 1:38–43, 1994.
20. Zhao, J., Z. Feng, F. E. Huggins, and G. P. Huffman. Organic acid treatment of ferrihydrite catalyst for improved coal liquefaction. *Energy & Fuels* 8:1152–1153, 1994.
21. Gray, D. Preliminary economics of coal/waste liquefaction. *Proceedings of the U.S. DOE Coal Liquefaction and Gas Conversion Contractors' Review Meeting,* Pittsburgh, September 7–8, 1994, pp. 87–94.
22. Gray, D., and G. Tomlinson. A techno-economic assessment of integrating a waste/coal coprocessing facility with an existing refinery. *Am. Chem. Soc. Div. Fuel Chem. Preprints* 40(1):20–23, 1995.
23. Warren, A., and M. El-Hawagi. Viability of co-liquefying coal and plastic wastes. *Am. Chem. Soc. Div. Fuel Chem. Preprints* 40(1):24–28, 1995.

PART TWO

PLASTICS, POLYMERS, TIRES, AND AUTOMOTIVE WASTES

CHAPTER
FOUR
ADVANCED RECYCLING TECHNOLOGIES FOR PLASTICS

Mark W. Meszaros

BACKGROUND

Plastics have provided consumers and industry with a wide range of benefits for over 50 years. In particular, benefits like light weight, breakage resistance, energy efficiency, and ease in shaping and design have made plastics an especially valuable and increasingly popular packaging material. This increase in popularity, however, has come at a time when the public's interest in recycling is also on the rise.

The perceived benefits of recycling, conserving natural resources, and reducing the need for additional landfill space are widely accepted by the public and have led to a surge in recycling programs. Key to the success of any recycling program is the availability of large, homogeneous, clean streams of high-value postconsumer material that is easily separated from the municipal waste stream. Examples of such high-value recyclables are source separated aluminum cans, clear glass, newsprint, polyethylene terephthalate (PET) soda bottles, and high-density polyethylene (HDPE) milk jugs. However, recycling can be difficult and expensive for other materials, like mixed plastics or paper [1].

With mixed, postuse plastics, there can be as many as six to eight different resins and thousands of different resin grades, blends, and composites. Hundreds of grades of polyethylene are used in packaging alone. Hundreds of additives, colors, and fillers may also be present. Each resin grade, or additive package, may have different properties, such as different melt flow rates and thermal stability. This vast number of resins, resin grades, and additives is needed because of the vast number of packaging requirements, including permeability, barrier protection, flexibility, rigidity, preservation of freshness, and need to retain carbonation, as well as filling the needs of the specific machines that make plastic packaging. However, the diversity of resins makes recycling incrementally difficult. Because of the wide range of properties and additives present in mixed plastics, it is virtually impossible to recycle these materials together into the types of products from which they came. Washing and compounding mixed, postuse plastics typically result in a poor-quality product with broad properties, inorganic impurities or contaminants, and unattractive colors.

A new approach to recycling postuse plastics, called advanced recycling, is under way in the United States and elsewhere and may offer a means of significantly increasing the overall quantity of plastics that can be recycled [2–6]. Advanced recycling technologies may help overcome some of the problems associated with conventional plastics recycling efforts, including the costly separation of different types of plastics, concerns abut the quality of end products, and finding reliable markets for products made from the recycled material.

Advanced recycling recovers the chemical or feedstock components of plastics, which are the building blocks from which new plastics and a variety of other products can be manufactured. This is achieved by converting plastics back to raw materials, either directly to monomers or to the petrochemical feedstocks that are used to make monomers and many other petrochemical products. Many advanced recycling processes accept most plastics and do not require them to be washed or sorted by color or type.

Advanced recycling also eliminates some of the grinding, shredding, and extruding processes used in conventional plastics recycling. Lastly, the chemical and feedstock products from an advanced recycling process are virtually indistinguishable from those made from virgin materials and will have wide market applicability.

Several advanced recycling processes are under development in Europe, Japan, Canada, and the United States and will supplement existing conventional recycling processes. Integration of these new recycling options with the existing plastics recycling infrastructure will provide a long-term solution for plastics by increasing the amount and types of plastics recycled (Figure 1). For example, several large-scale advanced recycling processes are already under construction in Germany that will increase the volume of plastics recycled in that country and result in lower overall recycling costs [7, 8].

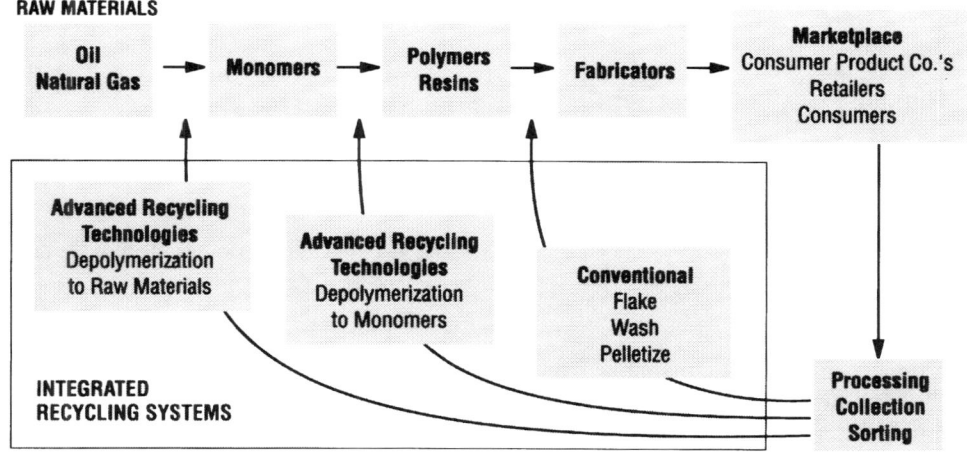

Figure 1 Integrating plastics recycling processes that recover polymers, monomers, and raw materials will greatly increase the amount and type of plastics recycled.

ADVANCED RECYCLING OVERVIEW

Advanced recycling reverses the polymerization process through a thermal (pyrolysis) or chemical (hydrolysis or methanolysis) step. Depending on the process and polymer, advanced recycling regenerates monomers or, in some instances, precursors of monomers, all of which have been derived from natural gas or crude oil. Converting a plastic into monomers creates a feedstock for reconstructing the same generic polymer. Conversion of a plastic into basic petrochemical feedstocks creates a product that can be used to produce new monomers, petrochemical products, or other products like gasoline, heating oil, and asphalt.

The appropriate advanced recycling technique depends on the type of plastic in the waste stream. Plastics such as polyesters (PET), polyamides, and polyurethanes are created via reversible reactions. It is feasible to convert these plastics back to their immediate starting materials, which are valuable and used to remake the same plastic. Other plastics such as HDPE, polypropylene (PP), and polyvinyl chloride (PVC) are created by irreversible reactions. These plastics can be converted into basic petrochemical components through thermal liquefaction or gasification. Further refining and purification take these components and isolate monomers, chemicals, and high-performing fuel ingredients.

The basic concepts of advanced recycling are illustrated in Figure 2. Plastics produced through reversible reactions can be recycled through route A or B; route A generally provides more valuable materials. Plastics produced via irreversible reactions must travel through route B. Regardless of the route, the new

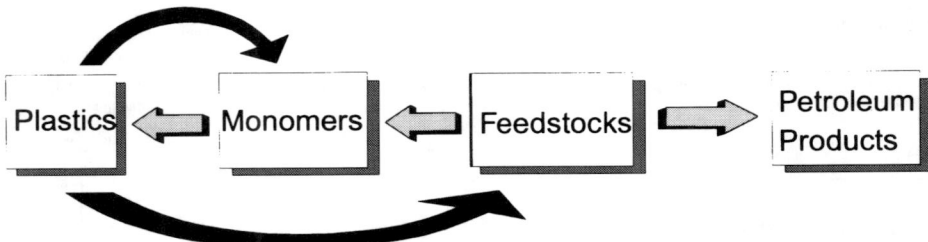

Figure 2 Plastics can be recycled into monomers (route A) or the feedstocks (route B) that are the precursors of monomers and many other petrochemical products.

products are indistinguishable from those produced from virgin raw materials. The following techniques are identified by the recycling route outlined in the figure: chemical and thermal depolymerization (route A) or pyrolytic liquefaction and gasification (route B).

Chemical Depolymerization

Alcoholysis-hydrolysis is one form of chemical depolymerization (route A). Step growth (condensation) polymers, which include polyesters, polyamides, polyurethanes, and polyethers, can be recycled in this manner. Forms of alcoholysis that include methanolysis and glycolysis effectively reduce condensation polymers to monomers or polymerizable oligomers by reversing their preparative chemistry (Figure 3). For years, polyester (PET) producers have recycled industrial PET plant scrap to monomer via methanolysis and glycolysis.

Stopping short of complete depolymerization, glycolysis cuts down long polymer chains (with typical repeat sequences of 150 units) into short-chain oligomers (repeat sequences of 2 to 10 units), which are then easily repolymerized to polymer. Shell uses this technology to produce REPETE, a recycled PET resin for use in packaging.

Complete depolymerization of most condensation polymers is accomplished with excess methanol and heat. For PET, the resulting product consists of dimethylterephthalate (DMT) and ethylene glycol. DMT is purified through several crystallization and distillation steps to yield a pure DMT monomer that is used to produce recycled PET resin, identical to virgin PET resin.

Hoechst Celanese, Eastman Chemical, and Du Pont have all demonstrated this process on a commercial scale [9, 10]. PET feedstock used in methanolysis can be of a lower quality than required for conventional recycling but the product, dimethylterephthalate, is virtually identical to virgin monomer. The DMT is repolymerized into a PET resin that is used for food contact and other applications. Coca-Cola and PepsiCo are now taking advantage of this recycling tech-

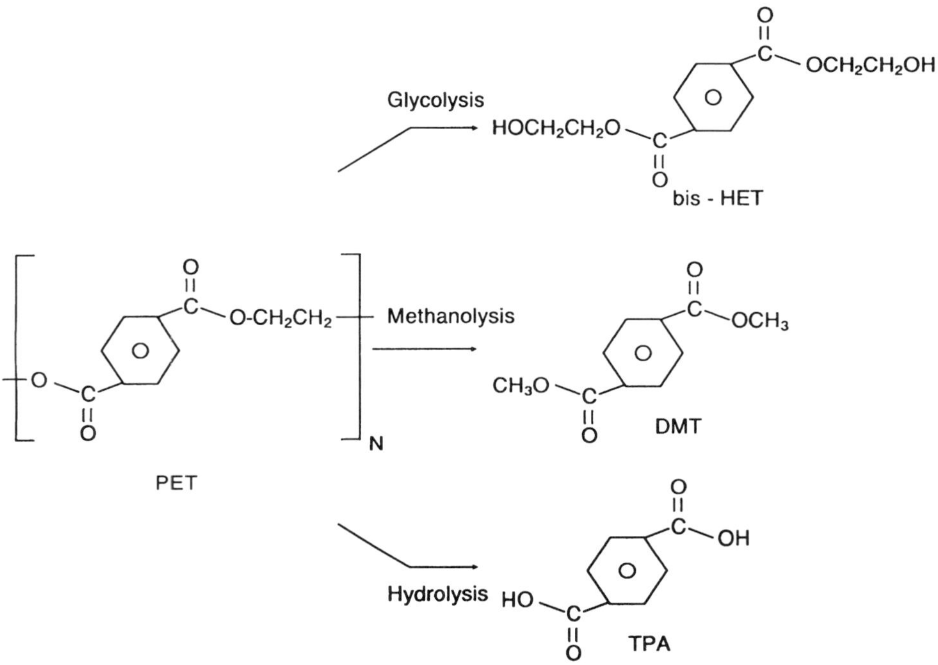

Figure 3 Polyethylene terephthalate (PET) is depolymerized by glycolysis, methanolysis, and hydrolysis.

nology to use recycled PET in soda bottles, having received letters of nonobjection from the U.S. Food and Drug Administration (FDA).

Hydrolysis can also be used to depolymerize PET. Amoco Chemical has a patented process using only water, heat, and pressure to completely depolymerize PET to terephthalic acid (TPA) [11]. The crude terephthalic acid is purified using Amoco's proprietary hydrogenation process to yield a recycled terephthalic acid that is virtually indistinguishable from virgin monomer.

Elsewhere, Innovations in PET, a Smorgon Swig joint venture (Australia), is developing a process that combines glycolysis and hydrolysis to produce terephthalic acid. This patented process uses several glycolysis steps to separate the PET from other polymers and impurities [12]. The resulting bis-hydroxyethylterephthalate (bis-HET) is purified and hydrolyzed to produce terephthalic acid. After further purification, the terephthalic acid is used to make recycled PET resin.

Du Pont is developing an aminolysis process to convert carpets made from either 6- or 6,6-polyamide to monomers. The process is being tested in a pilot plant and will require several years of development before commercialization.

BASF and Allied Signal have also announced plans for developmental projects to recycle polyamides into monomers.

Polyurethane producers have developed several chemolysis processes to recover monomers. Glycolysis, hydrolysis, and aminolysis all convert the polyurethane back to usable starting materials and are being investigated. Recycling polyurethanes is a little more complicated than recycling other condensation polymers because one of the polymer building blocks, the diisocyanate monomer, is not recovered by hydrolysis or alcoholysis. Hydrolysis of polyurethane produces the diamine and the polyol; the diamine can be converted back to the diisocyanate if desired. Nevertheless, polyurethane producers individually and through the Polyurethanes Recycle & Recovery Council (PURRC), a unit of the Polyurethanes Division of the Society of the Plastics Industry, are developing new advanced recycling processes for polyurethanes.

Some thermal depolymerization processes also fall under route A. Heating plastics such as poly(methyl methacrylate), polystyrene, and various acetal homopolymers in the absence of oxygen results in high yields of component monomers. The monomers are isolated, purified, and subsequently used to synthesize new polymers. This technique is in commercial use in India, where scrap poly(methyl methacrylate) is thermally depolymerized to methyl methacrylate, which is purified and then repolymerized.

Pyrolytic Liquefaction

Pyrolytic liquefaction falls under route B. The process is a form of thermal depolymerization but produces a liquid feedstock instead of individual monomers. This liquefaction must be conducted in the absence of oxygen to prevent production of oxygenated products.

The liquid products are similar to naphtha, a valuable refinery stream that serves as the foundation for many petrochemical and plastics feedstocks. The liquids produced by thermal liquefaction of plastics are particularly valuable because they contain a large fraction of hydrogen-rich aliphatic compounds. The aromatic fraction is also rich in benzene, toluene, and xylenes, also valuable feedstocks for the plastics and petrochemical industry.

Although pyrolytic liquefaction yields a wide variety of petrochemical products that are less valuable than the monomers produced under route A, it still has certain advantages. First, pyrolytic liquefaction treats all plastics and not just those that cannot be recycled under route A. Second, liquefaction process conditions can be selected to meet the overall thermal degradative requirements of most plastics. This makes prior separation of the different resins found in waste plastics unnecessary. Removal of food and paper impurities is also unnecessary, because they will be dehydrated and thermally degraded into other chemical products. Finally, this process converts waste plastics into liquid, ready for transport and use as a petrochemical feedstock.

Thermal degradation of plastics to liquid hydrocarbons is a simple process, similar to many common refinery cracking processes in use today. Waste plastics, however, are not crude oil and pose some intriguing challenges, including logistics, economics, and the ability to produce a marketable product. The logistics of getting a stream of solid, postuse plastics to a process facility is a reverse distribution and a materials handling challenge. Advanced recycling, however, is not unique, because it shares these challenges with other recycling processes and materials.

Favorable economics for pyrolytic liquefaction processes will be difficult to achieve because the products have to compete with existing petrochemical operations and virgin feedstocks. Economies of scale will be difficult to achieve because only 3 to 4% of crude oil is converted into plastics and only a fraction of these plastics will ever be brought back to any single facility. Products from advanced recycling processes must compete with virgin products, and at current crude oil prices ($18/bbl) the more valuable refinery streams such as naphtha and gasoline are worth only 7 to 9 cents per pound. To obtain this value, the product stream from an advanced recycling process should be relatively pure and free of nonhydrocarbon components such as halogens and metals. It must also be completely compatible and interchangeable with the current feed streams to the refinery or petrochemical process unit to avoid operating problems. Economics are further stretched if plastics require some preprocessing to facilitate their transportation and introduction into an advanced recycling process unit.

Despite these hurdles, several processes are moving forward on a pilot or commercial scale. These efforts have succeeded because their processes are simple and robust and can accommodate a variable feed stream while producing a consistent product, free of nonhydrocarbon impurities. The process technologies that are most often utilized are batch liquid-phase, fluidized-bed, and kiln processes. These processes are being located at refineries or petrochemical facilities to take advantage of existing infrastructure and markets and to reduce product transportation costs. They are also being evaluated in small industrial parks close to the source of waste plastics to reduce front-end transportation costs and for possible integration with conventional recycling facilities. Local solid waste management conditions, available petrochemical facilities, and local regulations will dictate which recycling technologies are most appropriate for different locales.

Pyrolytic Liquefaction Processes

BASF (Germany) has announced plans to invest DM300 million ($176 million) in a 300,000 MT/year full-scale plant at their Ludwigshaffen petrochemical facility [8, 13, 14]. The facility, planned to be on stream in 1996, will accept a preprocessed agglomerate of postconsumer plastics from Germany's Duales System Deutschland (DSD) Green Dot system. This agglomerate has most of the nonhydrocarbon components removed and is a free-flowing material, which greatly facilitates material movement. The BASF process utilizes a two-stage, batch liquefaction process. The first stage is a relatively low temperature (300°C) step that melts the plastics. More important, at this low temperature PVC will be dehydrohalogenated and release hydrogen chloride gas. The gas is captured and used elsewhere at Ludwigshaffen. The second stage, also a batch liquefaction step, depolymerizes the plastics at 400°–500°C to a mixture of gases (20–30%) and liquids (60–65%). The products are partially refined by distillation before being used on site as feedstocks. From each kilogram of plastics, the process is expected to produce some 900 g of petrochemical feedstocks. BASF will receive DM350 per metric ton from the DSD to recycle the plastics.

Veba Oel (Germany) is operating a waste plastics to feedstocks plant in Bottrop. The recycling plant, which was converted from a coal hydrogenation demonstration plant, has an initial capacity of 40,000 MT/year, but plans are under way to expand capacity to 120,000 MT/year by 1996 [6, 8]. Veba Oel, like BASF, accepts mixed plastics preprocessed into free-flowing agglomerates from the DSD Green Dot system. The plastics are prepared for the hydrogenation unit by using a liquid-phase process in which vacuum residue oil is mixed with the plastics and heated to 300°–400°C. At these temperatures, the plastics dissolve, dehydrohalogenate, and experience some cracking. After light hydrocarbons and hydrogen chloride are removed, the liquified plastics are hydrogenated in liquid-phase reactors at 450°–490°C under a hydrogen partial pressure of 150–250 psi. The organic products are passed through a hydrotreater unit operating at the same pressure but a somewhat lower temperature than the liquid-phase unit. The second hydrotreatment uses a fixed catalyst bed and produces a high-quality synthetic crude oil, free of halogens and other heteroatoms. Veba is receiving about DM757/MT from the DSD to recycle plastics but predicts that this cost could be decreased by half by 1996.

A consortium led by BP Chemicals (England) has developed a fluidized-bed process to recycle postconsumer plastics [6, 15]. DSM, Petrofina, Elf-Atochem, and Enichem are also members of the consortium and are contributing to the commercial development. The first step of the process is a low-temperature liquefaction step that removes some chlorine as HCl and also reduces the molecular weight of the plastics to assist in feeding them to the fluidized-bed unit. The fluidized-bed unit operates at 400°–600°C and contains sand and calcium oxide to absorb halogens. Plastic streams containing up to 10% PVC have been successfully processed. The product, a naphtha-like intermediate, is produced in 90% yield and can be fed directly into a steam cracker or fluidized cat cracker without further processing. BP estimates that a 25,000 MT/year unit will cost $30 million to $50 million and processing costs at this scale are about $300/MT.

BP Chemicals has demonstrated that the liquid resulting from the thermal cracking of mixed plastics is an excellent feedstock for steam crackers or cat crackers (Figure 4) [15]. In the steam cracker, the plastics-generated feed produced higher yields of ethylene (34% versus 28%), propylene (17% versus 15%), and butylene (12% versus 7%) than the traditional naphtha feed. In the cat cracker, the plastics-generated feed resulted in an 86% yield of a naphtha-grade product versus 62% yield from vacuum oil. These results clearly demonstrate that there is value in the products from the liquefaction of plastics.

At the Energy and Environmental Research Center (EERC; North Dakota), a project funded by the American Plastics Council (APC) and the U.S. Department of Energy (DOE) is also evaluating fluidized-bed reactors [16]. This process is similar to the BP process, but the cracking of plastics in a fluid bed is being evaluated without prior liquefaction. In recent trials, ground postconsumer plastics were successfully recycled using a 100 lb/h fluidized-bed reactor (600°C, CaO/sand bed).

Results from the EERC studies showed a synergistic effect from the thermal cracking of plastic mixtures; mixtures of plastics depolymerize at lower temperatures and yield more naphtha-range products than single resin streams. For example, a 50:50 mixture of polypropylene and polyethylene cracked at a lower temperature than polyethylene does alone and produced a narrower distribution of products than either polypropylene or polyethylene would individually. Polypropylene normally cracks at a lower temperature than polyethylene (Figure 5), and the consensus is that the reactive species from polypropylene depolymerization catalyzes the cracking of polyethylene, which in turn hinders further cracking of the polypropylene fragments to lighter products by chain termination steps. The result is that both polymers depolymerize at the polypropylene cracking temperature and produce more attractive products (Figure 6).

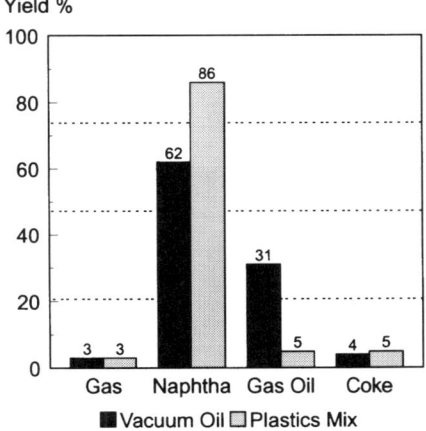

Figure 4 Steam cracking and catalytic cracking yields from polymer cracking product.

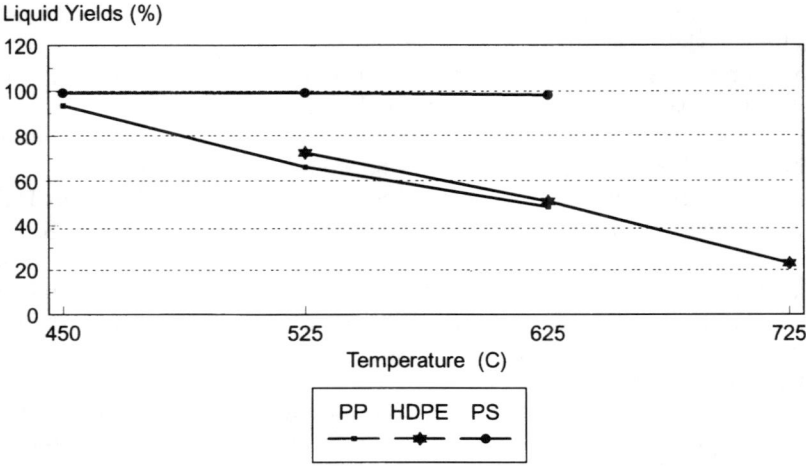

Figure 5 Liquid yields from pyrolytic liquefaction of plastics in fluidized bed.

Fuji (Japan) is evaluating small pyrolysis plants with a capacity of around 10,000 MT/year that can be used in light industrial parks [17]. Here, plastics are melt extruded and then further liquefied in a stirred tank at 200–300°C. HCl is also removed during these first two steps. The liquefied plastics are cracked at 400°C and passed through a guard bed to remove any remaining halogens. The last step uses a zeolite (ZSM-5) catalyst bed for catalytic cracking to lower

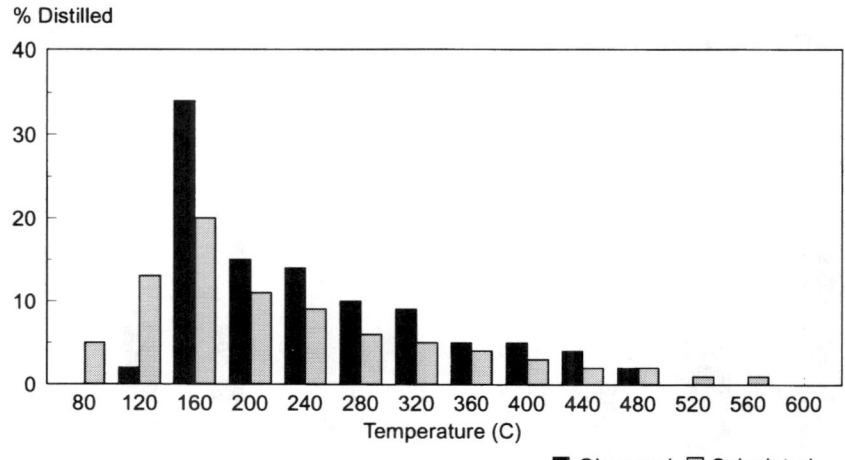

Figure 6 Observed and calculated yields from pyrolytic liquefaction at 525°C of a 50:50 mixture of polypropylene and polyethylene.

molecular weight aliphatics and olefins that are also converted into aromatics over the ZSM-5. The overall liquid yield is about 80%. The product is a mixture of gasoline (50%), kerosene (25%), and gas oil (25%). Two Fuji units are operational in Aioi and Okegawa, Japan. The process is also under evaluation in Canada and Germany.

Conrad Industries (Washington) is working with APC to demonstrate the feasibility of utilizing a tire recycling process to recycle postconsumer plastics [18, 19]. In the Conrad process, shredded plastics are added directly into an auger kiln reactor (500°–600°C) and depolymerized to gases that are condensed to a naphtha and gas oil range product in 80% yield. Noncondensable gases are recycled back into the process to generate heat for the process. The liquid product has been shipped to the Lyondell-Citco refinery and found to be an attractive feedstock for its petrochemical operations. Conrad recycling units are smaller (10,000 ton/year, $4–5 million capital costs) and consequently do not need to be sited at a petrochemical facility. (Additional details on Conrad and the results of a comprehensive parametric study are featured in a later section of this chapter.)

Pyrolytic Gasification

Pyrolytic gasification falls under route B. Gasification processes usually require harsh conditions, including high temperatures or catalysts, and produce an olefin-rich hydrocarbon gas or a synthesis gas product. Thermal or steam cracking of plastics at elevated temperatures produces ethylene and propylene in good yields but requires a sophisticated distillation to separate and purify the olefins. Synthesis gas, a mixture of carbon monoxide and hydrogen, is produced by the partial oxidation of plastics using high temperatures and a specific amount of oxygen. These gases are valuable feedstocks for the manufacture of ammonia, methanol, methyl *tert*-butyl ether (MTBE), and acetic acid.

Battelle (Ohio) has developed a pyrolytic gasification process that reduces a commingled waste stream of low- and high-density polyethylene, polystyrene, and polyvinyl chloride into a mixture of 40% ethylene, 27% methane, 17% hydrogen, and other fractions [20]. This process uses a circulating fluid bed to heat the plastics rapidly to around 850°C. Battelle claims that the short residence times of less than 2 seconds and the extremely high heating rates result in cleaner cracking of the plastics without rearrangements or recombination. Battelle has not announced any plans to develop this process further.

Several pyrolytic liquefaction processes also produce high yields of olefins when operated at elevated temperatures. The Energy and Environmental Research Center has produced good yields of ethylene and propylene in their fluidized bed process at 625 to 700°C from HDPE and PP [16] (Figure 7).

The difficulty with producing olefins (ethylene, propylene, and butenes) is that they are difficult to separate and transport. The U.S. petrochemical industry

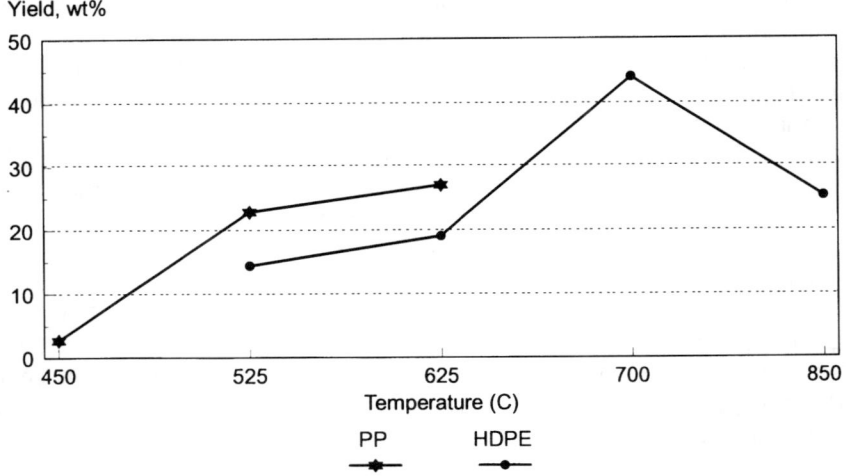

Figure 7 Olefin yields from high-temperature gasification using EERC fluidized bed.

has achieved a global competitive advantage in plastics manufacture partially because the majority (>94%) of the steam crackers to produce olefins and the cryogenic distillation units to separate and purify the olefins to polymer-grade monomers are located along the Gulf Coast. These facilities are interconnected with downstream polymerization facilities. Pyrolytic gasification of plastics to olefins will probably make sense in the United States only if it is done along the Gulf Coast and is integrated with an existing olefin production facility. In Europe, olefin and polymer production facilities are more evenly distributed throughout the continent and may provide more opportunities for pyrolytic gasification to olefins.

The Texaco Gasification Process is a commercially proven technology that has been used for more than 40 years to convert complex or heavy hydrocarbons into synthesis gas. Texaco recently performed successful gasification tests on several postconsumer plastics streams, including a sample from the DSD system and a sample containing 30% polyvinyl chloride (PVC) [21]. Conversion of these plastics to synthesis gas was in excess of 99%. Minimal preprocessing of the feed is required, as Texaco has developed a liquefaction step to prepare the plastics for the gasification process.

CONRAD ADVANCED RECYCLING PROCESS

The American Plastics Council (APC) [22] is working with Conrad Industries, Inc., of Chehalis, Washington [23] to demonstrate the viability of the Conrad process for recycling plastics back into liquid petrochemical feedstocks. The project is part of APC's overall program to increase the recycling rate for plastics and to develop or improve recycling technologies. The Conrad process was chosen because of its simplicity, versatility, and safety features. More important, because of its size and low cost it also allows a rapid progression from pilot-scale to commercial-scale process units.

The objective of the program is to determine the technical and economic feasibility of recycling plastics into liquid petrochemical feedstocks. To meet this objective, a series of parametric studies has been conducted to examine how various types of plastics perform under different operating conditions. Studies of various chlorine capture processes are also being conducted. Data acquired from these studies have been used to define process conditions for commercial-scale demonstrations, identify potential design improvements for future units, and develop preliminary process economics.

Conrad Recycling Process

The Conrad recycling process uses an auger kiln reactor that applies heat to plastics and/or tires in the absence of oxygen to produce liquid petroleum, solid carbonaceous material, and noncondensable gases [24] (Figure 8). The unit was originally designed to recycle tires but with minor modifications is well suited for plastics. Developmental studies on plastics and tires are ongoing at the Chehalis facility, where two Conrad process units are operational. Parametric studies of plastics have been performed primarily on a 200 lb/h pilot unit, but a 2000 lb/h commercial-scale unit is also available. Both units are fully automated and process data are downloaded to a computer on a continuous basis.

The feed system for the pilot unit is designed for granulated or pelletized plastics, consisting of a simple rotary air lock followed by an auger feeder. Feed systems to handle whole plastic bottles, films, and other plastic components are under investigation and will be incorporated in future Conrad recycling units. The horizontal auger kiln, or retort, is of a proprietary design and is well suited for the depolymerization of plastics to volatile products. The gases exit at the top of the discharge end of the retort and pass through several condensers. Gases that are not condensed are used within the system to generate heat for the process. Solid products, such as carbonaceous or inorganic materials, exit at the bottom of the discharge end of the retort to the solids collection drum.

The operating conditions for the auger kiln are adjusted to maximize liquid product yields while still providing enough gas to sustain fuel requirements for the heating of the retort. At current operating conditions, liquid products represents about 70 to 80% of the final output of this system. The carbon produced is of such a small yield (1–3%) and low quality due to contamination from inorganic impurities that it will have limited markets. The liquid product is

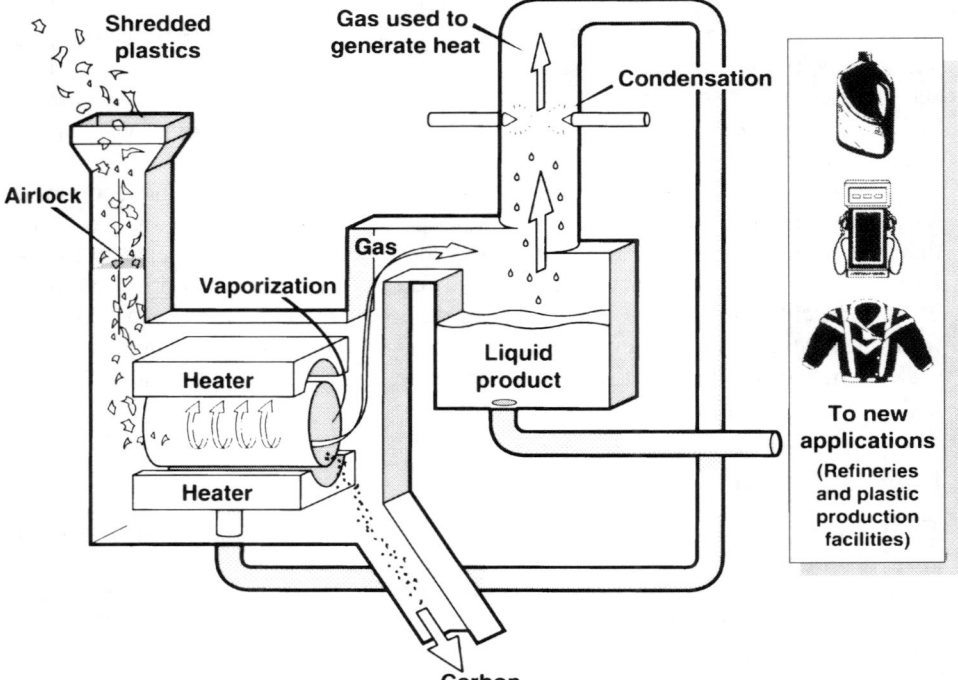

Figure 8 Plastics are recycled into liquid petroleum, carbon, and noncondensable gases in the proprietary Conrad recycling process.

shipped to refineries and plastics production facilities for conversion into feedstocks for products such as synthetic fibers, new plastics, and other petroleum-based derivatives.

Feed preparation is simplified in the Conrad process because minimal size reduction, washing, and removal of nonplastic contaminants are required. Plastics are introduced into the retort without oxygen through a rotary air lock if the plastics are granulated or by using a ram feeder (e.g., V-ram, reciprocating, or conical auger) if the plastic products are still in their original form. Inorganic contaminants do not disrupt the process and ultimately exit the unit with any unreacted materials or carbonaceous products. The plastics do not have to be fluidized or molten to enter the process, resulting in reduced preprocessing requirements and costs.

The Conrad process allows a variable residence time in the reactor, producing a more uniform product distribution. The plastics initially melt upon entering the hot retort, and the auger keeps the molten mass moving. The molten plastics continue to be heated until carbon-carbon bond cleavage (depolymerization) begins and will continue until volatile products are produced. The gaseous products are swept out of the reactor and nonvolatile fragments continue to be heated until they become volatile. Plastics that are easily depolymerized (e.g., polystyrene or polypropylene) are quickly cracked to volatile fragments and removed from the reaction vessel before undergoing further significant cracking, thus keeping the molecular weight of the products as high as possible. Plastics that require more energy to cleave their carbon-carbon backbone (e.g., polyethylene) are either catalyzed by reactive species (radical or carbonium ions) generated by the easily cracked polymers or remain in the retort until sufficient energy is applied to effect depolymerization.

The Conrad unit has also proved to be both a robust and efficient unit. During the parametric study, the system ran continuously for 5-day periods. It takes between 8 to 12 hours for the unit to warm up and line out, after which minor changes in feed rate, auger rotation, and oven temperatures are easily accomplished. Additional work on a versatile feed system is required, but once the plastics entered the reaction vessel, the auger kiln easily handled them.

Conrad Parametric Study

An 18-month parametric study has been under way to assess the Conrad recycling process [19]. The objectives of this study were to identify process bottlenecks, develop operating parameters, and begin to assess product value and markets. A base feed mixture of 60:20:20 high-density polyethylene/polypropylene/polystyrene (HDPE/PP/PS) was chosen as representative of the major nonrecycled constituents found in the postconsumer plastic stream and was used during the start-up and initial phases of the study. Off-spec resin pellets were used as feed material throughout the first phases of the study to eliminate feed composition variation. It also simplifies the preparation of mixtures and feeding the unit. The last phase of the program is focusing on using postconsumer plastics to gain an understanding of the effect that contamination and variable feed stream have on the process.

Base feed. Overall, liquid product yields and composition from the 60:20:20 HDPE/PP/PS base feed exceeded expectation. However, a delicate balance does exist between temperature, feed rate, residence time, and product yields (Table 1). Lower temperatures produce a high-quality naphtha-grade stream and high yields of paraffin waxes (which result from incomplete depolymerization of the plastics, especially polyethylene). Reducing the feed rate and slowing the auger rotation result in a reduction of wax formation and increased yields of liquid products. Better insulation and more efficient heating of the discharge end of the retort also substantially reduce the yield of waxes in the solids collection drum. Because of these challenges, reported liquid yields from the early phases

Table 1 Product yields from depolymerization of base feed[a] at various reaction temperatures

Oven temperature (°F)[b]	Liquid yield (wt %)[c]	Gas yield (wt %)	Solid yield (wt %)
1450	28	64	8
1300	48	51	1
1200	73	27	—
1100	79	21	—

[a]Base feed is 60% HDPE, 20% PP, and 20% PS by weight.
[b]Retort temperatures are about 200°F cooler than oven temperature.
[c]Combination of liquids and paraffin waxes from solids collection drum.

of the parametric study are the combined liquid product yields and paraffin wax products yield found in the solids collection drum. Paraffin wax yields were determined by xylene extraction from the products in the solids collection drum.

Part of the difficulty in determining the optimum combination of temperature, feed rate, and auger speed is inherent in the design of the pilot unit. The Conrad pilot unit is a single-pass retort with a relatively short kiln length (~6 foot heated zone). Commercial-scale Conrad units consist of two larger retorts in series, which substantially increases residence time. Nevertheless, valuable engineering and design data have been collected from the pilot unit that will undoubtedly improve the efficiency of the Conrad recycling process.

Gas chromatographic analyses of the liquid and gas products from the first phase of the parametric study are shown in Figure 9 and Table 2. The liquid products from the first phases of the parametric study were combined and 6000 gallons of product oil were shipped to the Lyondell-Citco Refinery in Houston, Texas and used as a feed in its coker unit. The coker unit was chosen because it minimizes risks to the refinery and the light-end gases from the coker go directly to several monomer production facilities, including Mobil's steam cracker.

LDPE, PS, and PET-rich feeds. To increase the understanding of process capabilities and product yields, studies were performed using the base feed spiked with either LDPE, PET, or additional PS. Again, only off-spec resin pellets were used to simplify the studies and reduce the variables. Liquid product yields are shown in Figure 10.

An LDPE-rich feed mixture containing 20:48:16:16 LDPE/HDPE/PP/PS behaved similarly to the base mixture and resulted in similar product yields and composition. Slightly higher yields of C_{12}–C_{15} aliphatics were observed and higher molecular weight aliphatics were decreased. No operating difficulties were observed.

A PS-rich feed mixture containing 48:16:36 HDPE/PP/PS also behaved similarly to the base mixture and resulted in similar product yields and com-

Figure 9 Gas chromatogram of liquid product.

position. In the liquid product, styrene yields almost doubled and slightly lower yields of higher molecular weight ($>C_{12}$) aliphatics were observed, as would be expected. Ethylbenzene yields did not increase significantly with the increased polystyrene level, suggesting that ethylbenzene may not result exclusively from polystyrene depolymerization but rather from the cyclization and dehydrogenation of various aliphatic compounds at elevated temperatures.

A PET-rich feed mixture containing 20:48:16:16 PET/HDPE/PP/PS did not behave like the base mixture. It was difficult to run the mixture at the lower temperatures (1100–1200°F) because of extensive production of solids. Upon analysis, the solids were identified as >95% terephthalic acid. The remaining 5% of the solids were found to be mono- and bis-hydroxyethyl esters. The solids were easily filtered, but because of the design of the pilot unit a substantial portion of the solids remained in the cooling tower, product tanks, and piping. Terephthalic acid yields were estimated at around 15 mol % at 1200°F. At higher temperatures, the PET and/or terephthalate moieties decarboxylate to produce CO_2 and aromatic products. There must be enough water encapsulated in the PET and other resins to supply the water for hydrolysis, and at the lower tem-

Table 2 Gas chromatographic analysis of the oil and gas fractions from the depolymerization of base mixture[a]

Oven temp. (°F)	1300	1200	1100
Auger temp. (°F)[b]	1095	980	895
Partial oil analysis (wt %)			
Benzene	7.0	2.3	1.1
Toluene	20.0	11.0	9.1
Ethylbenzene	9.0	5.7	4.8
Styrene	20.0	17.0	14.7
C_{10}	10.0	9.4	6.8
C_{11}–C_{15}	14.5	16.6	13.1
C_{16}–C_{20}	8.6	9.9	14.5
C_{21}–C_{25}	3.0	3.5	6.4
C_{26}–C_{30}	1.7	1.7	4.5
C_{31}–C_{40}	1.1	1.5	4.2
Partial gas analysis (vol. %)[c]			
H_2	4.0	2.6	2.2
Methane	18.0	13.6	10.5
Ethylene	19.4	18.1	13.8
Propylene	21.4	21.8	21.0
Isobutylene	5.0	6.1	6.8

[a] Paraffin waxes found in the liquid product tanks are included in this analysis. Waxes from the solids handling drum are not part of this analysis.

[b] Auger temperature was measured via thermocouple inserted inside the auger shaft at the midpoint of the retort.

[c] Normalized after subtracting out N_2 and O_2.

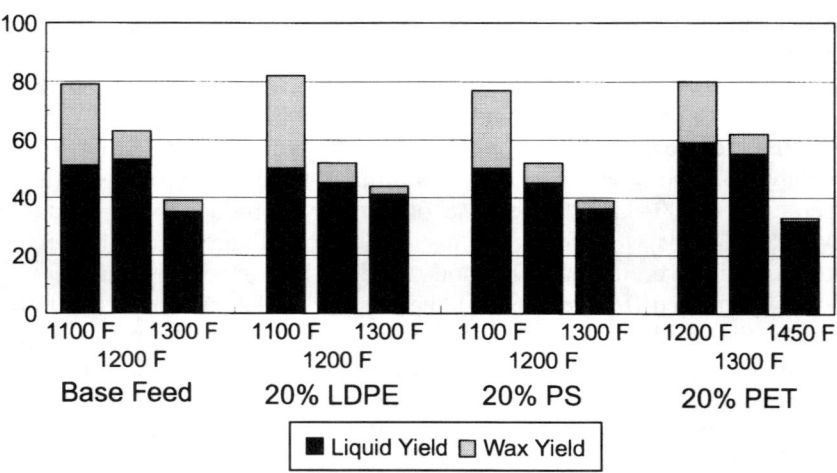

Figure 10 Product yields from Conrad parametric study.

peratures the terephthalic acid sublimes out of the retort before it undergoes decarboxylation. Another possibility is the PET undergoes an ene reaction to produce an acid group and a vinyl ester.

Polypropylene feed. Polypropylene (100%) was depolymerized at a retort temperature of 950°F to give high yields of aliphatic (48%) and aromatic (40%) products (Table 3, PP). Single unsaturated, low-molecular-weight ($<C_{15}$) aliphatics were a major component of the liquid products, but ethylbenzene (19.1%) was the major individual product. High levels of propylene (23%), isobutylene (21.4%), and C_5 gases (13%) were found in the gas products (Table 4, PP). The mechanism for converting branched aliphatics to ethylbenzene is unclear, but ethylbenzene is a major product in almost all cracking runs with polyolefins.

Table 3 Gas chromatographic analysis of various depolymerization runs

Component	PP[a]	PS[b]	PVC[c]	PVC/PET[d]	PCR-1[e]	PCR-2[f]
Aliphatics	47.9	7.7	41.3	42.3	44.4	22.7
$<C25$ saturated	4.3	2.5	6.7	6.7	9.6	6.0
$<C25$ unsaturated						
One unsaturation	25.0	2.9	23.5	23.1	23.3	10.5
Two unsaturations	10.3	1.8	5.3	6.7	5.9	4.2
$C25+$ Aliphatics	6.9	0.5	4.7	4.4	4.0	1.2
Aliphatics by carbon number						
$<C_{10}$	22.1	0.7	18.2	19.8	19.4	12.0
C_{11}–C_{15}	10.1	1.4	10.9	9.2	13.7	5.2
C_{16}–C_{20}	4.8	3.8	5.1	6.1	4.9	2.9
C_{21}–C_{25}	3.5	1.4	2.3	2.8	2.2	1.4
C_{26}–C_{30}	2.4	0.3	1.6	1.8	1.3	0.4
C_{31}–C_{35}	1.5	0.1	1.0	1.0	0.8	0.2
C_{36}–C_{40}	1.3	0.0	0.8	0.6	0.7	2.0
C_{40+}	1.4	0.0	0.9	0.3	0.7	0.2
Aromatics	40.4	91.0	49.2	54.9	49.9	73.1
Benzene derivatives						
Benzene	2.0	0.7	2.0	4.9	2.9	7.6
Toluene	1.3	10.0	7.9	837.0	7.1	22.0
C_2-benz.	22.5	67.6	28.1	26.9	25.0	27.4
C_3-benz.	2.8	7.7	5.4	5.9	6.4	6.9
C_{4+}-benz.	6.6	0.9	1.4	3.9	4.3	3.9
Naphthalenes	2.4	0.5	1.5	2.7	2.1	3.2
Phenanthrenes	2.7	3.6	1.4	1.8	2.1	2.1
Unidentified compounds	11.8	1.3	9.5	2.8	5.8	4.2

[a] 100% polypropylene.
[b] 100% polystyrene.
[c] 3% PVC in a 60:20:20 HDPE/PP/PS base mixture.
[d] 1% PVC, 5% PET in a 60:20:20 HDPE/PP/PS base mixture.
[e] Postconsumer plastic containers; retort temperature was 977°F.
[f] Postconsumer plastic containers; retort temperature was 1058°F.

Table 4 Gas chromatographic analysis of gas products[a]

Component	PP	PS	PVC	PVC/PET	PCR-1	PCR-2
Hydrogen	5.3	10.5	5.2	5.3	7.1	8.3
Carbon monoxide	1.1	39.5	1.1	0.7	2.0	2.8
Carbon dioxide	0.2	16.1	0.0	0.0	0.3	0.7
Methane	11.9	9.4	11.9	14.9	13.0	16.1
Ethylene	14.7	9.6	15.4	17.6	15.1	15.7
Ethane	9.3	1.2	4.0	8.7	8.8	8.5
Propylene	19.9	2.9	21.7	20.6	19.0	17.9
Propane	3.2	0.0	4.9	2.5	2.9	2.2
C_{4s}	21.4	2.9	24.2	20.3	20.5	16.8
C_{5s}	4.1	0.9	4.8	1.7	2.1	1.5
C_{5+}	8.7	7.0	6.9	7.8	9.3	9.5

[a] Volume %, normalized after subtracting N_2 and O_2. All samples are the same as in Table 3.

Polystyrene feed. Postconsumer and postindustrial polystyrenes have been used as a feed and produced very high yields of liquids that are rich in styrene. Postindustrial polystyrene (8173 pounds over 2 days) were fed to the Conrad pilot unit, resulting in 7495 pounds of liquid product for a 91.7% overall liquid yield. Only 70 pounds of solids were collected and the noncondensable gas yield was estimated to be 280 pounds. Gas chromatographic (GC) analysis of the liquid and gas products from this run is shown in Tables 3 and 4 as PS.

Postconsumer food service polystyrene was evaluated and gave slightly lower liquid yields but similar styrene yields, presumably due to the higher levels of contamination. The product mixture is being evaluated to determine its value as a feed for styrene production.

Chlorine capture studies. The second phase of the parametric study focused on halogen capture. Any postconsumer plastic stream will contain some halogens in the form of polyvinyl chloride, polyvinylidene chloride, brominated flame retardants, halogenated additives, food waste, or salt. Therefore, two issues must be addressed. First, the gas stream resulting from the depolymerization of plastics must be scrubbed to remove any halogenated gases (HCl, HBr, Cl_2, Br_2, etc.) to satisfy emission controls. Second, halogens in the liquid product must be minimized to increase its value and marketability.

Parallel APC studies at the Energy and Environmental Research Center (EERC) [16] and Energy and Environmental Research Corporation (Irvine, CA) [25] showed that feeds containing as little as 3% PVC produced up to 10,000 ppm chlorine in the liquid products. Most of the organochloride formation is thought to occur during the condensation step, not during depolymerization. Also, the aliphatic carbon-chlorine bond is fairly weak and is probably facile at reaction temperatures. Therefore, if the chlorine can be captured prior to condensation, the organochloride level in the liquid product should be minimized.

The first chlorine capture process studied attempted to capture the chlorine from the hot effluent gas prior to the condensers. Hot calcium oxide fixed beds placed between the retort and the condensers were evaluated. Initial tests on a bench-scale auger kiln showed that a plastics mixture containing 3% PVC produced liquid products with <25 ppm chlorine. However, difficulties arise when PET is present, because the terephthalic acid and CO_2 produced during PET depolymerization react with the calcium oxide and lead to premature plugging of the calcium oxide beds.

A second approach captured the chlorine while it was still in the retort. Calcium oxide or hydroxide (hydrated lime) was added directly into the retort along with the plastic feed. This approach was effective even when PET was present in the feed. Several months of the parametric study were devoted to optimizing this approach, and the results are shown in Table 5. Total chloride concentrations in the retort effluent, heavy oil product, and light oil product were measured by Dohrmann microcoulometer.

Mostly through trial and error, a calcium hydroxide loading of 10 wt % was found to give good results over a wide range of feed compositions. If only PVC is present, a lower loading is effective, but PET can consume large quantities of calcium hydroxide. Total chlorine levels of the liquid products were consistently less than 50 ppm and frequently less than 20 ppm.

Postconsumer plastics. Several postconsumer plastic streams have been evaluated as feedstock for the Conrad process. The first stream used as a feed was from a curbside collection program. The plastics were analyzed and found to be 8–10% PET, 0.5–2.0% PVC, 75–80% HDPE, 8–10% PS, and 5–7% PP. Plastics were fed into the system at a rate of 100 lb/h and hydrated lime was added at a rate of 10 lb/h. Oven temperature was set at 1250°F and the auger temperature was between 1050° and 1100°F.

Table 5 Summary of chlorine capture runs

Run	PVC[a] (%)	PET[a] (%)	Feed rate (lb/h)		Temp (°F)	Cl (ppm)
			Feed	Ca(OH)$_2$		
25	3.0	0.0	80	0	1215	2500–3500
34	0.5	2.0	134	10	925	4–19
35	1.0	5.0	120	12	950	4–12
36	1.0	10.0	120	12	975	6–13
41a	3.0	0.0	120	9	900	90–140[b]
41b	3.0	0.0	120	9	900	120[c]
41b′	3.0	0.0	120	9	900	80[d]
41c	3.0	0.0	120	12	900	60–70

[a]Percent PVC/PET in base mixture.
[b]Average from run.
[c]Individual sample from run, analyzed unwashed.
[d]Same sample as 41b, but washed to remove inorganic chlorides.

Two GC analyses of the postuse plastics are shown in Tables 3 and 4 (see PCR-1 and PCR-2). The auger temperature for PCR-1 was 977°F and the liquid products were fairly evenly split between aliphatic and aromatic compounds. At an auger temperature of about 1060°F, the component mix changes dramatically to aromatic compounds with toluene and benzene more dominant products. Unfortunately, the Conrad pilot plant is not equipped to measure liquid yields over a short period of time, and liquid yield cannot be correlated with this abrupt change in product mix. These results, however, are consistent with results from the fluidized-bed pyrolysis program at EERC that also showed abrupt increases in aromatic products in this temperature range [16].

Economics. The economics for the Conrad recycling unit are still unclear. Economic driving forces such as labor rates, scale of economy, transportation costs, and feedstock costs all dramatically affect the overall economics. Small-scale (<20 MM lb/year), stand-alone units may require tipping fees around 5 cents per pound to break even. Integrating a Conrad recycling unit with a conventional recycling facility, however, may allow existing recyclers to lower overall costs while increasing revenues. Existing recyclers may be able to accept a lower quality bale of plastics, extract resins that have good market value via conventional recycling, and recycle the remainder through a Conrad unit. Feedstock, landfilling, and transportation costs may be reduced and revenues increased by the sale of products from the Conrad process.

CONCLUSION

Advanced recycling processes are inherently uneconomic because their products have to compete with virgin products that are now produced in efficient, large-scale facilities using inexpensive feedstocks. Nevertheless, advanced recycling processes may ultimately provide the most economical means of handling large volumes of mixed plastics. Germany, which has a federal mandate to recycle 60% of all plastic packaging, is leading the way in developing processes for recycling plastics into feedstocks. The German projects are all large-scale processes, integrated with existing refinery or petrochemical facilities. Because of the framework of the petrochemical industry in the United States, smaller process units that convert plastics into liquids on a local level may be more attractive. Detailed studies of the Conrad advanced recycling project have led to a better understanding of the viability of advanced recycling and the technical and economic hurdles that remain. Both advanced and conventional plastics recycling are viewed by the plastics industry as important elements of an integrated solid waste management system that will introduce new products to commerce, conserve resources, and reduce landfill-bound wastes.

REFERENCES

1. Liedner, J. *Plastics waste, recovery of economic value.* New York: Marcel Dekker, 1981.
2. Menges, G., R. Fischer, and V. Lackner. *Intern. Polymer Processing VII,* 4:291, 1992.
3. Leaversuch, R. D. *Modern Plastics,* July:40, 1991.
4. Ehrig, R. J. Development and Commercialization of Tertiary Plastics Recycling. Spectrum report 26-1, Decision Resources Inc., Burlington, MA, July 1992.
5. Randall, J. C., M. W. Meszaros, A. A. Adams, and J. E. Lohr. *Modern Plastics Encyclopedia,* mid-December:54, 1992.
6. *Chemical and Engineering News,* October 4:11, 1993.
7. Mapleston, P. *Modern Plastics,* November:58, 1993.
8. *European Chemical News,* April 25:37, 1994.
9. Naujokas, A. A., and M. Ryan. U.S. Patent 5 051 528, 1991.
10. *Food and Drug Packaging,* February:49, 1991.
11. Rosen, B. I. U.S. Patent 5 095 145, 1992.
12. West, S. M. WO Patent 93/23465, 1993.
13. *ChemicalWeek,* March 2:20, and April 27:24, 1994.
14. *Chemical and Engineering News,* March 28:19, 1994.
15. *European Chemical News,* February 15:27, 1993.
16. Thermal Recycling of Plastics, Final Report to the American Plastics Council. Energy & Environmental Research Center, Grand Forks, ND, March 1994.
17. *Japan Chemical Week,* May 31:6, 1990.
18. Powell, J. *Resource Recycling,* May:52, 1993.
19. Meszaros, M. W. Advances in plastics recycling: Thermal depolymerization of thermoplastic mixtures. Presented in part at the 208th National Meeting of the American Chemical Society, Washington, DC, 1994.
20. Paisley, M. A., and R. D. Litt. U.S. Patent 5 136 117, 1992.
21. Curran, P. F., and K. A. Simmonsen. Gasification of mixed plastic waste. Presented at the 8th Annual Recyclingplas Conference, Washington, DC, June 1993.
22. The American Plastics Council is a ChemStar program as a joint initiative with the Society of Plastics Industry, Inc. and comprises 25 major resin producers. For more information, contact the American Plastics Council, 1275 K Street, NW, Washington, DC 20005 (202-371-5319).
23. For more information on Conrad Industries and the Conrad recycling process, contact Bill Conrad at Conrad Industries, Inc., 121 Melhart Rd., Chehalis, WA 98532 (206-748-4924).
24. Oeck, R. C. U.S. Patent 4 412 889, 1982.
25. Kryder, G. (EER), and G. Mackey (Dow), personal communication, 1994.

CHAPTER
FIVE

RECYCLING OF POLYMERS FROM AUTOMOBILE SHREDDER RESIDUE

B. J. Jody, E. J. Daniels, and A. P. S. Teotia

INTRODUCTION
Background

In 1991 the automobile shredder industry supplied more than 10 million tons of recovered ferrous scrap for use in the iron and steel industry. The single largest source of that raw material is obsolete automobiles, although other source materials (such as obsolete refrigerators, washing machines, and other appliances, commonly referred to as "white goods") also contributed to the supply of recovered metals. On the basis of available data and vehicle statistics [1–3], we estimate that in 1991 the recycling potential from shredded automobiles was about 10.3 million tons of ferrous scrap, 0.8 million tons of nonferrous metals, 0.8 million tons of plastics/composites, and 2.1 million tons of other materials. Because only metals were recycled, the shredder industry was left with about 3 million tons of residual material requiring disposal. This is in addition to a smaller quantity of waste that resulted from recovering metals from shredded white goods. Almost all of this waste was disposed of in landfills. The cost of landfilling, which has increased dramatically over the past few years, combined with more stringent regulations, long-term liability concerns, and shrinking landfill space, dictates that viable disposal alternatives be developed to maintain a healthy metals recycling industry. The impact of the landfill constraints and cost will also be exacerbated by the increasing nonmetal content of vehicles, primarily plastics. For example, the consumption of plastics/composites in domestic automobiles increased from 176 lb in 1978 to 243 lb in 1992 (Figure 1). This contributed to the decline in the total weight of the average passenger car from 3494 lb in 1978 to 3136 lb in 1992. Of the five main categories of construction materials used in vehicles (Figure 2), only plastics/composites showed increased market penetration over the same period. The share of plastics/composites in domestic cars increased from 5.0% in 1978 to 7.7% in 1992 (Figure 3). By 2000 the average automobile is expected to weigh about 3000 lb and contain 300 lb of plastics. The weight ratio of metals to automobile shredder residue (ASR) is estimated to decrease from about 4.1 for model year 1978 to 3.4 for model year 1992. For many shredder operators, the profit margin on vehicles is declining rapidly [4, 5]. In February 1990, Waxman Metals Group used fluff disposal fees of $125/ton for a 1979 passenger car [4, 6]. The net value of a typical 3150-lb hulk in eastern Canada was estimated to be $48.75, based on the following assumptions: 2250 lb of scrapped steel at 6 cents/lb, 150 lb of nonferrous metals at 12 cents/lb, 750 lb of fluff with disposal fee of 6.25 cents/lb, freight of $10 per car, and processing cost at 15 cents/lb [4, 6]. Field and Clark [6], using the Waxman data, calculated the constant-dollar value of the average scrapped car for model years 1976 to 1989, assuming fixed proc-

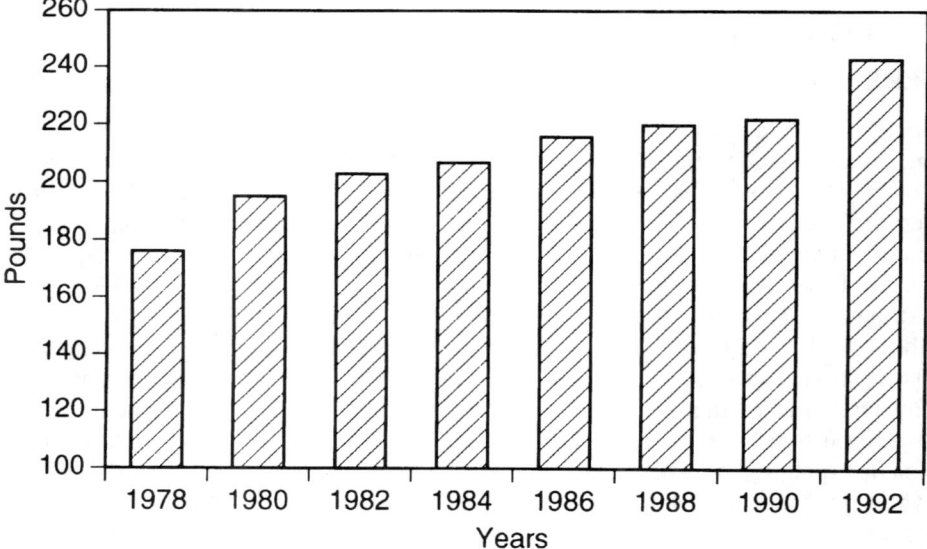

Figure 1 Average plastics/composites in domestic cars. (Developed in part from data published in Reference 2.)

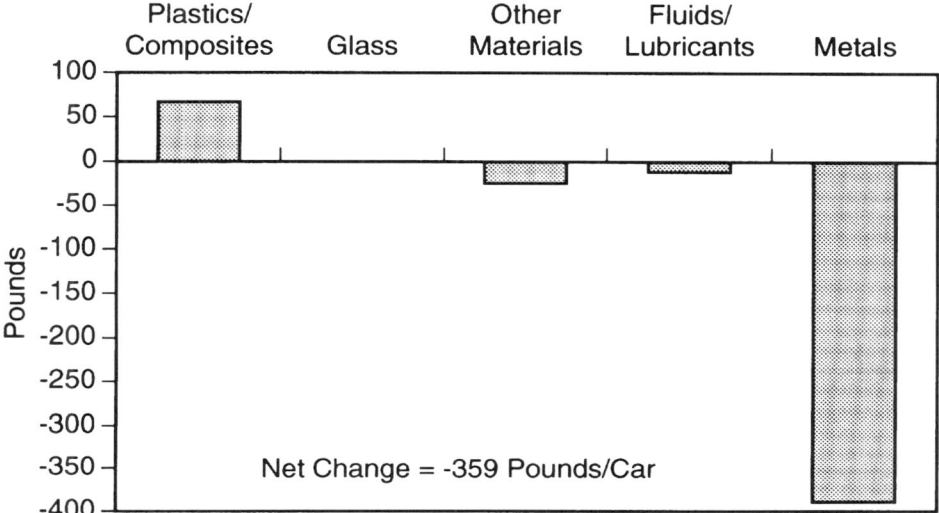

Figure 2 Changes in average material consumption in domestic cars, 1978–1992. (Developed in part from data published in Reference 2.)

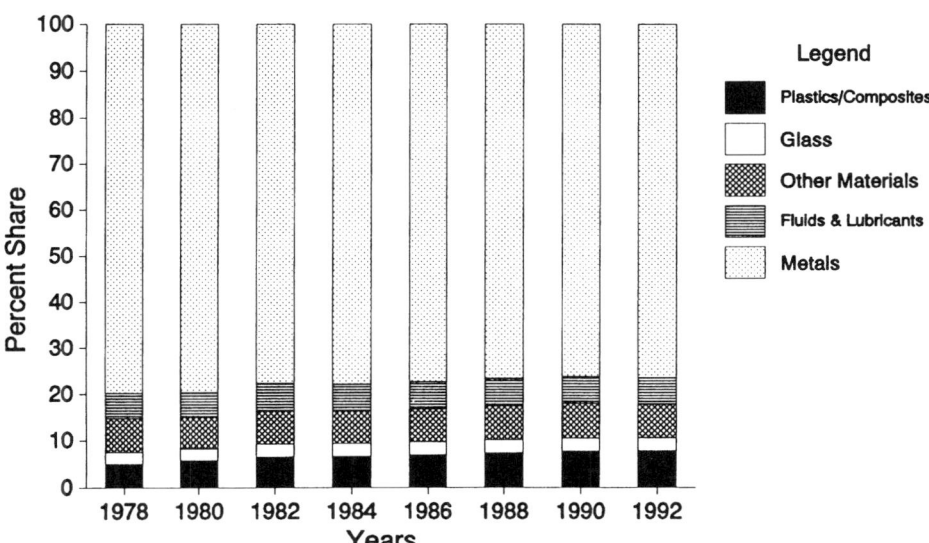

Figure 3 Average material consumption share in domestic cars. (Developed in part from data published in Reference 2.)

essing cost at 15 cents/lb, freight of $10/car, and fluff disposal fee of 6.25 cents/lb. On the basis of the average automobile material contents, the value of a scrapped car in constant 1990 dollars was estimated to be $60.57 for model year 1976, $55.01 for model year 1979, and $51.51 for the model year 1989. Field and Clark stated that this simplistic analysis illustrates that the value of the scrap car is falling as the material content changes [6]. The handling of automotive plastic waste is becoming a global problem, according to a study by Euromotor Reports Limited [7]. Concern about car scrapping is even greater in Europe than in the United States; about 14 million cars are junked annually in Europe, compared with 9 million cars in the United States [2].

Characterization of ASR

The development of technology for recycling ASR is complicated by the fact that ASR is a heterogeneous waste material; its composition, density, and moisture content change from site to site, and from day to day at the same site, as different types of source materials and different models and years of obsolete automobiles and white goods are shredded [8]. As shown in Table 1, ASR contains a large number of intermingled species, including plastics (thermoplastics, thermosets, and composites), rubber, fibers, fabrics, paper, glass, wires, wood, tar, dirt, rocks, sand, rust, automotive fluids, moisture, and small amounts of metals and metal oxides that are not recovered in the shredding operation. Approximately 40–55% of the ASR is combustible material. The heating value of ASR may vary from approximately 4000 to 6000 Btu/lb and averages approximately 5400 Btu/lb [9]. Furthermore, in some cases, polychlorinated biphenyl (PCB) contamination can result from the inadvertent shredding of capacitors contained in old white goods (such as obsolete refrigerators, washing machines, and dishwashers). In addition, ASR will contain heavy metals, such as lead (from unremoved batteries and other sources), cadmium (used as a coloring agent in some old plastics and fabrics), and mercury (from mercury switches). The moisture content of ASR may vary between 5 and 40% by weight (wt %), depending

Table 1 Composition of ASR

Recyclables	Energy value	Inerts
Thermoplastics	Paper	Glass
Foams	Brake fluid	Dirt
Fibers	Engine oil	Sand
Metal chunks	Transmission oil	Gravel
Rust	Grease	Moisture
Wires	Wood chips	
	Thermosets	
	Rubber	
	Tar	

on the type of shredding operation (i.e., wet or dry) and the degree to which the ASR is exposed to rain while in inventory. An approximate composition of ASR (including that generated by the shredding of white goods) is shown in Figure 4.

ASR constituents are also heterogeneous with regard to density and shape. For example, the polyurethane foam (PUF), with absorbed moisture and oils, might constitute about 10% of the ASR mass but could account for over 30% of its volume. On the other hand, fines ($<\frac{1}{4}$ inch) might constitute about 45% of the mass but less than 10% of the volume. The volume ratios of the large and small PUF pieces could be as high as 250:1. The physical variability of ASR is in itself problematic in terms of attempting to recycle it.

An expected composition of the plastics portion (based on automotive plastics use in model year 1981 and developed in part from data published in Reference 10) of the ASR is as follows [10].

Polyurethane foam	22.6%
Reinforced polyesters (glass or fiber)	21.9%
Bulk molding compound	
Sheet molding compound	
Polypropylene	19.2%
Polyvinyl chloride	15.5%

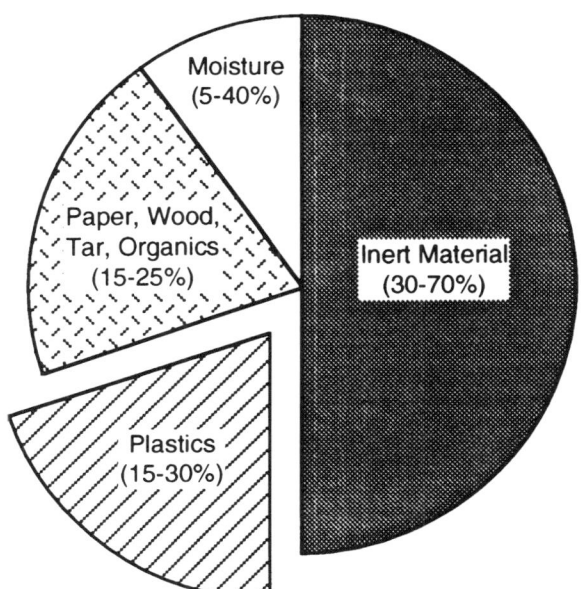

Figure 4 Approximate composition of ASR.

Acrylonitrile butadiene styrene	7.3%
Nylons	3.7%
Acrylic	2.5%
Phenolic	2.1%
Other	5.2%

Figure 5 illustrates the use and the expected growth of some of the leading plastics in the automotive section [11].

Historical Perspective

Before the introduction of the automobile shredder in the early 1960s, the most common method of recycling obsolete automobiles involved open-air combustion of the automobile hulk to burn off plastics and other combustibles. Open burning is no longer practiced because better recovery technology (namely the shredding process) was developed and because open burning is a gross violation

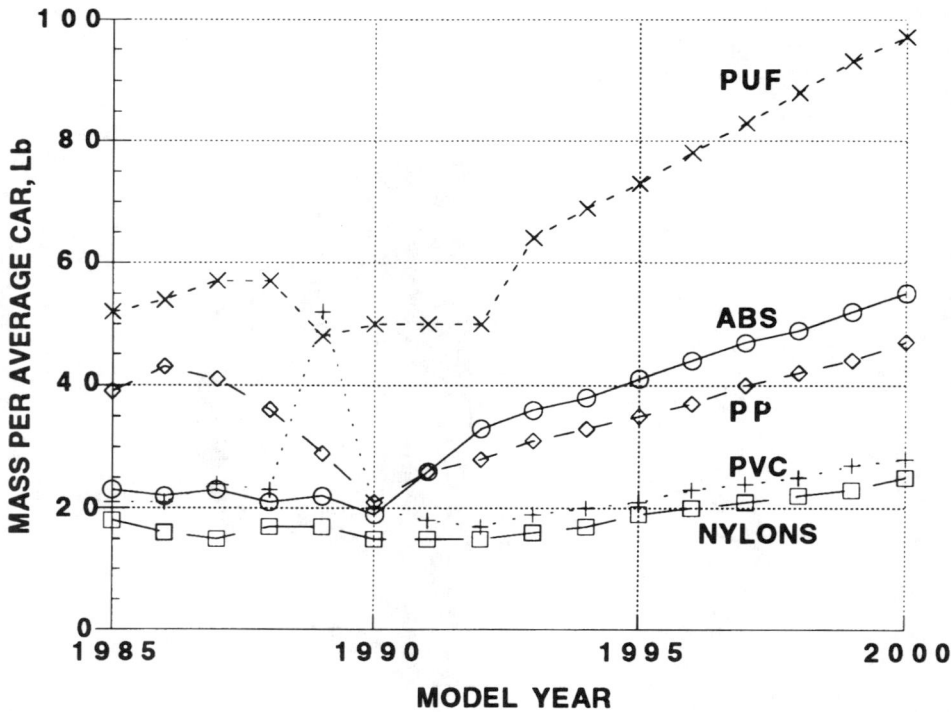

Figure 5 Use and expected growth of leading thermoplastics in the automotive industry. (Developed in part from data published in Reference 11.)

of environmental (Clean Air Act) regulations. More controlled incineration techniques were explored in the late 1950s and early 1960s [12–14].

With the introduction of the shredder and the incorporation of better separation techniques (such as multiple magnetic separation stages, air classification, screens, water wheels, and wet and dry cyclones in the early 1960s [12]), the number of obsolete automobiles processed increased dramatically, and so did the amount of recovered metals. Several publications that appeared in the late 1960s and early 1970s addressed a variety of topics on the subject, including the production of quality scrap, improved separation methods, the impact of increasing plastics content of automobiles, process economics, and environmental concerns associated with the process [15–21]. The 1970s also brought increased public interest in the processing and recycling of plastics waste [22–25]. This interest was stimulated primarily by increased public awareness of the magnitude of the plastics waste problem in general and by increased disposal costs [7, 8]. During the 1970s, two methods of ASR disposal were practiced: landfilling and incineration. During the 1980s and 1990s, concern over solid waste contamination of land and water resources, together with more stringent regulations and higher disposal costs, resulted in increasing interest in recycling as an alternative to landfilling, which became virtually the only practiced method. Lack of treatment/recycling/disposal technology for ASR could result in larger and larger piles stored on site, creating real fire hazards. Such fires were found to emit large quantities of hazardous and toxic pollutants, including volatile organic compounds, such as benzene, chlorobenzene, ethylbenzene, toluene, and xylene, and metal aerosols, including lead, cadmium, and copper [26].

DISPOSAL/TREATMENT METHODS

This section describes some of the disposal/treatment methods that are used with or could potentially be applied to ASR.

Landfilling

Landfilling continues to be the most widely practiced disposal technique, and in many areas it is an inexpensive disposal method for ASR. However, in most parts of the United States, tipping fees have increased dramatically in the past 10 years, and the ASR has had to be trucked ever greater distances to the nearest landfill available. If this trend continues, landfilling of ASR may soon be a cost-prohibitive technique in many parts of the country, if it is not banned altogether. Many states already require treatment of ASR to fix and immobilize heavy metals before landfilling. As discussed earlier, Waxman Metals Group used a fluff disposal fee of $125/ton in its estimation of the net value of a typical 3150-lb hulk in Eastern Canada in 1990. Disposal costs in landfills, including transportation cost, in most parts of the United States are between $15 and $75 per ton, with an average of about $35 per ton. For a disposal fee of $35 per ton of fluff, the net value of the hulk would have been estimated to be about $83, compared with about $49 under the disposal fee of $125 per ton. The analysis illustrates that the value of the scrap car is very sensitive to disposal costs. The presence of heavy metals (lead, cadmium, and mercury), automotive fluids, and low levels of PCBs in the ASR also creates a potential future liability risk when this material is landfilled.

Incineration

Incineration could be a cost-effective technique for disposing of ASR because its heating value is nearly half that of coal; consequently, ASR will burn without the assistance of a supplemental fuel [9]. Several studies have addressed this topic, and many ASR incinerators have been built by scrap processors [9, 12–14, 27]. Dean et al. [12] estimated that at least 23 such units were built by the end of 1973. To the best of our knowledge, none of these units are in operation today, although incineration can achieve more than a 50% reduction in the weight and over a 75% reduction in the volume of ASR. This lack of acceptance of incineration results from a combination of economic, environmental, and logistical problems; for example,

1. Landfilling is still less expensive in most locations than incineration.
2. Gas scrubbers are required to control HCl emissions.

Although the technology for scrubbing HCl exists, it is expensive, especially when the HCl is present in high concentrations, which may be the case for ASR. Hubble et al. [9] reported chlorine concentrations as high as 16.9% (the average concentration was 3.4%) on the basis of samples from 12 shredding sites. Moreover, the high chlorine content, in particular, makes the flue gas very corrosive, which increases the materials and maintenance costs of the incinerator. High chlorine concentrations are also suspected of increasing the risk of producing dioxins and benzofurans during the combustion process [9]. Therefore, we anticipate that operation of incinerators will be subject to a complicated permitting process that may call for extensive, costly chemical analysis. Incineration is even more expensive if the ASR contains PCBs. Government regulations require that PCBs be incinerated at very high temperatures (about 2000°F). High-temperature incinerators are costly to build and maintain. Supplementary fuel might also be required to attain such temperatures, which, of course, increases operating costs. Moreover, the location of incineration facilities can be made difficult by the "not in my backyard" (NIMBY) syndrome, which further complicates the permitting process.

The cost-effectiveness of incineration depends on the site, the type of incinerator, and local landfill costs, as well as on the variability of the composition of the ASR, including heavy-metal composition. Woodruff, Mechel, and Albertson [28] of Energy Products of Idaho reported that "preliminary results of one and one-half years of pilot plant test work indicate that the fluidized bed combustion process does offer an economical alternative for the disposal of automobile shredder residue. Another incineration process, using a proprietary Austrian technology, was reported in the Euromotor Reports Limited of London [7]. A pilot project using this technology is scheduled to be completed in the United Kingdom in 1994 [7]. The plant is designed to generate 2.5 MW of electric power, and it is claimed that it would consume about 90% of the ASR, leaving a small residue for landfilling [7]. We believe that the ash content of the average ASR is about 50% by weight; therefore, we doubt that 90% reduction will be achieved on a regular basis.

Recycling

Because ASR has a high hydrocarbon and plastics content, it could conceivably be recycled by all four recycling modes: primary, secondary, tertiary, and quaternary. The greatest emphasis on ASR recycling has been on secondary and quaternary recycling, relying on the plastics content of the ASR. The applicability of these recycling procedures to ASR is discussed next.

Primary recycling. Primary recycling generally refers to the recycling of uncontaminated waste material in the industrial manufacturing sector. Primary recycling could also apply to the separation of individual components within ASR, with the intention of using those components as a blend with virgin material. Dean et al. [12] reported on a set of experiments conducted by the Bureau of Mines in which heavy-medium separation of the individual plastics in ASR was attempted. In these experiments, the float and middling fractions from a water elutriator were sent to a gravity plastics separator comprising a series of brine cells, each maintained at a different specific gravity. Although 100% separation was not achieved, the tests showed that certain plastics could be concentrated in the various fractions (Table 2). Concentrating the various fractions of ASR

Table 2 Results of heavy-medium separation test

Fraction	Type of plastic in fraction (wt %)						
	ABS	Acrylic	PE	PP	PVC	Rubber	Other[a]
1	0	0	5.4	67.0	2.0	18.6	7.0
2	69.5	1.2	0.5	0	4.5	2.6	20.2
3	48.5	1.1	0	0	5.7	20.1	20.1
4	0	34.9	0	0	3.3	31.9	27.0
Sink	0	0	0	0	16.5	50.5	12.9

[a]Does not include nonplastic materials.
Source: Developed in part from data published in Reference 12.

might improve the viability of any recycling opportunity. However, we do not believe that heavy-medium separation can produce materials from ASR that will be adequate for primary recycling. A process that may produce close to primary-grade recycled plastics is selective dissolution and purification of individual solvents [29–37]. Because this method involves solvents and chemical processing, we discuss it under tertiary recycling.

Secondary recycling. The most comprehensive research regarding secondary recycling of ASR was that funded by the Energy Conversion and Utilization Technologies (ECUT) Branch of the U.S. Department of Energy [8, 38–40]. Production of secondary products from ASR has been pursued basically because of the thermoplastics content of the ASR. Thermoplastics, representing approximately 70–80% of the total plastics in ASR, can be heated and remolded into different products without the need to achieve a high degree of separation of impurities. Such products as park benches, lampposts, road traffic furniture, shingles, and other construction materials can be made from waste-containing plastics by using state-of-the-art extrusion and molding equipment. Many such products were identified in Reference 8. The presence of hard objects in the waste material (such as metals and glass) could, however, damage the molding equipment and/or increase its maintenance cost.

The key limitations of this type of recycling have been (1) the small market for such products, (2) the cost of making such secondary products, and (3) the greater attractiveness of less contaminated, commingled waste plastic streams, such as curbside plastics and manufacturing waste plastics, for producing such products (because they are richer in plastics and require less pretreatment than does ASR). In general, secondary products are, at best, only marginally competitive with their counterparts made of virgin materials (such as wood, sand, and gravel).

Recent experience in using recycled plastics from the municipal solid waste (MSW) stream in manufacturing fence posts and guardrail posts for highway agencies in Florida is encouraging. Smith and Ramer [41] tested such products for important characteristics and properties, such as durability, soil microorganisms, water absorption, environmental exposure, and flexural and tensile strength and concluded that such products could potentially be used in transportation applications. Specifically, flexural testing indicated that some posts were stronger than concrete. The U.S. Department of Energy program [8] also evaluated the use of ASR as an additive for "polymer concrete." In general, the conclusions appear to indicate a lower compressive strength for the ASR-polymer concrete. Boeger and Braton [42, Appendix A*] report on another concept for ASR, in which it would be physically separated into three fractions: metals, mill fuel,

---

*Boeger, K. E., and N. R. Braton. Mill fuel and mill cover recycled from shredder fluff. Appendix A in *Automobile Shredder Residue: The Problem and Potential Solutions* [42].

and mill cover. The mill cover is the nonmetallic fraction under 2 inches, and it is being tested as a landfill cover. The fraction greater than 2 inches is reported to have a calorific value of almost 11,000 Btu/lb and could conceivably be used as a fuel for quaternary recycling.

Other factors to be considered in secondary recycling are as follows:

1. The heavy metals in ASR should be locked in the product once the plastics are melted and remolded. However, many plastics tend to degrade slowly in an acidic or basic medium. Oil, PCBs, and other organics that may be present in the ASR could also become exposed and leachable.
2. The quality of products that can be derived from ASR can be upgraded by partial separation of undesired components (such as glass, metals, and gravel) and by the addition of virgin plastics and other additives [8, 39]. The appearance of the product can also be enhanced by sandwiching the recycled product between thin layers of virgin material.
3. The continuously changing composition of ASR demands that ample tolerances be provided for in the design of the process in order to meet product specifications regularly and to prevent obsolescence.

Equipment that has been developed or is under development for the secondary recycling of plastics in MSW streams could be adopted for the secondary recycling of ASR. This equipment includes Mitsubishi's Revezer, the Klobbie, the FN machine, the Flita System, the Remaker, the Regal Converter, and Kabor's K board.

Because ASR contains, in addition to the plastics, natural fibers (wood chips and paper) and synthetic fibers (carpet materials), it should be a good candidate for making composite secondary products after a mechanical/physical separation to concentrate these species in the separated product [43–46]. The fibers could help reduce the "creep" and weight problems from which many potential secondary products suffer. However, such products will still be difficult to market, because they are likely to be more expensive than conventional products. For example, it was estimated that the cost of plastic lumber (for outdoor use) to wood suppliers is about twice the cost of treated wood [47]. This is, at least partly, because the cost of raw materials in such products is only a small portion of the total cost of the finished product. The plastic lumber is likely to be even more expensive if the plastics are derived from ASR, because of the extra processing required to separate the plastics.

Another interesting aspect of recycled plastic lumber used in marine applications is pointed out by Weis et al. [48], who found that leachates from formulations of recycled plastics are far less toxic than those from chromated copper arsenate–treated wood. They also reported that styrene-containing recycled plastic lumber resulted in the leaching of plastisizers (phthalates), long-chain alkanes, phenol, and alkylated phenols, the source of which could have been the antioxidants that are commonly added to many types of plastics.

Tertiary recycling. Tertiary recycling or chemical processing could be used to produce value-added products (such as monomers, solvents, light hydrocarbons, and/or solid, liquid, and gaseous fuels) from the hydrocarbon-based fraction of ASR (plastics, rubber, paper, automotive fluids, and wood) [8, 25, 40, 49–53]. Plastics, however, will be the main source of such products. Processes that may be employed for this purpose include pyrolysis, hydrolysis, selective dissolution, hydrogenation, and gasification.

Pyrolysis, the thermal decomposition of organic material in an oxygen-deficient environment, is a well-known technique for the production of fuels and chemicals from such feedstocks as wood, coal, tires, and municipal waste. The type and relative quantities of different products are generally functions of the operating conditions, primarily temperature. This property enables the operator to change the product mix when market prospects change. The applicability of pyrolysis to plastics has also been demonstrated. Banks et al. [52] reported that the major products of this process, using polyethylene as a feedstock, were hydrogen, benzene, methane, ethylene, and propane; the major products from the use of polyvinyl chloride were benzene, acetylene, styrene, and hydrogen chloride; and the major products from the use of polystyrene were styrene, benzene, toluene, and methylstyrene.

To the best of our knowledge, no commercial-scale plastics pyrolysis reactors are in operation in the United States, and no data are reported on the pyrolysis of ASR. However, some limited testing has been done by more than one organization [7].*

Pyrolysis of ASR could also yield a mixture of products similar to those produced by the pyrolysis of polyethylene, polyvinyl chloride, and polypropylene. More work is required to establish the operating conditions necessary for the production of a desired product mix. We expect that the variability of the ASR's composition might present a control problem for the system. The potential presence of PCBs in the ASR, along with finely divided heavy metals–containing particles, could render the liquid products useless, if not a hazardous waste. The high chlorine content would also result in the formation of HCl, thus reducing the H/C ratio in the products. This tends to produce more heavy and fewer light products.

Some of the problems encountered in pyrolyzing plastics-containing materials are [25]:

1. Pyrolysis of plastics requires more time than pyrolysis of other hydrocarbons because of the poor heat transfer characteristics of the plastics.

*Private communications with various shredder operators.

2. Carbon residue produced by plastics pyrolysis has a tendency to stick to the walls of the reactor.
3. Some plastics, when heated, produce a high-viscosity material that is difficult to pump.

A low-temperature (500°–600°F) pyrolysis process employing a double-screw extruder has been tested on several waste materials, including MSW and a single sample of ASR that was dug up from a landfill [54, 55]. The material is fed at one end of the screw extruder/mixer, and coal-like products leave from the other end [54]. The limited results on ASR indicated that a small portion (< 20 wt %) of the ASR was volatilized and collected as condensable and noncondensable vapors. Organic vapors and CO_2 dominated the gases collected. This indicates that actual pyrolysis occurred, even at the low operating temperature of this process. It may be, however, that the pyrolysis took place at localized hot spots that resulted from the shear forces applied to the ASR.

Pyrolysis tests were conducted on ASR and ASR-MSW mixtures in a proprietary pyrolysis system to produce a gaseous fuel that can be transported to nearby industries [56]. The tests produced a low-Btu gas. Further tests are planned with ASR and with proposed ASR, in which the polyurethane foam and the fines (<0.25 inch) are removed. The material that was preprocessed by the Argonne process produced about 40 wt % more gas than did the original material, primarily because the Argonne process separates a substantial portion of the inerts.

Hydrolysis involves reactions with H_2O, generally at elevated temperatures. Its applicability to ASR has been investigated primarily for the treatment of polyurethane foam [49, 50, 57]. Hydrolysis is also applicable to the treatment of polyesters, polycarbonates, and polyamides. Mahoney et al. [50] showed that the hydrolysis of PUF produces two main products: polyols and amines. Braslaw and Gerlock [49] developed a process to separate and purify the product mix to produce very high quality polyols. Experiments reported by Braslaw and Gerlock used (separately) clean foam as a material input to the process and also as-received foam from a shredder facility. They then tested the physical properties of their seat foam formulation by blending the recovered polyols with virgin material. A blend with up to 50% of the polyols recovered from the clean foam waste produced a foam with physical properties not "significantly different" from those of all-virgin foam. When the polyols from the as-received foam were blended, physical properties consistent with all-virgin foam were achieved only when less than 10% of the recovered polyols was used with 90% virgin material. Braslaw and Gerlock [49] estimated a $275,000 after-tax cash flow on a $1.1 million plant operating one shift per day with an annual capacity of about 1000 tons per shift. They also pointed out that the technology would be applicable only for processing dirty PUF, because there is a market for clean industrial PUF waste among carpet underlay manufacturers. When the market value of the clean foam is taken as an operating cost for the process, the positive cash flow becomes negative.

Brown et al. [7] reported that, in addition to the U.S. research on seat foams and hydrolysis, work was carried out in Germany. The results using seat foam were found to be poor. However, it was stipulated that the results could have been improved by using higher quality foams, such as uncontaminated manufacturing waste. The hydrolysis of clean foam was reported to be potentially more cost-effective than the landfilling option [7]. They also reported that Mazda Company in Japan has developed a promising technology, using a new hydrolysis catalyst, to break down some polymers [7]. This hydrolysis process reduces polymers to either monomers or oil and gases more efficiently than other processes. Another interesting development, reported by *Plastics Recycling Update*, is that Toyota Motor Company has been able to achieve the removal of paint layers from painted plastic bumpers by using its high-pressure, high-temperature hydrolysis process [58].

Selective dissolution of plastics, from a stream that is rich in mixed plastics, by using solvents is a technique that could be used to recover plastics in a usable form that might qualify for primary recycling [28–35, 38, 59]. Solvent selectivity and operating temperatures, for both dissolution and separation, are essential to produce products that can be substituted for or mixed with virgin materials without reducing the quality of the final product [60]. Argonne National Laboratory [29–35] has been conducting research to recover thermoplastics from ASR by using solvents. Our process is briefly described in this document. The approach emphasized at ANL is to use several solvents, one at a time when possible, to extract individual plastics.

Hydrogenation of the ASR—or parts of it, such as the "oils"—can also be used to remove chlorine and sulfur and produce more saturated hydrocarbons. An advantage of hydrogenation in this application is that the metal compounds present may be reduced to their elemental state, which increases their value. However, hydrogenation processes are generally carried out under high pressures and elevated temperatures, which means much costlier equipment, especially when the feedstock contains about 50% inerts. The cost of hydrogen also needs to be considered carefully in this case, because it may be substantial. Hydrogenation experiments on waste plastics to produce petroleum-like "oil" using an alumina silicate catalyst were conducted by M. Taghiei at the University of Kentucky [61, 62]. In these experiments, the mixed plastic waste was treated at 400°–500°C and at ~800 psia for about 1 hour.

Gasification could also be used to convert the hydrocarbon material in the ASR into a low-Btu gas containing carbon monoxide, hydrogen, and other light hydrocarbons. The product gas can be used as either a gaseous fuel or a feedstock to produce liquid products, such as methanol. High-temperature gasification tests were conducted by using the gasifier developed by Voest-Alpine Industrielanlagenbau, GmbH, of Austria [42, Appendix E]. In these tests, ASR was blended with mixed plastics, waste oils, and fuel oil. The net calorific value of the blended material was about 17,000 Btu/lb, which would indicate that the amount of actual ASR used relative to the oils and mixed plastics was less than 25%. Analysis of the gasification products showed that several dioxin species were present in minute quantities. The product gas, which consisted of nitrogen (58 wt %), hydrogen (13%), carbon monoxide (17%), and carbon dioxide (5%), with the balance water vapor, had a calorific value of about 100 Btu/ft^3. This product would be considered a low-Btu industrial fuel gas. Brown et al. [7] noted that Dow Europe and Voest-Alpine have been working to develop a high-temperature gasification process using plastic scrap from ASR as a feedstock. The ingredients are heated to 1600°C, yielding essentially hydrogen, carbon monoxide, and a solid residue. After cleanup, the gaseous fuel can be used as a heat source for on-site power generation. Van Stolk et al. [63] discussed an "electro-chemical" gasification process for polymers and for ASR. In this process, gasification takes place in an electric-arc furnace to produce CO and H_2. The first demonstration plant was built in Italy [63].

Gasification plants are generally very capital intensive and may be marginally economical only at very large scale. As discussed earlier, ASR cannot be supplied in large enough quantities for the desired scale of gasification plant. Furthermore, we doubt that combustion of some of the ASR alone can achieve the high temperatures required for efficient gasification. Therefore, although ASR may be blended with other solid and liquid fuels, such as coal, MSW, and waste oil, for gasification, we doubt that gasification of ASR by itself can be economical or capable of producing a quality fuel.

Other techniques applicable to the processing of scrap plastics, such as methanolysis and glycolysis, can also be applied to some automotive plastics when they are separated from the ASR. However, these processes are generally difficult to justify economically when applied to plastics, especially to a stream containing dilute concentrations of plastics, and to the best of our knowledge none has been commercialized. The additional cost for separating the plastics from ASR makes these processes less economically attractive.

Quaternary recycling. Essentially, quaternary recycling is recovery of the energy value of waste via incineration, with heat recovery from the combustion gases being used to produce steam or steam and electricity. In a study sponsored by the U.S. Department of Energy, the potential of harvesting some of the energy released in the incineration process was evaluated [9]. During field tests, samples of ASR from 12 different facilities were burned in a rotary-kiln incinerator. Different scenarios for utilization of the released thermal energy were also analyzed. The major findings of the study are summarized and discussed below:

1. The average heating value of the ASR material was about 5400 Btu/lb, but this could vary by ±50%.

2. The chlorine content of the ASR samples varied between 0.7 and 16.9 wt %. This large variation is likely to result in an overdesigned scrubbing system in order to deal with the worst-case scenario. We believe the high value (16.9%) is not typical and might have resulted from the sampling procedure adopted. (It is difficult to obtain a representative sample from this heterogeneous material.)
3. The moisture content of the samples varied between 5 and 35 wt %. The water in the ASR consumes a substantial portion of the heat released in the incineration process, resulting in a lower flame temperature than could otherwise be obtained. As a result, the specific thermodynamic availability or useful energy is reduced. Water in the ASR also lowers the dew point of the flue-gas stream. This lower dew point will increase the chance that condensation will occur during the heat recovery process. Condensation will be detrimental to the ductwork because of the highly corrosive nature of the flue gas. The study demonstrated that the flue gas can be scrubbed to comply with environmental regulations by using state-of-the-art equipment.
4. The ash remaining after incineration is about 25–70 wt % of the starting material (average value is about 50 wt %). This proportion of ash nearly doubles the concentration of metals in the ash; as a result, the ash could be classified as hazardous, which would then require fixing and/or postprocessing for recovery of the metals. For example, it was reported [42] that Puremet Corporation had developed a hydrometallurgical process that would be applicable for recovery of the metals from ASR incinerator ash.
5. When the heat is used to produce steam, the amount of steam generated as a result of incinerating the ASR material far exceeds the on-site heat requirements of a shredder. Therefore, unless there is a customer for the heat nearby, it will be uneconomical to recover and/or transport this heat.
6. On-site electricity generation was also evaluated. In general, the amount of electricity that could be produced from the ASR available at most shredders exceeds the on-site electricity requirements of the shredder; the excess electricity can be sold back to the utility. The sale of electricity depends on many factors, including compliance with Public Utility and Regulatory Policy Act (PURPA) requirements in order to qualify as a cogenerator. The economic competitiveness of such a venture will also depend on such factors as the cost of electricity, the rate at which the local utility will purchase excess electricity, and the consistency of ASR production rates, which will affect the design of the equipment relative to its utilization. Under the assumptions of the study, a "median" plant generating 4.7 million kWh of electricity per year from 60,000 tons of ASR per year would yield a before-tax cash flow of about $105,000 per year on an investment of about $1.9 million. A large-scale plant (180,000 tons of ASR per year) would produce a cash flow of $665,000 on an investment of $4.6 million. The fact that most, if not all, shredders operate only about 8 to 10 hours a day and shut down the rest of the time makes this scenario less economically attractive.

A study [64] conducted by the U.S. Environmental Protection Agency (EPA) concluded; "The highest PCDD [polychlorinated di-benzo-P-dioxins] emissions concentrations appear to be associated with low temperature combustion processes whose function is to recover energy of other resources (e.g., metal values) by combustion of waste materials." This study also reported that the highest PCDD emissions were measured at sites containing plastics in the feed and that total organic halogens (TOX) showed a stronger association with PCDD emissions than with total chloride content. Thus, any incineration system will likely require a high-temperature afterburner to prevent such emissions; this would, of course, increase the cost of the incineration process.

Combustion of ASR in fluidized beds has also been investigated. Tabery [65] reported that fluidized-bed incinerators are suited for ASR and that they could be economically attractive for the larger applications in areas where the disposal cost is high. He estimated that for a hypothetical case of a shredder generating 62,500 tons of ASR per year, and paying $40 per ton disposal cost, a rate of return of about 35.5% could be realized, ignoring depreciation and tax considerations. The heat generated from burning the ASR can produce as much as six times the power requirement of the shredder; the excess electricity can be a revenue generator. In the analysis, 62.5% of the total revenue was attributed to avoided ASR disposal cost and 31.9% was attributed to electrical sales to the grid. These data indicate that the analysis is very site specific.

THE ARGONNE NATIONAL LABORATORY PROCESS

Argonne National Laboratory has developed a process for the recovery of most ASR constituents, including thermoplastics [29–35]. The concept of recovering the plastics for reuse as plastics is intuitively appealing, because (1) the plastics content of shredder fluff is expected to increase, (2) plastics use in automobiles is a growing market, and (3) the plastics recycling industry, although in its infancy, is growing. We expect that more than 60% of the total thermoplastics content of shredder fluff might be recoverable by dissolution of the plastic in solvents. Conceptually, given differences in the solubilities of certain plastics and their differences in susceptibility to specific solvents, the intent of the ANL process is to extract selectively specific plastics or groups of compatible plastics from shredder fluff by using solvents. A four-step process was developed: (1) drying, (2) mechanical separation, (3) extraction of the thermoplastics, and (4) regeneration of the solvents for reuse.

Drying

The drying process is by indirect heating (i.e., steam coil) only, because flame or sparks could set the material on fire. Temperatures in excess of about 105°C (220°F) should be avoided in the presence of air. Localized smoldering was observed on occasions, but it could not be reproduced when the material was dried in a regular oven at higher temperatures. Not enough data exist at this time to conclude that drying at lower temperatures is completely safe. Therefore, careful monitoring and control of the dryer are necessary to avoid smoldering fires. Drying at lower temperatures also minimizes the evaporation of organic species along with the moisture, thus helping to avoid a potential environmental problem that might require scrubbing or wastewater treatment. Furthermore, because of the potentially high water content of some ASR material (up to 40 wt %), the drying process can be both energy intensive and time consuming. The problem with low-temperature drying, apart from the time it takes, is that it could result in poor drying of the inner material, unless the material is agitated or remixed frequently during the drying process. This drying process may be eliminated in cases in which the moisture content of ASR is low (<10 wt %). However, the moisture must then be separated later, during the "oils" washing process.

Physical Separation

Mechanical separation of the polyurethane foam and the fines (<0.62 cm, or 0.25 inch, in size) from ASR is necessary for the following reasons:

1. To recover the foam as a potential product.
2. To concentrate the plastics in the remaining fraction. This results in smaller, less expensive equipment for the solvent extraction operation and in fewer solvent losses, because smaller quantities of the nonplastic materials that can absorb solvent (foam) or be wetted by it (dirt) will be present.
3. To avoid the need for costly filtration treatment. The presence of the fine nonplastic particles will contaminate the extracted plastics, unless extensive and costly filtration of the fine particles from the solution is performed.

Attempts to separate the shredder fluff by using small, commercially available multideck vibrating screens were not successful. The wires in the shredder fluff caused plugging of the openings in the upper deck screen in a short period of time. Small pieces of plastics and nonplastic materials were trapped in the fuzz and PUF and could not be shaken loose by the vibration of the screens. A laboratory classification column was built, fitted with several screens, and equipped with a variable-output air blower in the bottom that can be cycled on and off to provide agitation of the shredder fluff as it is being separated (Figure 6). This resolved the plugging problem to a large extent; it also pushed the light PUF to the top of the top screen and liberated some of the entrained dust and fines content of the PUF. No material was leaving the column during the agitation and separation process except for some heavy fines that dropped out of the bottom of the column. Therefore, the apparatus was operated in a batch

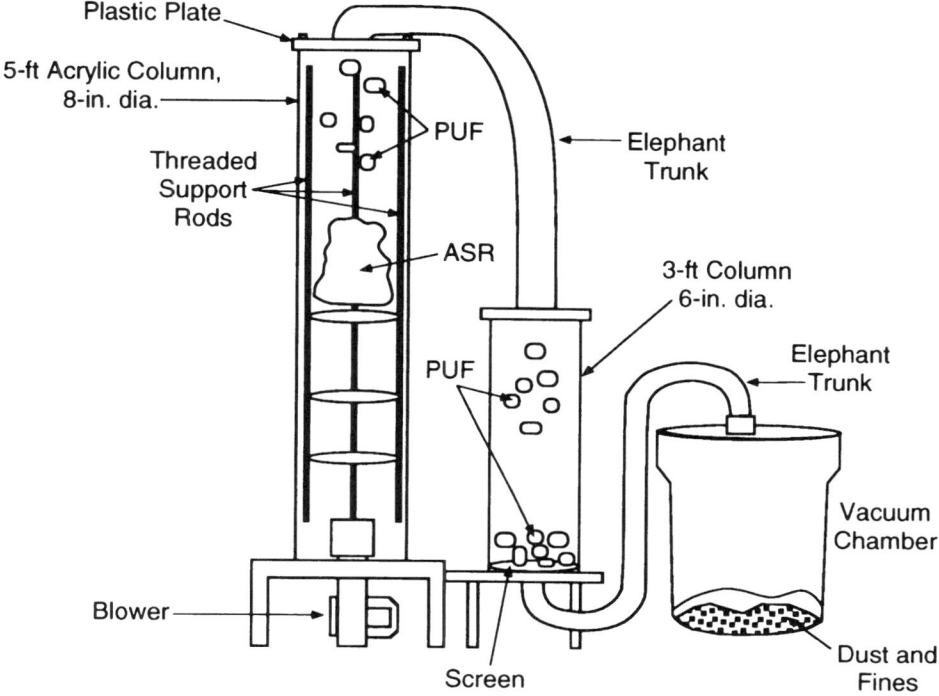

Figure 6 Laboratory apparatus for physical separation of PUF, fines, and plastics-rich shredder fluff fractions.

mode. An "elephant trunk," connected to a vacuum system that may be activated when the blower is in the off mode, was attached to the top of the column. This resulted in the separation of the foam from the top of the top screen. Some features of this system are being incorporated at present into a large-scale system to separate the foam and the fines, as part of a demonstration project to separate and recover the foam.

The collected foam constituted about 10 wt % of the ASR. A process to clean the foam by using organic solvents and water-detergent solutions was developed and tested in the laboratory. About 30–40 wt % of the dirty foam was separated as "oils," or automotive fluids and moisture. A pilot plant for full-scale demonstration of the foam separation and cleaning processes is under construction. Preliminary economic analysis indicates that the foam separation and cleaning process could pay for itself in about 1 year.

The fines, which constituted about 40 wt % of the ASR (values as low as 20% and as high as 55% were observed for different samples), were separated further by using a magnet to produce two different fractions: a magnetic fraction and a nonmagnetic fraction. The magnetic fraction was about 40% of the total fines. The heating values of a sample of the magnetic and a sample of the nonmagnetic fines were determined and found to be 3844 and 2355 Btu/lb, respectively. This is an indication that some of the brittle plastics, mainly thermosets, and wood chips were also broken down into smaller pieces and were screened with the fines. However, because of the significant changes encountered in the composition of the fines from shredder to shredder, the heating values reported here should be interpreted carefully and may not be typical values. Masswise, the organics represented about 20–30 wt % of the total fines and about 10–20 wt % of the magnetic fines. Elemental analysis of the inorganic compounds was performed on several samples, after the samples were combusted to burn off the organic compounds. Average data for samples from six shredders are shown in Table 3.

We are pursuing using the fines as raw materials (a source of iron oxides and silica) for the cement-making process. Further conditioning of the fines may be necessary to produce a consistent product for such an application.

Solvent Extraction of the Plastics

We identified polyurethane foam (PUF), polypropylene (PP), polyvinyl chloride (PVC), and acrylonitrile butadiene styrene (ABS) as potential candidates for

Table 3 Approximate elemental analysis of metals in the different fractions of fines derived from ASR

Metal	Percent by weight in different fines fractions		
	Total fines	Magnetic	Nonmagnetic
Boron	0.1	0.04	0
Chromium	0.07	0.09	0.045
Zirconium	0.04	0.02	0.03
Cobalt	0.01	0.01	0.008
Barium	0.3	0.18	0.35
Potassium	0.25	0.11	0.2
Manganese	0.18	0.27	0.12
Nickel	0.15	0.17	0.03
Lead	0.25	0.19	0.22
Titanium	0.4	0.22	0.37
Aluminum	1.5	0.65	1.5
Calcium	3.5	2.2	4
Iron	16	35	15
Copper	0.05	0.05	0.1
Magnesium	1.2	0.3	1.5
Sodium	2.2	0.45	2
Silicon	11	5	12
Zinc	2	1.6	1.2

recovery from ASR for recycling. Polyurethane foam, ABS, and PP are the fastest-growing construction materials in the automobile industry [11]. ABS was also targeted because of the high market value of the virgin resin and because of its good solubility characteristics in several mild organic solvents at low and moderate temperatures. PVC was selected because its chlorine content could be an obstacle if the stream is to be processed for its energy or chemical value (incineration, pyrolysis). In addition, reasonably clean PVC can be used for making numerous products with flexible specifications. PVC, however, is susceptible to degradation upon thermal cycling. Among the first signs of degradation is the appearance of a blackish color. PP was targeted for recovery because it is present in large quantities and its use by the automotive industry is on the rise. Moreover, because it is soluble only at high temperatures, it can be recovered with little contamination except for the PE, which dissolves under similar conditions. In this document, we did not address the potential for dismantling of some of the plastics prior to shredding. Even though we believe that it can be technically feasible, it is unlikely to be economical, at least until cars that are designed for dismantling reach the end of their useful life, which might be 15–25 years away. Hock and Maten [66] estimated (based on studies sponsored by the American Plastics Council) that the "cumulative costs associated with recovery, by selective dismantling, through reclamation of the automotive plastic parts . . . fall between $0.36–$1.08 per lb and average $0.72 per lb."

To minimize the contamination of the recovered plastics, the first step in the extraction process was to use a mild solvent, such as hexane, to extract the automotive fluids, at ambient temperature and pressure, without dissolving the targeted plastics. The "oils" recovered in the process, including those recovered from the polyurethane foam, constituted about 5 wt % of the ASR and had heating values between 16,000 and 18,000 Btu/lb. However, the hexane extracts also contained the PCBs that may be present in the ASR. Therefore, the recovered oil may require disposal as a hazardous waste, instead of being considered as an energy source. Table 4 shows the results of the analysis conducted on one composite sample prepared from oils derived from ASR and obtained from several shredders. We are also testing a biodegradable solvent that has a flash point comparable to that of mineral spirits, and it is showing promising results.

After the oils were extracted, the extraction of the plastics at ambient pressure was tested in two methods. In the first method, selective solvents were used to extract individual plastics in series. In the second method, a solvent was used to dissolve all of the plastics of interest at atmospheric pressure and elevated temperatures, and then the mixed plastics were separated using different solvents. Table 5 shows some of the solvents for the plastics of interest. Each of these methods has advantages and disadvantages. For example, in the first method, the ABS and the PVC can be recovered without being exposed to the high temperature required for the recovery of PP and PE. However, the extraction steps all take place in large reactors, because all of the plastics-rich stream will be used in the extraction process. In the second method, only one extraction has

Table 4 Approximate analysis of a composite sample of oils recovered from ASR obtained from several shredders

Component	Presence (wt %)
Silicon	0.082
Iron	0.059
Calcium	0.039
Copper	0.030
Aluminum	0.028
Titanium	0.026
Zinc	0.025
Lead	0.024
Magnesium	0.021
PCBs	
Aroclor 1016 (ppm)	230
Aroclor 1254 (ppm)	86

to be done in a large vessel. The separation of the mixed plastics can be conducted in smaller reactors. In addition, this method enables more efficient cascading of the heat between the stages to minimize the energy requirement of the process. However, all of the plastics will be exposed to the high temperature and will experience a higher number of thermal cycles. Potential degradability of some of the products is higher. Products generated by both methods are under evaluation at present.

Solvent Regeneration

After a thermoplastic is dissolved in a solvent, the plastic is recovered by one of several different methods, depending on the characteristics of the solution. These methods include cooling, using an antisolvent, and evaporation of the solvent. When antisolvents are used, the binary liquid mixture remaining after the plastics are recovered is distilled to recover both the solvent and the anti-

Table 5 Solubility of some plastics in organic solvents

Thermoplastic	Solvents
ABS, PVC, ABS-PVC alloys	Chlorinated hydrocarbons, ketones, tetrahydrofuran, and aromatics
Polyethylene and polypropylene	Hot aromatic and chloroaromatic hydrocarbons
Polystyrene	Aromatic and chlorinated hydrocarbons
Polyvinyl chloride	Esters, ketones, chlorinated hydrocarbons
Nylons	Phenols
Acrylics	Aromatic and chlorinated hydrocarbons, ketones, and esters

solvent. In all three cases, the solvent is reused. We have regenerated the same solvent over 10 times without any measurable degradation in its performance.

Residual Material

The residual material from which the thermoplastics have been recovered is a dry mixture that could be used as solid fuel. Low in ash and chlorine, it should be easy to pelletize or cube because it is dry and its foam content has been separated. Samples of such material have been sent for testing in incineration and pyrolysis test facilities. Results are not available at this time.

Economic Analysis of the ANL Process*

The following analysis refers to a conceptual process based on scaling up the results of the Argonne bench-scale and pilot-scale experiments to a full-scale operation that accepts 183 million lb of ASR per year. The process is designed to separate, clean, and recover PUF, as well as to separate and recover ABS, PVC, and the polyolefins PE and PP. The facilities are intended to operate continuously for 330 days per year. The scoping estimate of required capital investment includes all the equipment unique to the operations (inside battery limits), as well as an electric substation, steam generation, cooling water system, and site preparation (outside battery limits)—in summary, a complete, operable unit at 1993 costs. Four operators are used per shift. The process is designed to recover annually 9.5 million lb of clean PUF and over 16 million lb (total) of thermoplastics.

Table 6 summarizes the estimated process economics, showing the capital investment ($7.8 million) and the annual balance sheet in thousands of dollars. Avoided tipping fees are evaluated at $30/ton. Revenues from sales of the recovered products are based on what are believed to be realistic prices (see note b, Table 6). Slightly more than $1 of annual revenue is received per dollar of capital invested. The operating expenses are itemized below. The second largest cost item is the disposal of hazardous oil waste at 12.5 cents/lb—at $1.275 million, this item alone is about 25% of the total annual operating expenses. Recycle and recovery of the PUF cleaning solvent and the plastic extraction solvent are a predominant part of the utilities, estimated at nearly $1.3 million. Gross profit is the difference between revenue and expenses. The payback (PB) period is calculated by dividing the required capital investment by the sum of after-tax profit (60% of gross) and depreciation. We estimate that the PB is 3.0 years for this conceptual process.

Obviously, the economic merit is extremely sensitive to market price and demand for the recovered plastics and PUF. The disposal cost of oily waste is

*This section was prepared by N. F. Brockmeier, Argonne National Laboratory (1994).

Table 6 Economic analysis of Argonne plastics recovery process

Date of development	1993
Product form	Foam and granules
Process type	Wash and precipitation
Feed, 10^6 PPY[a]	183
Plastic output, 10^6 PPY	26
Capital investment, 10^6	7.8
Revenues, 10^3/yr	
Avoided tipping fees	1990
Polyurethane foam[a]	2370
Precipitated plastics[a]	3582
Subtotal	7942
Expenses, 10^3/yr	
Chemicals and solvents	490
Oil waste disposal	1275
Utilities (electricity, steam)	1290
Operating labor	720
Other fixed costs	1132
Subtotal	4907
Gross profit, 10^3	3035
Less depreciation	780
Annual realization, 10^3	2255
Payout time, yr	3.0[b]

[a] PPY = pounds per year.
[b] Based on values for cleaned foam = $0.25/lb, PVC = $0.25/lb, ABS = $0.40/lb, and PP and PE = $0.05/lb.

a very important factor—one that further work will focus on in an attempt to reduce it. It may also be possible to streamline the process to reduce the estimated fixed costs. By itself, the separation and cleaning of PUF is extremely attractive, with a PB estimated to be 1.2 years.

CONCLUSIONS

In the past three decades, several treatment and disposal methods for ASR have been investigated. Landfilling is still the most common practice. However, increasing transportation, tipping, and (in some states) metal-fixation costs, and concern over long-term liabilities associated with landfilling of waste in general, are reducing the economic attractiveness of this approach. This has resulted in increased interest in ASR recycling and in particular in recycling its polymers content because of the potential marketability of these polymers. One of the problems associated with attempting to recycle ASR is the variability in its composition (i.e., plastics contents, nonplastics combustibles, and inert compounds). All four types of recycling have been investigated for the recovery of recycled materials from ASR. Using the ASR as-is for making even low-grade products does not appear technically feasible or economical—nor is it recommended, because of the potential presence in the ASR of lead, cadmium, mercury, PCBs, and sharp objects. Therefore, preprocessing of the ASR to concentrate and clean the potentially useful components appears necessary. For example, work done at Argonne National Laboratory indicates that separation and cleaning of the polyurethane foam found in ASR is technically feasible, using commercially available equipment, and the process could pay for itself in about 1 year. This process is being demonstrated by using near-full-scale equipment and it is close to commercialization. Even though the cleaned foam constitutes about 5–10 wt % of the ASR, it constitutes a substantial portion of its volume.

Methods for separating thermoplastics from ASR have also been investigated by different workers and appear technically feasible. However, the economic competitiveness of such processes is yet to be proved. The solvolysis processes could yield plastics good enough to be mixed with virgin materials to produce primary products. However, our preliminary cost estimate for the solvolysis process that we have been developing indicates a potential payback period of more than 3 years. Tertiary recycling of ASR to produce chemicals and liquid and gaseous fuels or feedstock, investigated by many researchers and organizations, has been shown to be technically feasible. Such processes can result in significant reduction in the weight of materials going to the landfill. However, the relatively high cost of the needed equipment is one of many problems such technologies must overcome before they can produce profitable and marketable products. For example, gasification of coal is a developed technology, yet the resultant cost of the product gas cannot compete with the current cost of natural gas. Because of its relative capital intensity, we do not believe that gasification economics would be significantly improved by using ASR, without a fee for accepting the ASR. We expect similar economics for pyrolysis and hydrogenation, although pyrolysis might be less costly than either gasification or hydrogenation.

Ultimately, the most cost-effective approach for recycling ASR may be a hybrid system that first incorporates physical separation to concentrate components into two or more fractions, then uses the recycling technology that can best be applied to the individual fractions.

ACKNOWLEDGMENTS

This work was supported by the U.S. Department of Energy, Assistant Secretary for Energy Efficiency and Renewable Energy. This chapter is from Argonne National Laboratory, operated by the University of Chicago for the U.S. Department of Energy under contract no. W-31-109-Eng-38.

REFERENCES

1. Wards Communications, Inc. *Wards automotive yearbook.* Detroit, MI: Wards Communications, 1992.
2. Motor Vehicle Manufacturers Association. *MVMA Motor Vehicle Facts & Figures '92.* Detroit, MI: MVMA 1993.
3. Davis, S. C., and S. G. Strang. Transportation Energy Data Book: Edition 13. Oak Ridge National Laboratory Report ORNL-6743, Oak Ridge, TN, 1993.
4. Keebler, J. Popular, durable plastic faces dilemma on recycling. *Automotive News,* March 5: 2 and 37, 1990.
5. Henstock, M. E. The impacts of material substitution on the recyclability of automobiles. In *Resources, conservation and recycling, 2,* 69–85. Amsterdam: Elsevier Science Publishers, 1988.
6. Field, F. R., and J. P. Clark. The recycling dilemma for advanced materials use: Automobile materials substitution. *Materials and Society* 15(2):109–147, 1991.
7. Brown, S., S. Steinstra, and K. E. Ludvigsen. *Closing the loop, the car recycling challenge.* London: Euromotor Reports Limited, 1993.
8. Secondary Reclamation of Plastics Waste, Phase 1 Report—Development of Techniques for Preparation and Formulation: Automobile/Appliances Shredder Residue, Mixed Industrial Waste, Curbside Separated Consumer Waste, Availabile from Plastics Institute of America, Inc., at Stevens Institute of Technology, Castle Point, Hoboken, NJ, 1987.
9. Hubble, W. S., I. G. Most, and M. R. Wolman. Investigation of the Energy Value of Automobile Shredder Residue. U.S. Department of Energy Report DOE/ID-12551, 1987.
10. McClellan, T. R. *Modern Plastics* 59(2):50–52, 1983.
11. Curlee, T. R., C. G. Rizy, and S. M. Schexnayder. Recent Trends in Automobile Recycling: An Energy and Economic Assessment. Oak Ridge National Laboratory Report ORNL/TM-12628, March 1994.
12. Dean, K. D., J. W. Sterner, M. B. Shirts, and L. J. Froisland. Bureau of Mines Research on Recycling Scrapped Automobiles. U.S. Department of Interior Bulletin 684, 1985.
13. Bilbrey, J. H., Jr., J. H. Sterner, and E. G. Valdez. Resource recovery from automobile shredder residue. *Conservation and Recycling Journal,* 1979.
14. Ellsworth, R. D., E. P. Ballinger, and R. B. Engdahl. Preliminary Survey of Development of an Incinerator for Removal of Combustibles from Scrapped Automobile Bodies. Final Report prepared by Battelle Memorial Institute for Institute of Scrap Iron and Steel, August 30, 1957.
15. Kaiser, E. R., and J. Tolciss. Smokeless Burning of Automobile Bodies. Coll. Eng. Tech. Report 764.2, New York University, June 1961.
16. Dean, K. D., and J. W. Sterner. Dismantling a Typical Junk Automobile to Produce Quality Scrap. U.S. Bureau of Mines Report of Investigations 350, U.S. Government Printing Office, Washington, DC, 1969.
17. Luntz, R. A. Separation of non-ferrous metals from hammermill fragmentizer. *Scrap Age* 13: October 1973.
18. Dean, K. D., J. W. Sterner, and E. G. Valdez. Effect of Increasing Plastics Content on Recycling of Automobiles. U.S. Bureau of Mines, U.S. Government Printing Office TPR 79, Washington, DC, December 1974.
19. Daellenbach, C. B., W. M. Mahan, and J. Drost. Utilization of automobile and ferrous refuse scrap iron in cupola iron production. *Proceedings of the Fourth Mineral Waste Utilization Symposium,* Chicago, May 7–8, 1974.
20. Auto shredder reduces junk cars in Michigan. *Scrap Age* 107: December 1974.
21. Sawyer, W. *Automotive scrap recycling: Processes, prices, and prospects.* Baltimore: Resources for the Future, 1974.
22. Sterner, J. W., D. K. Steele, and M. B. Shirts. Hand Dismantling and Shredding of Japanese Automobiles to Determine Material Contents and Metal Recoveries. U.S. Bureau of Mines Report of Investigations 8855, 1984.

23. Warner, A. J., C. H. Parker, and B. Baum. Solid Waste Management of Plastics. Prepared by DeBell and Richardson, Inc., for the Manufacturing Chemists Association, Washington, DC, 1970.
24. Mack, W. Recycling plastics at a profit. Twenty-Eighth Annual Western Conference of Society of Plastics Industry, May 1971.
25. Leidner, J. *Plastics waste: Recovery of economic value.* New York: Marcel Dekker, 1981.
26. Rayan, J. V., and C. C. Lutes. Characterization of Emissions from the Simulated Open-Burning of Non-Metallic Automotive Shredder Residue. U.S. Environmental Protection Agency Final Report EPA-600/R-93-044, March 1993.
27. Bilbrey, J. H., K. C. Dean, and J. W. Sterner. Design and Operation of an Automobile and Railroad Car Incinerator. American Institute of Mechanical Engineers Reprint A 74-88, 1974.
28. Woodruff, K. L., B. D. Mechel, and D. M. Albertson. Development of a fluidized bed combustion system for auto shredder residue. Appendix C of Reference 44.
29. Bonsignore, P. V., B. J. Jody, and E. J. Daniels. Separation techniques for auto shredder residue. Presented at the 1991 SAE International Congress and Exposition, Detroit. published as SAE Technical Paper No. 910854, pp. 59–63, 1991.
30. Daniels, E. J., B. J. Jody, and P. V. Bonsignore. Alternatives for recycling of auto shredder residue. *Journal of Resource Management and Technology* 20(1):14–26, 1992.
31. Jody, B. J., E. J. Daniels, and P. V. Bonsignore. A process of recycle shredder residue. DOE Case No. S-76, 923, Patent Application S.N. 972,426, 1994.
32. Jody, B. J., and E. J. Daniels. Recycling of plastics in automobile shredder residue. *Proceedings of the 25th IECEC Conference,* Reno, NV, August 1990.
33. Daniels, E. J., B. J. Jody, and P. V. Bonsignore. Automobile shredder residue: Process development for recovery of recyclable constituents. *Proceedings of the 6th International Conference on Solid Waste Management and Secondary Materials,* Philadelphia, December 1990.
34. Jody, B. J., E. J. Daniels, P. V. Bonsignore, and E. L. Shoemaker. Chemical and mechanical recycling of shredder fluff. *Proceedings of the Pollution Prevention Conference for Iron and Steel Industry,* Chicago, October 14–15, 1992.
35. Jody, B. J., E. J. Daniels, P. V. Bonsignore, and N. F. Brockmeier. A process to recovery plastics from obsolete automobiles using solvents at ambient pressure. *Proceedings of the 206th American Chemical Society National Meeting,* Chicago, August 22–27, 1993.
36. Lynch, J. C., and E. B. Nauman. Separation of commingled plastics by selective dissolution. *Proceedings of SPE-RETEC,* Charlotte, NC, October 30, 1989.
37. Lynch, J. C., and E. B. Nauman. Recycling of commingled plastics via selective dissolution. *Proceedings, 10th International Coextrusion Conference,* 1989, pp. 99–110.
38. Maximizing the Life Cycle of Plastics. Final Report, Plastics Institute of America, February 1980.
39. Program on recycling auto scrap, DOE/ECUT and PIA. *Inside R&D* 13(2): January 11, 1984.
40. Curlee, T. R. *The economic feasibility of recycling: A case study of plastic wastes.* New York: Praeger, 1986.
41. Smith, L. L., and R. M. Ramer. Recycled plastics for highway agencies. *Transportation Research Record* 1345:60–66, 1992.
42. Schmitt, R. J., Automobile Shredder Residue: The Problem and Potential Solutions. Center for Metals Production Report No. 90-1, Carnegie-Mellon Research Institute, Pittsburgh, January 1990.
43. Felix, J., and P. Gatenholm, Interphase Design in Cellulose Fiber-Polypropylene Composites, *Proceedings, American Chemical Society Meeting, Division Polymer Materials Science and Engineering,* 67:315, 1992, Washington, D.C.
44. Klason, C., J. Kubat, and P. Gatenholm. New Wood-Based Composites with Thermoplastics. In *Cellulosic utilisation,* ed. H. Inagahi. New York: Elsevier, 1989.

45. Felix, J., and P. Gatenholm. The effect of compatibilizing agents on the interfacial strength in cellulose-polypropylene composites. *Proceedings, American Chemical Society Meeting, Division of Polymer Materials Science and Engineering,* Atlanta, April 1991.
46. Klason, C., J. Kubat, and P. Gatenholm. Wood fibre-reinforced composites. In *Viscoelasticity of biomaterials,* ed. W. Glasser. Washington, DC: American Chemical Society, 1992.
47. Leaversuch, R. HDPE recycle: It's a big volume opportunity waiting to happen. *Modern Plastics* 65(8):44–47, 1988.
48. Weis, P., J. S. Weis, A. Greenberg, and T. J. Nosker. Toxicity of construction materials in the marine environment: A comparison of chromated-copper-arsenate-treated wood and recycled plastics. *Arch. Environ. Contam. Toxicol.* 22:99–106, 1992.
49. Braslaw, J., and J. L. Gerlock. Polyurethane waste recycling 2, polyol recovery and purification. *Industrial and Engineering Chemistry* 23:552, 1984.
50. Mahoney, L. R., S. A. Weiner, and F. C. Farris. Hydrolysis of polyurethane foam waste. *Environmental Science and Technology* 8(2): 1974.
51. Dean, K. C., C. J. Chindgren, and E. G. Valdez. Innovations in Recycling Automotive Scrap, Presented at the Institute of Scrap Iron and Steel, Annual Meeting, Washington, DC, January 15–18, 1972.
52. Banks, M. E., W. D. Lusk, and R. S. Ottinger. New Chemical Concepts for Utilization of Waste Plastics. U.S. Environmental Protection Agency Report SW-1GC, 1971.
53. Huang, C. J., and C. Dalton. Energy Recovery from Solid Waste. National Aeronautics and Space Administration Report NASA-CR-2526, 1975.
54. Jones, F., Cogen Designs, East Lansing, MI, private communications, September–October 1993.
55. Loomans, B. A., J. E. Kowalczyk, H. A. Lange, and J. W. Jones. Method of continuously carbonizing primarily organic waste material. U.S. Patent 5,017,269, May 21, 1991.
56. Taylor, R., General Conservation Crop., private communication, 1993.
57. Valdez, E. G., K. C. Dean, and J. H. Bilbrey. Recovering Polyurethane Foam and Other Plastics from Auto Shredder Reject. U.S. Bureau of Mines Report of Investigations 8091, 1975.
58. Automotive Plastics Recovery, *Plastics Recycling Update* 6(8), Resource Recycling, Portland, OR, August 1993.
59. Tesoro, G. Recycling of synthetic polymers for energy conservation—The state of the art. *Polymer News* 12:1987.
60. DuBois, J. H., and F. W. John. *Plastics.* 5th ed. New York: Van Nostrand Reinhold, 1967.
61. *The Wall Street Journal,* Forget recycling, turn plastics back into oil, Sept. 29, 1993, Dow Jones & Co., Inc., New York.
62. *Automotive Plastics Newsletter,* August:A-7, 1993. Market Search, Inc., Toledo, OH.
63. Van Stolk, A. G., A. C. Lewis, and C. A. Snavely. A new gasification process for auto fluff-depolymerization of auto shredder residue into hydrogen and carbon dioxide. *Proceedings of the 49th Annual SPE Technical Conference (ANTEC): In Search of Excellence,* 37:2, 142–2, 145, May 2–9, 1991.
64. National Dioxin Study Tier 4—Combustion Sources, Engineering Analysis. U.S. Environmental Protection Agency Report No. EPA-450/4-84-014h, prepared by Radian Corp., September 1987.
65. Tabery, R. S. Fluidized bed incineration of automobile shredder residue with energy recovery. Presented at the Shredder Operators Conference, Austin, TX, November 29, 1990.
66. Hock, H., and M. A. Maten. A Preliminary Study of the Recovery and Recycling of Automotive Plastics. SAE Special Publication No. 966, pp. 59–71, 1993.

CHAPTER
SIX

PYROLYTIC REPROCESSING OF SCRAP TIRES INTO VALUE-ADDED PRODUCTS

Michael A. Serio, Marek A. Wójtowicz, Hsisheng Teng, and Peter R. Solomon

INTRODUCTION

Scrap tires present formidable disposal problems. The same properties that make them desirable as tires, notably their durability, also make their disposal and reprocessing difficult. Tires are well known to be virtually immune to biological degradation. Landfilling of the 280 million tires that are generated each year in the United States is becoming an unacceptable solution [1]. In addition to the continuous flow of waste tires, approximately 2–3 billion tires are already stored in piles through the United States; illegal dumping is also becoming a serious problem. The tires take up large amounts of valuable landfill space, provide breeding sites for mosquitoes and rodents, and present fire and health hazards. Recently, a large mountain of tires caught on fire in Canada, with widespread environmental consequences resulting from the oils and gases generated from the decomposing rubber. Such fires are difficult to contain because of the high flammability of tire material and because of the pockets of air present in the piles. One fire in Huntington, Virginia, burned for 9 months.

The most desirable solution from an environmental and economic standpoint is to thermally reprocess the tires into value-added products [2]. Large-scale efforts employ tires either as a fuel (Oxford Energy Corporation, Modesto, CA)

or as a filler for asphalt (Rubber Asphalt Producers, Phoenix, AZ). These two technologies annually consume about 28 million tires [3]. However, tire burning has had repeated problems with feeding the tires and slagging, and the rubber asphalt costs 40% more than conventional material. RW Technology, Cheshire, Connecticut, has also tried to convert tires into other plastic products, but the market niche seems to be small.

The success of a tire-processing technology is strongly dependent on process economics and environmental performance (emissions). The two most important factors affecting process economics are the tipping fees charged for tire disposal and the selling prices of the products [4]. The tipping fees have been steadily increasing over years and this trend is expected to continue. Depending on the geographic region, the fees vary from $35 to $108 per ton of whole tires delivered in mass quantities [5]. Selling prices of the products depend on the process, with fuel- and energy-producing technologies yielding rather low returns on investment. Thus, it would be desirable to develop a process based on the recovery of value-added products such as carbon black, activated carbon, and valuable chemicals (e.g., benzene, toluene, and xylene). Pyrolysis of scrap tires appears to be a logical choice, the more so that apart from minor fugitive sources and equipment leaks, virtually no emissions are produced by this process [4, 5].

Nippon Zeon estimated that the break-even cost of its tire pyrolysis pilot plant was $0.25 per tire [6]. According to a different estimate [7], a plant processing 81,000 tons of scrap tires per year could be made profitable, based on sales of reclaimed products. Economic analysis of several pyrolysis units, however, has shown negative cash flows [4], presumably due to insufficient value of the products. Significant improvement of the economic leverage can be accomplished through upgrading of the primary products to obtain secondary products of higher value. This constitutes the essence of the approach advocated in this study.

The proposed solution is to thermally reprocess the used tires into activated carbon, other solid carbon forms (e.g., carbon black, graphite, or carbon fibers), and liquid and/or gaseous fuels. The process is based on pyrolysis of tires, and the key to its successful development is understanding and controlling the chemistry of low-temperature carbonization so that the yield and physical properties of the products can be optimized.

Pyrolysis has been widely used to convert solid fuels, such as coal, into liquid and gaseous hydrocarbons, a process that results in a solid char residue. Used automotive tires contain polymeric structures that, in some respects, are similar to those of coal. Not surprisingly, typical elemental compositions of both materials are also similar (rubber: ~88% C, ~8% H, ~2% O, ~0.5% N, ~1.5% S [8–10]; and coal: 65–95% C, 2–7% H, 3–30% O, 0.2–2% N, 0.4–10% S). Coal pyrolysis has been extensively studied [11–13], but investigations of tire pyrolysis are rarely reported in the open literature.

The most commonly used vulcanized tire rubber is a styrene-butadiene-copolymer (SBR) containing about 25 wt % styrene [8]. A typical composition of tire rubber is 60–65% SBR, 29–31% carbon black, 1.9–3.3% zinc oxide, 1.1–2.1% sulfur, ~2% extender oil, and ~0.7% additives (wt. %, as received) [8, 14]. In most cases, tire pyrolysis studies were performed under inert conditions [8, 15, 16]. Pyrolysis may also be carried out in mildly oxidizing atmospheres, such as steam and carbon dioxide, to improve the quality of pyrolytic products [16–18].

Tire pyrolysis experiments have usually been conducted in the temperature range 773–1173 K [8, 14, 15]. As with coal pyrolysis, the principal products of thermal degradation of tires are gases, liquid oils, and solid carbon residues. The following yields of tire pyrolysis are typical (on an as-received basis): 33–38 wt % char, 38–55 wt % oil, and 10–30 wt % gas. The product yields are affected by the pyrolysis conditions, such as pyrolysis temperature and heating rate. The literature on the analysis of products from tire pyrolysis is summarized in the following sections.

Gas Analysis

Gases produced from tire pyrolysis are mainly hydrogen, carbon dioxide, carbon monoxide, methane, ethane, and butadiene, with lower concentrations of propane, propene, butane, and other hydrocarbon gases [8]. The gas has a calorific value between 170 and 2375 Btu/ft^3 [5], with an average of 835 Btu/ft^3 [19] (natural gas averages 1000 Btu/ft^3).

Oil Analysis

The yield of oil from tire pyrolysis is high (~50 wt % of initial tire rubber), reflecting the potential of tire rubber to act as a substitute for fossil fuel and chemical feedstocks. The oils have high aromaticity, low sulfur content, and are considered relatively good fuels [18]. The molecular weight range for the oils is up to 1600 g/mol, with an average molecular weight between 300 and 400 g/mol [8]. Infrared analysis of the oils indicates the presence of alkanes, alkenes, ketones or aldehydes, and aromatic, polyaromatic, and substituted aromatic groups [8]. Tire-derived oils may be an important source of refined chemicals as they contain high concentrations of valuable chemical feedstocks, such as benzene, toluene, and xylene [8].

Carbon Residue Analysis

The carbon residue could become a marketable product if its properties were similar to those of activated carbons. In fact, the main reason why pyrolysis of scrap tires has not been commercially successful is the problem with upgrading the quality of char. As mentioned before, tire pyrolysis performed in an inert environment can produce 33–38 wt % carbon residue. It has been reported that the char yield increases with decreasing pyrolysis temperature and decreasing heating rate [8]. The surface area of the tire char also depends on pyrolysis temperature and heating rate. For a tire char produced by pyrolysis in an inert gas, surface area usually ranges from 30 to 90 m^2/g [8, 14, 15].

There are two main uses of tire chars: as a reinforcing filler and as an adsorbent. Commercial carbon black is usually used for filling polymers and vulcanizates. The use of carbon black derived from tire char in the tire and printing-ink industries has been reported to be unsuccessful [8, 15], mainly due to the high ash content. Chars from tire pyrolysis contain as much as 15 wt % ash, with the majority being zinc oxide [15]. Removing the ash from tire char is an important issue in the process of producing useful carbon black from the solid waste-tire residue. An alternative approach, which is advocated in the current study, is to use the solid residue to produce activated carbon for which the ash content is less critical. High-quality carbon black can be made from the liquid products, which are absolutely ash free. Finely divided carbon can also be obtained from the CO produced during char activation.

Carbon as an adsorbent is usually evaluated by its surface area, and this is why an activation process is necessary to produce salable activated carbons from tire char. To develop a high internal surface area, char can be activated by mild oxidation with steam or carbon dioxide at high temperatures. The slow kinetics of carbon gasification in steam and CO_2 allow gas molecules to diffuse into the micropores and enlarge the existing surface area. The activation process usually follows hydrocarbon pyrolysis performed in an inert environment, but it is possible to accomplish pyrolysis and activation in one stage by pyrolyzing under mildly oxidizing conditions [14].

Ogasawara et al. [14] carried out the pyrolysis and activation of tires in one stage. The carbon residue from 1 hour of steam activation at 1173 K was found to have a surface area of 1260 m^2/g, whereas pyrolysis in helium yielded char with a surface area of 87 m^2/g. The carbon residue produced from this "wet method" is as good as the commercial activated carbon in terms of surface area, but the carbon yield was only 9 wt % of the starting tire material. Therefore, a method for increasing char yield from tire pyrolysis is one of the most important issues in making activated carbon from waste tires. In a recent study [9], a high-surface-area activated carbon (>800 m^2/g) was produced in relatively high yields by pyrolysis at up to 1173 K, followed by CO_2 activation at the same temperature. The surface area of this carbon is comparable to that of commercial activated carbons.

The objective of the present study is to propose and carry out preliminary assessment of a process leading to the conversion of waste tires into marketable products such as activated carbon or carbon black. In addition, the pyrolysis-based process would also produce some quantities of gaseous and liquid fuels. Process flexibility, in terms of the type and amounts of particular products, is a desirable feature in view of high variability of markets.

PROCESS DESCRIPTION AND DISCUSSION

The concept of pyrolytic reprocessing of waste tires into value-added products is presented in Figure 1. Characteristics of the reactants and products, process conditions, and the yields of pyrolytic products are based on an extensive literature review and on the data collected in our laboratory [9]. As shown in Figure 1, the primary products of tire pyrolysis are oils, char, and pyrolytic gas, also known as pyro-gas. Tire pyrolysis yields substantial quantities of oils and char, which undergo further processing toward secondary, high-value products. According to the proposed scheme, char upgrading is implemented in a closed-loop activation step, which, in principle, eliminates undesirable by-products and emissions. A stream of carbon dioxide is used for high-temperature char activation, which is followed by CO_2 recovery through the reverse Boudouard reaction. Char upgrading results in producing high-surface-area activated carbon and finely divided carbon, which will be referred to as the Boudouard carbon. Ash-free oils are turned into carbon black using the well-known furnace process. As an alternative, oils can be separated into valuable chemical feedstocks by means of distillation. It is expected that product upgrading will greatly improve the economic leverage of scrap-tire pyrolysis.

A schematic of the proposed process is shown in Figure 2. A stream of scrap tires enters the pyrolyzer, where it comes into contact with a recycle stream of product gas that acts as a carrier gas for pyrolysis. Mixing used tires with other waste material and solid-waste pretreatment are optional features of the process and are discussed below. An alternative arrangement involves recycling part of the pyrolysis oil into the pyrolyzer to sweep the volatile products out of the reactor. This option is represented by a dashed line in Figure 2. It is expected that the recycled oil will undergo (partial) cracking, leading to an altered gas composition at the reactor outlet. System flexibility allows for wide variations in the ratio of gas to oil recycling, depending on product characteristics and requirements. The pyrolyzer operates at approximately 1173 K, and the product streams are fuel gas, which also contains pyrolysis liquids, and char. The pyrolysis liquids are separated from the volatiles in a condenser, and they are subjected to partial combustion in an oil furnace to produce carbon black. The high-Btu gas is combusted to supply the process heat requirements. The solid product does not have sufficient surface area to make it commercially attractive and thus an activation step is needed. Char activation is carried out using CO_2 at approximately 1173 K, which results in CO_2 reduction to form carbon monoxide. The consumed CO_2 is regenerated in a Boudouard reactor with simultaneous formation of finely divided carbon. The net result of the process is the conversion of used tires into three marketable products: activated carbon, carbon black, and Boudouard carbon. The particular elements of the process scheme are described in more detail below. Pyrolysis, char activation, and integration of these steps constitute the heart of the process. The production of carbon black

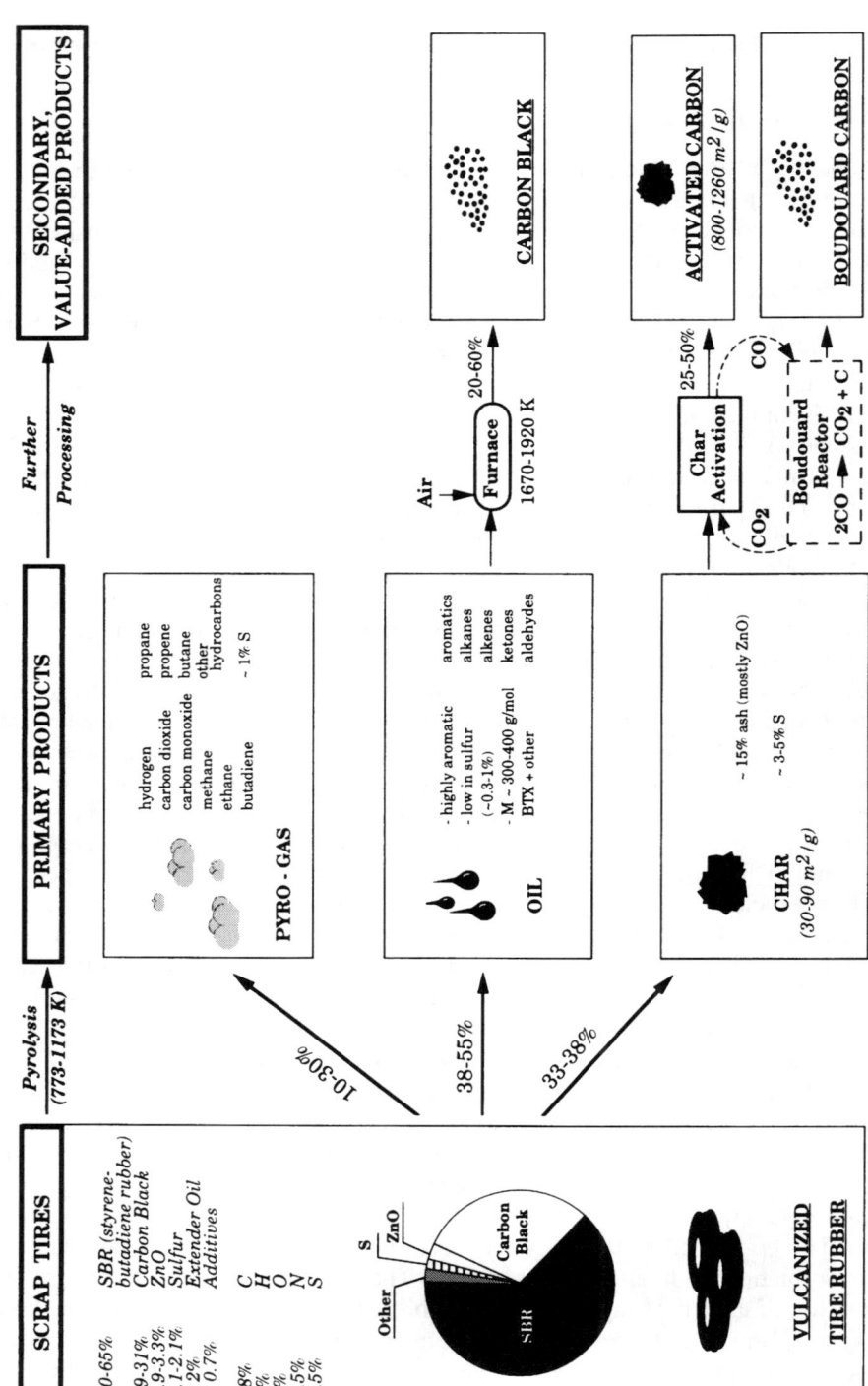

Figure 1 The concept of pyrolytic reprocessing of scrap tires into high-value products.

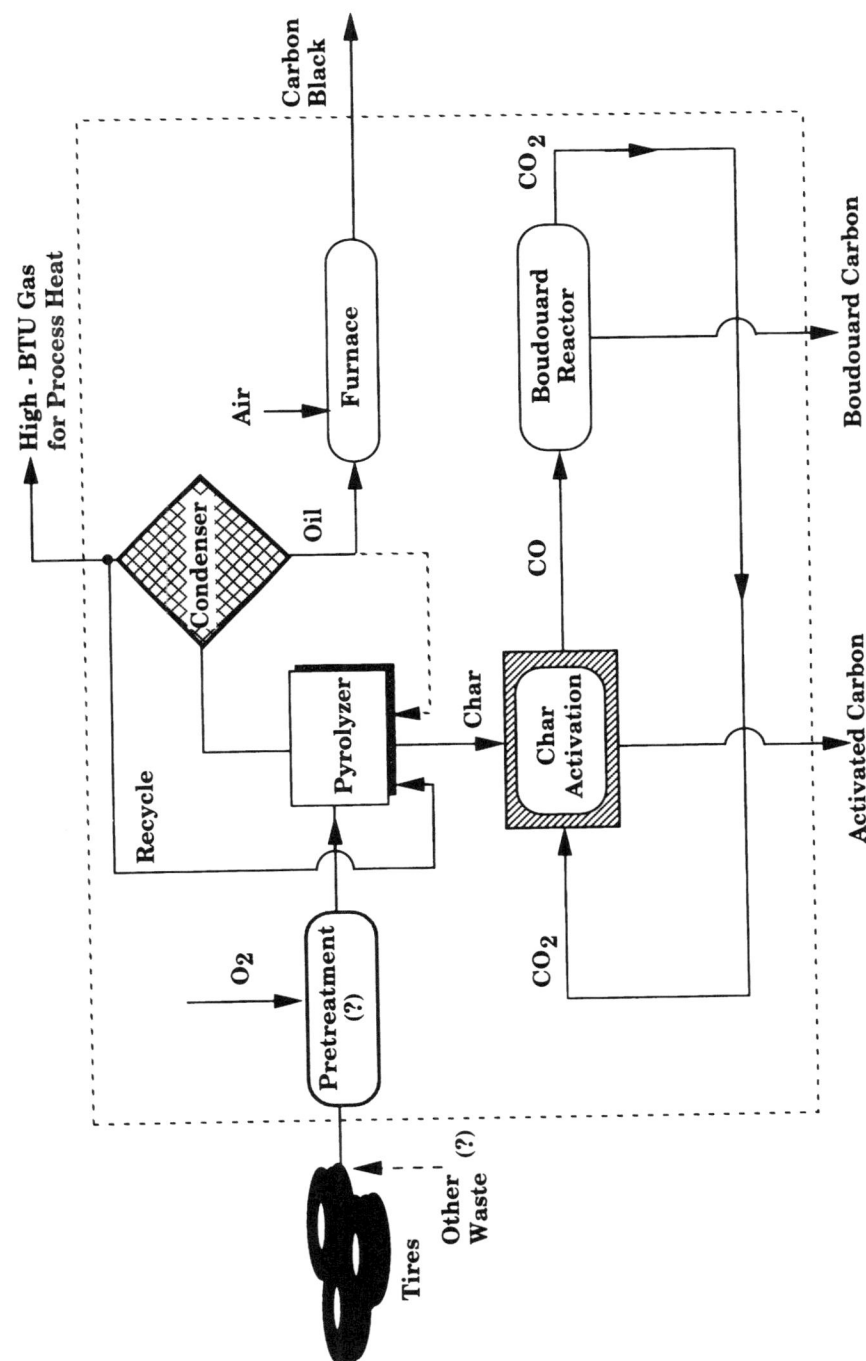

Figure 2 A schematic diagram of the proposed tire pyrolysis process.

and Boudouard carbon is very much a novelty in the proposed scheme, although it will be integrated into the process by adapting existing technologies.

Feed Stream Preparation

The tires are first shredded and then possibly combined with another waste stream. The use of mixed wastes would improve the appeal and the number of sites where a plant could be installed. This feature, however, increases the complexity of the design as well as the operating costs. The amount of size reduction required for scrap tires does not appear to be excessive. In fact, higher char activation efficiencies were reported when coarser waste material was used ($\sim$170 mg pieces versus -50 mesh particles) [9]. The size of scrap tire material that would be optimal for the process is still to be determined. Because the cost of shredding increases with decreasing particle size, there is an economic incentive to make the tire pieces as large as possible. For smaller particle sizes, pretreatment with O_2 was found to increase char yields and surface areas [9].

Pyrolysis and Char Activation Units

A rotary-kiln reactor has been used successfully for pyrolysis of tires, but this is not necessarily the optimal solution for a high-throughput system [15]. A kinetic analysis of tire pyrolysis indicates that the process is relatively fast and can be completed in under a second at 873 K. This fact would allow the use of an entrained-flow reactor, which is quite appropriate for high-throughput applications. Unfortunately, only relatively small particles can be fed into such a reactor, mainly due to heat transfer limitations. For somewhat larger pieces of tire material, a fluidized-bed reactor could be used. In either case, a hot cyclone is desirable to separate char particles from the volatiles. According to the scheme presented in Figure 2, char activation by CO_2 takes place in a separate reactor. This has the advantage of allowing the oxides of carbon to be cycled between the activation step (producing CO) and a disproportionation step (producing CO_2 and carbon). In this way, CO_2 is periodically consumed (activation) and regenerated (disproportionation); both CO and CO_2 stay relatively pure in this scheme.

Another possibility is to use a moving-bed countercurrent system. Although solids throughput would necessarily be lower, this disadvantage would be offset by combining the pyrolysis and char activation stages in a single reactor. In this scheme, tire pieces would be fed into the reactor at the top and CO_2 at the bottom (Figure 3). A temperature of about 1173 K would prevail at the bottom of the reactor so that the char would be progressively pyrolyzed and then activated as it moves through the reactor. Such a design, shown schematically in Figure 4, simplifies the process of feeding the tire pieces and eliminates the requirement for a hot cyclone, because the volatiles and char come out of opposite ends of the reactor. The moving-bed concept has been employed in the

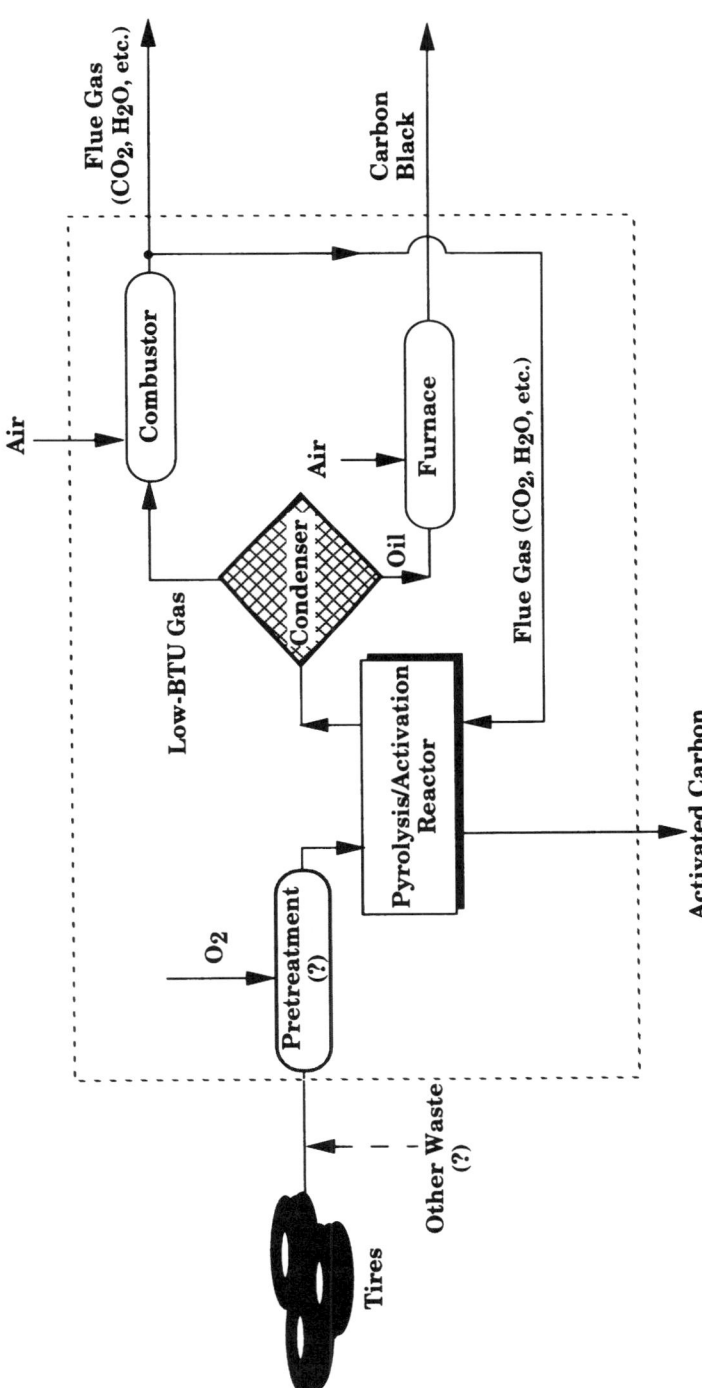

Figure 3 The concept of a countercurrent moving-bed reactor that combines tire pyrolysis and char activation in a single unit.

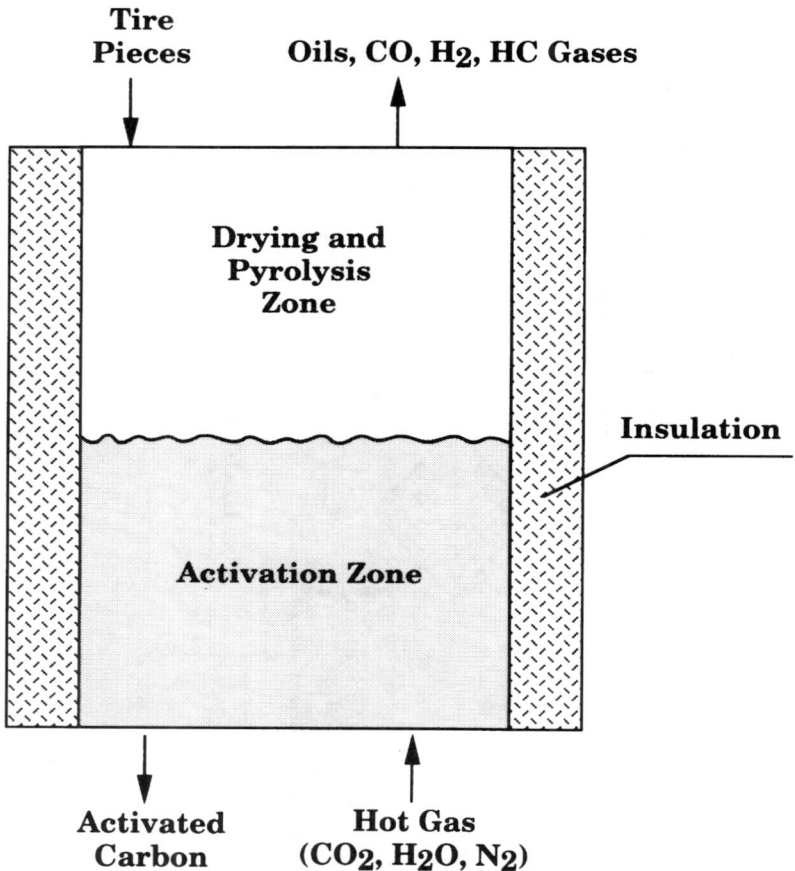

Figure 4 Schematic representation of the proposed tire pyrolysis process involving a single pyrolysis/char activation reactor.

large-scale gasification of coal at a plant in North Dakota that is producing synthetic pipeline gas [20]. There are some differences between the two schemes, however. The tire-processing unit would operate under atmospheric pressure, because there is little or no advantage in producing activated carbon at an elevated pressure. Second, the coal gasification system is internally heated by introducing oxygen at the bottom and partially combusting a portion of the coal. In the tire-processing reactor, the pyrolysis gases that come off the top of the unit would be combusted externally and the hot exhaust gases (mostly CO_2 and H_2O) would be introduced into the bottom of the reactor for char activation. The reactor would be insulated to prevent heat losses. The disadvantages of the single-reactor scheme shown in Figure 3 are that (1) the pyrolysis gas has a lower heating value due to dilution with the flue gas and (2) if Boudouard carbon is to be produced, a fairly complex separation step is required to recover pure CO from the pyrolysis gas (Figure 5). Previous work has demonstrated that the final activated carbon product is not very sensitive to the conditions under which pyrolysis is carried out [9]. This allows greater design flexibility and the possibility of considering several feasible schemes.

It has been shown that activation using an 8% CO_2-He mixture at 1173 K for 3–10 hours was sufficient to obtain high-surface-area chars [9]. This residence time could be reduced by raising the temperature, increasing the CO_2 partial pressure, or by using steam rather than CO_2. The char activation chemistry can be represented by the following two reactions:

CO_2 activation: $CO_2 + C \rightleftharpoons 2CO$ (the Boudouard reaction)

Steam activation: $H_2O + C \longrightarrow CO + H_2$

Both reactions are endothermic and the equilibrium becomes more favorable as the temperature increases. The use of steam would offer advantages in terms of a higher reaction rate and a higher surface area product. The use of CO_2 would increase the production of CO from the activation step. The CO could then be subjected to the reverse Boudouard reaction for the production of finely divided carbon [21–24].

Carbon Black Production

The oils resulting from tire pyrolysis can be used to produce carbon black, because the properties of tire-derived oils are known to be similar to those of the petroleum fraction used in carbon black production [18]. In the oil furnace process, a highly aromatic feedstock is converted to carbon black by partial combustion and pyrolysis at 1673–1923 K in a refractory-lined steel reactor. The carbon black properties that are important in reinforcement material applications (e.g., in tires) are the particle size and structure (degree of agglomeration into three-dimensional networks). These properties are controlled by the nozzle design, reaction chamber geometry, temperature, residence time, and turbulence intensity [25]. As an alternative to carbon black production, tire pyrolysis oil can be used for its fuel value, although further processing to remove aromatic components would be required. The conversion of the oil stream to carbon black is a more attractive option for a number of reasons: (1) a solid product is easier to store and handle; (2) the value of the solid product is higher than that of any possible fuel; (3) little or no upgrading of the material is required; (4) possible transportation fuel uses would require upgrading because of the relatively high aromatic content of oil; and (5) since carbon black is required to make tires, the production of carbon black from the oil is a form of tire recycling and seems

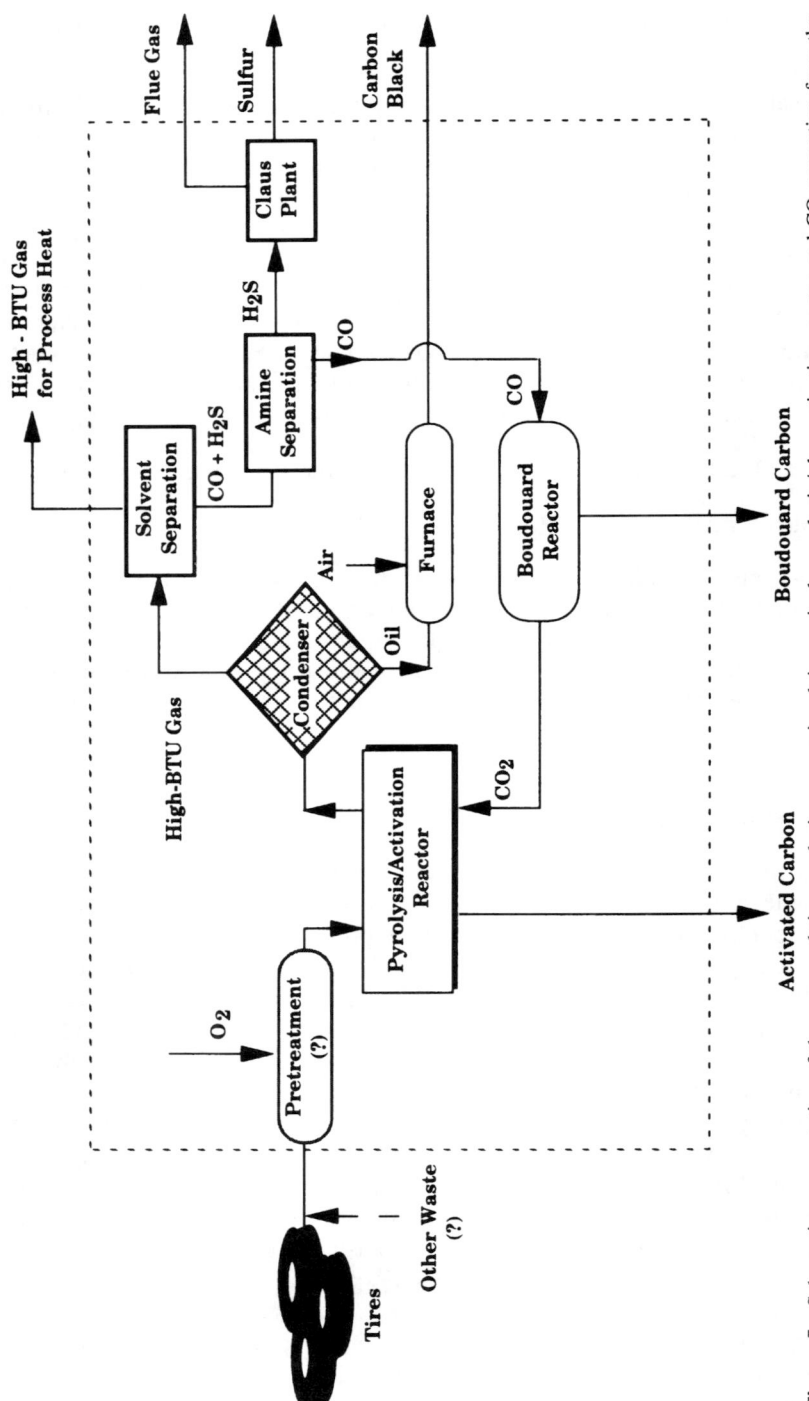

Figure 5 Schematic representation of the proposed tire pyrolysis process involving a single pyrolysis/char activation reactor and CO separation from the pyrolysis gas.

to be more practical than trying to recover the original carbon black from the char.

Boudouard Carbon Production from CO

This process step was discussed by Walker [21] as a part of his coal-processing scheme. The temperature of the Boudouard reactor is maintained at 773–800 K and the carbon product is ash free and of a particle size much smaller than can be obtained by grinding [26]. In fact, the particle size of the Boudouard carbon is small enough to burn completely in a diesel engine [27]. The Boudouard carbon is similar to carbon black (except for having zero hydrogen content) and in some cases can probably be substituted for carbon black. The potential applications of this product include the manufacture of colorants and lubricants.

The production of finely divided carbon from CO has never been commercialized, presumably because of the relatively high value of CO and uncertainties associated with marketing of the Boudouard carbon. During World War II, however, the reaction was used in Germany to produce a substitute for carbon black [28]. Although some information on the reaction kinetics and product characteristics has been available for a long time [29, 30], more research and development is certainly needed to make the process commercially viable. In particular, it is expected that the following topics will have to be addressed: (1) product quality, (2) the determination of reaction rates at various CO/CO_2 ratios in the feed gas, (3) reactor type selection and design, (4) separation of the Boudouard carbon from the iron catalyst, and (5) exploration of noncatalytic reaction pathways.

Energy and Mass Balances

The fuel gases produced during pyrolysis can be used to provide process heat for the pyrolysis and activation stages. Some heat can also be recovered from the manufacturing of carbon black by partial combustion of the oil and from the production of Boudouard carbon. Preliminary estimates show that the process outlined in Figure 2 will be self-sufficient from an energy standpoint. Existing tire pyrolysis units (e.g., the one operated by Conrad Industries in Centralia, WA) usually generate more pyro-gas than is needed for process heat, and excess gas is often burned in an outside flare. On-site energy savings could be realized to improve process economics.

The net yields of the products depend on the degree of activation required in the production of activated carbon. A high degree of activation would correspond to a larger amount of Boudouard carbon produced from the CO stream. Estimated yields are shown in Figure 1.

Process Economics

A preliminary cost analysis for the process indicates that the proposed tire-reprocessing scheme shown in Figure 2 could be profitable. The following formula is used to evaluate process economics:

$$P = F + R - C - T - S - D$$

where P is the profit, F is the tipping fee collected for tire disposal, R is the revenue received from the sale of products, C is the processing cost for operating the facility, T is the cost of transportation of tires, S is the cost of tire shredding, and D is the cost of disposal of waste products. Furthermore,

$$R = R_1 + R_2 + R_3$$

where R_1, R_2, and R_3 are selling prices of activated carbon, carbon black, and the Boudouard carbon, respectively. The value of fuel gas is neglected, although proper utilization of this source of energy can improve process economics. All the foregoing terms can be expressed in dollars per tire, and the main features of the cost analysis are discussed below.

Pyrolysis yields of 35% char, 20% pyrolytic gas, and 45% oils are assumed as well as a 50% char burn-off during activation. The following additional assumptions are made:

1. The average weight of a tire is 20 lb.
2. The carbon-black yield in the furnace process is 40%.
3. The most economical tire-processing capacity is 100 tons/day (i.e. ~4 million tires per year) [19].
4. The capital cost of such a plant is $20 million (two times larger than the capital investment estimated in Reference 19 because more equipment is required).
5. Tipping fee charged by the plant: $0.50/tire.
6. Selling price of activated carbon: $0.30/lb (activated carbon for wastewater treatment sells for about $0.68/lb [31]).
7. Selling price for carbon black: $0.16/lb (high-quality carbon black sells for $0.28/lb [5]).
8. Selling price of the Boudouard carbon: $0.13/lb (comparable to the price of low-grade carbon black derived from scrap-tire char [5]).
9. Cost of plant operation: $0.50/tire (comparable to the cost of operating the tire incineration plant in Modesto, CA [5]).
10. Tire transportation: $0.20/ton/mile (this cost usually ranges from $0.15 to $0.20 ton/mile for transportation of whole tires over a distance of 100 miles [5]).
11. Average distance over which tires need to be transported: 100 miles.

12. Tire shredding cost: $0.40/tire (the typical cost charged by shredding companies was $0.19–$0.75/tire in 1989 [5]; shredding costs vary depending on the fineness of the product).
13. Waste-product disposal cost is negligible.

Simple calculations based on the above assumptions lead to the following figures ($/tire): $F = 0.50$, $R_1 = 1.05$, $R_2 = 0.58$, $R_3 = 0.45$, $C = 0.50$, $T = 0.18$, $S = 0.40$, and $D = 0$. This is equivalent to the net profit $P = \$1.50$/tire, an annual gross income of $6 million, and a payback period on the investment of approximately 3.3 years.

Of course, more detailed economic analysis needs to be performed, but even this simplified and fairly conservative treatment demonstrates the high commercial potential of the proposed approach. It should be pointed out that the profitability of the process is expected to increase in time because of increasing tire disposal charges.

For comparison, the cost of scrap-tire incineration without heat recovery is about $0.02–0.03/lb [32] (i.e., about $0.40–0.60/tire). The cost of power generation from waste tires is 1.5–5.1 times higher than the corresponding cost for coal, depending on whether whole or ground tires are used [32]. Depolymerized scrap rubber (DSR) can also be used as liquid fuel but, in general, depolymerization is difficult and involves extensive high-pressure treatment. Clearly, reprocessing of waste tires into fuels does not seem to be an economically attractive option.

As demonstrated above, pyrolysis processes with value-added products can be made profitable and competitive with respect to incineration plants. In addition, despite its reasonably good environmental record, tire incineration is facing increasing opposition from communities concerned about potential emissions. This noneconomic barrier is expected to be significantly lower in the case of scrap-tire pyrolysis. Asphalt rubber is probably the most competitive tire-derived product currently on the market. Although nearly twice as expensive as regular pavement, asphalt rubber has already demonstrated superior performance and durability. According to legislation passed in 1991, the use of asphalt rubber may be required in 5% of new pavement as early as in 1994, with a projected increase to 20% by 1977 [5]. Unfortunately, the estimated total demand for this product is still at about 2% of the amount of scrap tires available [32]. Therefore, it can be concluded that a large market exists for other technologies targeted at scrap-tire reprocessing into useful products.

Product Utilization

Tire-derived activated carbons could be used, for example, in wastewater treatment, stabilization of landfills, and recovery of organic solvents and vapors. The

Boudouard carbon can have a variety of uses that are yet to be explored; they include the manufacture of colorants and lubricants. The liquid stream can be utilized in the production of carbon black, fuel oil, or valuable chemical feedstocks. The oil-derived carbon black can be reused in tire manufacturing, which forms a recycle loop for this material. Pyrolytic gas can be used for process heat generation.

CONCLUSIONS

A preliminary process design was developed in which scrap tires are used as the input and in which activated carbon, carbon black (or fuel oil), Boudouard carbon, and fuel gas are produced as the output. The proposed technology has the potential to convert an unmitigated waste stream of tires into marketable products (activated carbon, Boudouard carbon, and carbon black). Economic feasibility of the process has been demonstrated. The typical feedstocks for activated carbon and carbon black are either coal or petroleum, and the proposed technology would obviate the need to deplete these resources. The process allows a high degree of flexibility in the relative amounts of each product to reflect changes in the feed stream and market conditions. Another advantage of the system is the production of both activated carbon and carbon black from used tires, with inorganic material ending up in the product with a high tolerance for this component. Finally, following additional research, the proposed technology should be able to find applications in reprocessing other polymer wastes found in American industry (e.g., end cuttings from automobile hoses or products that do not pass quality control). Further research and development of the process and study of product utilization are under way.

ACKNOWLEDGMENTS

The financial support of the National Science Foundation under grant III-9215045 is gratefully acknowledged. The Project Officer is Dr. Edward Bryan.

REFERENCES

1. *New York Times.* Editorial, May 9, p. D1, May 9, 1990.
2. Schulman, B. L., and P. A. White. Pyrolysis of scrap tires using the Tosco II process: A progress report. In *Solid wastes and residues: Conversion by advanced thermal processes,* ed. J. L. Jones and S. B. Radding, 274. ACS Symposium Series 76. Washington, DC: American Chemical Society, 1978.
3. U.S. Environmental Protection Agency. Office of Solid Waste. Markets for Scrap Tires. Report No. EPA/530-SW-90-074B, 1991.

4. Dodds, J., W. F. Domenico, D. R. Evans, W. Fish, P. L. Lassahnn, and W. J. Toth. *Scrap tires: A resource and technology evaluation of tire pyrolysis and other selected alternate technologies.* Washington, DC: U.S. Department of Energy, 1983.
5. U.S. Environmental Protection Agency, C. Clark, K. Meardon, and D. Russell, *Scrap tire technology and markets.* Park Ridge, NJ: Noyes Data Corporation, 1993.
6. Saeki, Y., and G. Suzuki. Fluidised thermal cracking process for waste tyres. *Rubber Age* 108: 33, February 1976.
7. Bracker, G. P. Pyrolytic resource recovery. *Conserv. Recycl.* 4(3):161, 1981.
8. Williams, P. T., S. Besler, and D. T. Taylor, The pyrolysis of scrap automotive tyres. *Fuel* 69: 1474, 1990.
9. Teng, H., M. A. Serio, R. Bassilakis, P. W. Morrison, Jr., and P. R. Solomon. Reprocessing of used tires into activated carbon and other products. *ACS Div. Fuel Chem. Prepr.* 37(2):533, 1992.
10. Ohio Air Quality Development Authority. Air Emissions Associated with the Combustion of Scrap Tires for Energy Recovery. Prepared by Malcolm Pirnie, Inc., 1991. Cited in Reference 5.
11. Howard, J. B. Fundamentals of coal pyrolysis and hydropyrolysis. In *Chemistry of coal utilization,* second supplementary volume, ed. M. A. Elliott, 665. New York: Wiley, 1981.
12. Gavalas, G. R. *Coal pyrolysis.* Amsterdam: Elsevier, 1982.
13. Solomon, P. R., M. A. Serio, and E. M. Suuberg. Coal pyrolysis: Experiments kinetic rates and mechanisms. *Prog. Energy Combust. Sci.* 18:133, 1992.
14. Ogasawara, S., M. Kuroda, and N. Wakao. Preparation of activated carbon by thermal decomposition of used automotive tires. *Ind. Eng. Chem. Res.* 26:2552, 1987.
15. Petrich, M. A. Conversion of Plastic Waste to Valuable Solid Carbons: Final Report of a Project in the Innovative Concepts Program. U.S. Department of Energy, 1991.
16. Torikai, N., T. Meguro, and Y. Nakamura. Pore size distribution of activated carbons made from tires which contain carbon blacks differing in particle size. *Nippon Kagaku Kaishi* 11:1604, 1979.
17. Funazukuri, T., T. Takanashi, and N. Wakao. Supercritical extraction of used automotive tire with water. *J. Chem. Eng. Japan* 20:23, 1987.
18. Merchant, A., and J. M. Torkelson. *Pyrolysis of scrap tires.* Evanston, IL: Chemical Engineering Department, Northwestern University, 1990.
19. Wolfson, D. E., J. G. Beckman, J. G. Walters, and D. J. Bennett. Destructive Distillation of Scrap Tires. Report No. 7302, U.S. Department of Interior, Bureau of Mines, 1973.
20. Penner, S. S. Coal Gasification: Direct Application and Synthesis of Chemicals and Fuels. Report No. DOE/ER-0326, U.S. Department of Energy, 1987.
21. Walker, D. G. *ACS Div. Fuel Chem. Prepr.* 36(3):1129, 1991.
22. Watanabe, T. The nature of the carbon produced by the catalytic decomposition of carbon monoxide with iron. *Bull. Inst. Phys. Chem. Research (Tokyo)* 8:288, 1929.
23. Donald, J. H. *An annotated bibliography of the literature and patents relating to the production of carbon by the decomposition of carbon monoxide.* Pittsburgh: Mellon Institute of Industrial Research, 1956.
24. Walker, D. G., and L. Hadley-Coates. 1988. Boudouard carbon, an alternative to gasoline and diesel oil. *International Journal of Energy Research* 12:243, 1988.
25. Austin, G. T. *Shreve's chemical process industries.* 5th ed. New York: McGraw-Hill, 1984.
26. Soehngen, E. E. The Development of the Coal Burning Diesel in Germany. Report FE/WEPO 3387-1 (Purchase Order No. 3387), U.S. Energy Research and Development Administration, 1976.
27. Essenhigh, R. H. Coal combustion. In *Coal conversion technology,* ed. C. Y. Yen and S. Lee, 171. Reading, MA: Addison-Wesley, 1979.
28. B.I.O.S. Final Report No. 1399, Item No. 22. Cited in Reference 29.
29. Walker, P. L., Jr., J. F. Rakszawski, and G. R. Imperial. Carbon formation from carbon monoxide-hydrogen mixtures over iron catalysts. I. Properties of carbon formed. *J. Phys. Chem.* 63:133, 1959.

30. Walker, P. L., Jr., J. F. Rakszawski, and G. R. Imperial. Carbon formation from carbon monoxide-hydrogen mixtures over iron catalysts. II. Rates of carbon formation. *J. Phys. Chem.* 63:140, 1959.
31. Petrich, W., private communication, August 1993.
32. Paul, J. Rubber reclaiming. In *Encyclopedia of polymer science and engineering,* Vol. 14, ed. H. F. Mark, N. M. Bikales, C. G. Overberger and G. Menges, 787. New York: Wiley, 1988.

CHAPTER
SEVEN
COPROCESSING OF COAL WITH WASTE TIRES AND POLYMERS: IN SITU ELECTRON SPIN RESONANCE INVESTIGATIONS

M. M. Ibrahim and M. S. Seehra

INTRODUCTION

Proper disposal or benefaction of waste tire rubbers and waste polymers is of great concern, both environmentally and for economic viability. As a good alternative route to disposal, Farcasiu and Smith [1] have investigated the coliquefaction of coal with waste tire rubbers. There is an increasing interest in promoting coliquefaction of coal not only with waste tire rubbers but also with waste polymers and waste oils and in establishing catalytic and thermal liquefaction routes to get better yields and quality of oils [2]. In this chapter we report on the use of electron spin resonance (ESR) spectroscopy in understanding the coliquefaction of Blind Canyon coal with waste tire rubbers and polymers.

The free radicals in coals, as monitored by electron spin resonance spectroscopy, are an integral part of the coal structure, as their characteristic properties vary systematically with coal rank and coal macerals [3, 4]. When coals are subjected to pyrolysis, new free radicals are generated through coal depolymerization whose intensities vary in a systematic fashion with pyrolysis temperature in bituminous coals [5–7]. In the process of direct coal liquefaction, hydrogenation of the products of coal pyrolysis must occur, resulting in capping of the free radicals by transferred hydrogen [8–10]. Therefore, a decrease in the free radical density is indicative of some process involving hydrogen transfer. Consequently ESR spectroscopy is an important experimental technique for coal conversion studies if the experiments can be carried out in situ under the realistic conditions of high temperatures and pressures used in coal liquefaction. We have developed such an ESR apparatus in which in situ experiments at temperatures up to 600°C can be carried out under various atmospheres, such as vacuum and flowing gases, and H_2 pressures up to 700 psi [11]. We have also modified the ESR sample tube so that waste tire rubber or polymers could be added during a reaction and we could observe in real time the effect of the added material. In the work reported here, we have employed this apparatus to investigate the coprocessing of coals with waste tire rubbers and we provide definite evidence of significant enhancement of hydrogen transfer facilitated by the tire polymers. These results are supportive of the increased coal liquefaction yield in the pres ence of waste tire rubbers reported first by Farcasiu and Smith [12] and more recently by Liu et al. [13]. Similar experiments on the coprocessing of coal with polyethylene and polystyrene show a rapid increase in the free radical densities beyond ~350°C indicating enhanced cracking of coal with these polymers. These results are described below.

APPARATUS FOR IN SITU ESR MEASUREMENTS

We have fabricated a microwave cavity system in which in situ ESR spectroscopy can be carried out at temperatures up to 600°C with gaseous flow or pressures up to 700 psi (pressures may be increased above 700 psi with the use of sapphire tubes instead of Pyrex tubes). For gas flow experiments, we have modified the sample tube so that we could incorporate tire rubber or polymers during an experiment to study the effect of adding the material in real time.

The high-temperature ESR cavity system consists of three primary units: (1) a rectangular TE_{102} cavity in which the field modulation coils are located outside the cavity and a circular hole through the center of the cavity that allows insertion of the quartz Dewar and the sample; (2) a heating unit external to the cavity, purchased from Wilmad Glass Co., in which flowing N_2 gas over a voltage-controlled chromel heater is used to deliver heat to the sample without heating the microwave cavity; and (3) a flow chamber unit, shown in Figure 1, that enables the sample to be exposed to flowing gas (such as N_2 or H_2)—a description of this is given in Reference 11. The ESR sample tube can also be modified (Figure 1 inset) by providing a side tube for some specific experiments to add material during the course of reaction.

The high-temperature quartz Dewar that is inserted in the hole of the TE_{102} cavity has an evacuated double wall ensuring only minimal heat transfer from the sample to the cavity. Because inserting a thermocouple inside the cavity distorts the microwave signal, the thermocouple is placed just at the entrance of

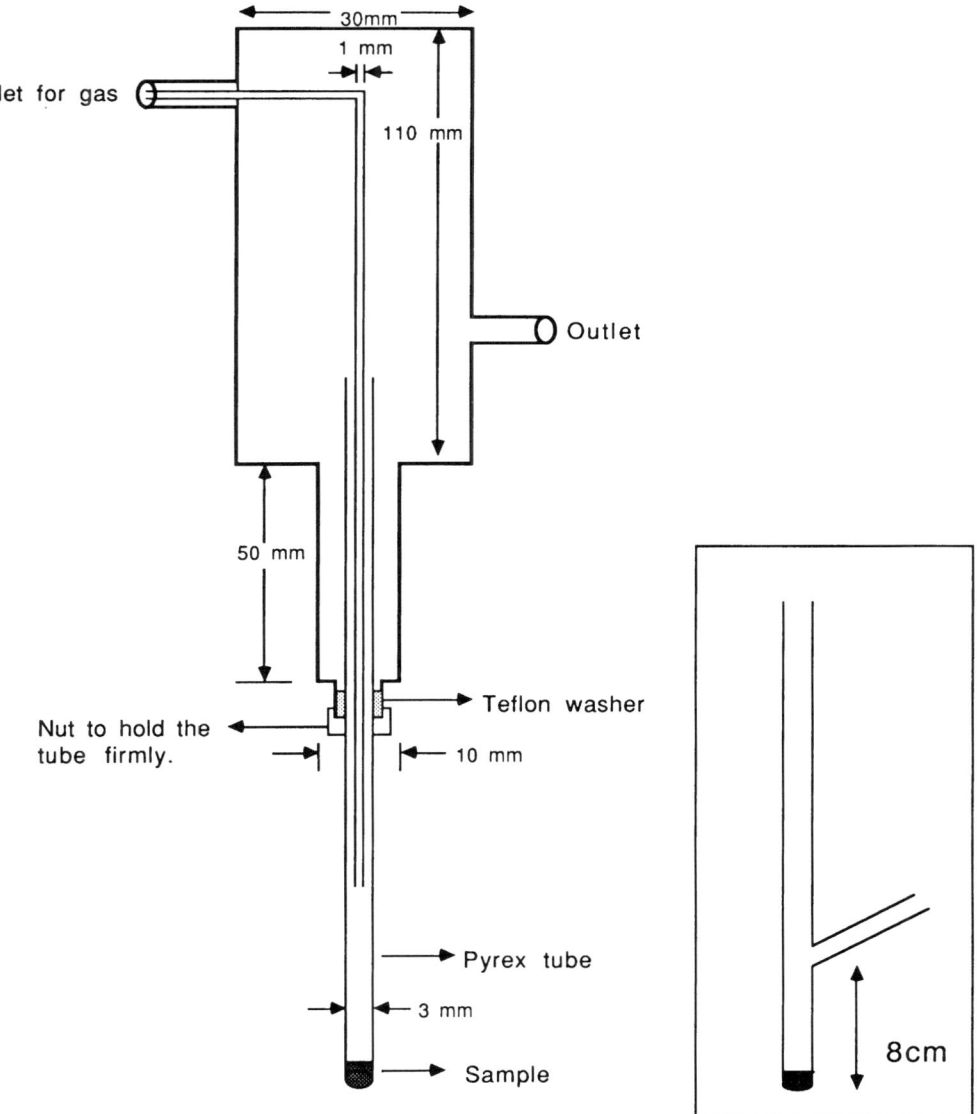

Figure 1 Flow chamber for gas flow through Pyrex sample tube. Inset shows the Pyrex tube with a side tube for adding samples in situ.

the cavity. The temperature is calibrated by an initial run with an additional thermocouple inside the sample, and the experiment is carried out under identical conditions. In the heater unit, the temperature is controlled by the voltage input to the heater, controlled by a variable transformer. For each data point, at least a 10-minute interval is allowed to ensure temperature stability. At each temperature, the cavity coupling is adjusted to a fixed value, because change in the coupling and quality factor Q of the cavity affects the signal intensity [14] and may introduce error in the g values if the samples are highly conducting [15]. However, our earlier experiments with bituminous coals diluted with silica gel [6] showed that free radical densities up to 500°C are equivalent for the highly diluted and undiluted samples. Consequently all the experiments reported here are for undiluted samples. The free radical densities are calculated from the area under the ESR signal obtained by integration of the standard derivative signal [6, 11].

EXPERIMENTAL RESULTS AND DISCUSSION

The ESR studies reported here were carried out on a Blind Canyon coal obtained from the Penn State Coal Bank. The analysis of this coal and details of the X-band high-temperature ESR cavity system have been published [11]. The tire rubber used in these experiments was chipped from a used tire rubber (Goodyear Vector) and another sample of tread of a Michelin tire was obtained from Dr. Farcasiu of the Pittsburgh Energy Technology Center (PETC). The tire rubber sample was shredded to millimeter-size pieces and mixed with the coal in a 1:1 ratio by weight. Two samples of Goodyear tire rubber were cut from the central portion and from the rim of the tire and are referred to as tire rubber(C) and tire rubber(R), respectively. Polyethylene (spectrophotometric grade) and polystyrene (molecular weight ~2500) were obtained from Aldrich Chemical Company. Polymers were mixed in a 1:1 ratio with coal for ESR measurements.

For the ESR experiments, samples were vacuum sealed in the ESR tube at room temperature followed by measurements at elevated temperatures or measurements were done in flowing H_2 gas [10, 11]. Only a single ESR line is observed from the coal sample and no ESR signal was observed from the tire rubber or polymers as determined in a separate experiment up to 500°C, indicating that the free radicals seen are all associated only with the coal. In Figure 2 we show the temperature variation of the free radical density N for the coal and the coal-tire rubber(R) mixture in a sealed configuration. For measurements in hydrogen flow the plotted values of N are determined from the equation $N = (N_T T/298)/m_T$, where N_T is the measured free radical density at any temperature T, ($T/298$) corrects for the Curie law variation of the free radical density relative to room temperature (298 K), and m_T is the mass of the coal at any temperature measured in a separate thermogravimetry experiment under identical conditions

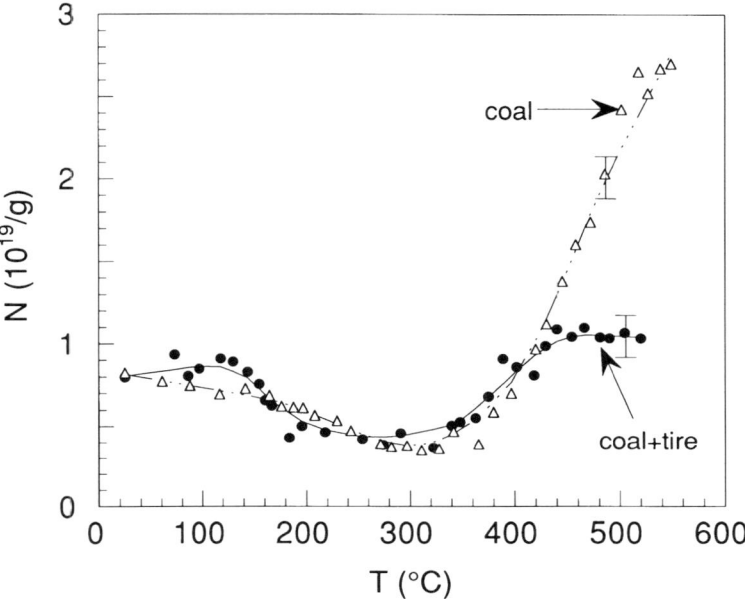

Figure 2 Variation of free radical density N with temperature for coal and coal + tire rubber(C) samples evacuated and sealed in the Pyrex tube.

of hydrogen flow (this corrects for the mass loss in coal due to devolatilization during heating). For the ESR measurements in the sealed configuration only the Curie law corrections are applied. From Figures 2 and 3, for measurements in the sealed and hydrogen flow configurations, it is clear that above ~400°C the tire has considerable effect in reducing the coal free radical densities. At the lower temperatures of about 100°C, there is a slight increase in N due to the effect of tire rubber. Above 400°C, the N values in the presence of tire rubber are suppressed by a factor of 2 to 3 and the presence of H_2 appears to provide additional suppression of the free radical density. Similar experiments were done using a Michelin tread with Blind Canyon coal. The results of coprocessing of the coal with Michelin tire rubber are similar to those for the Goodyear tire rubber and show a decrease in free radical densities above 400°C (Figure 4), although the magnitude of N at all temperatures is higher than for coprocessing with Goodyear tire rubber.

ESR spectroscopy measurements were also done for the coal mixed with polyethylene and polystyrene, 1:1 by weight. Measurements were done for samples evacuated and sealed in a Pyrex tube and also under flowing hydrogen. Figure 5 shows the variation of free radical densities with temperature for the coal mixed with polyethylene and polystyrene for measurements in flowing H_2. Here, above 350°C there is a very rapid increase of free radicals compared with

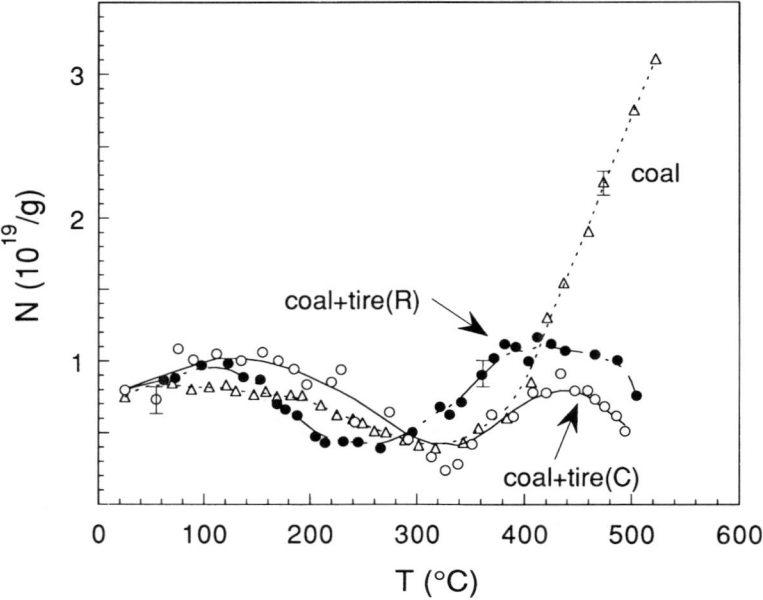

Figure 3 Variation of free radical density N with temperature for coal and coal + tire rubber samples for measurements under flowing hydrogen.

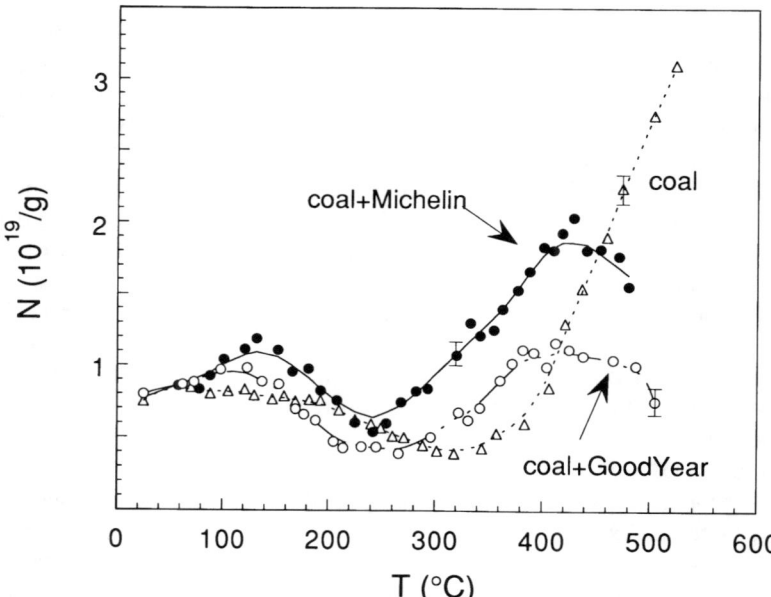

Figure 4 Variation of free radical density N with temperature for coal coprocessed with two different tire rubbers: Goodyear Vector and Michelin tread.

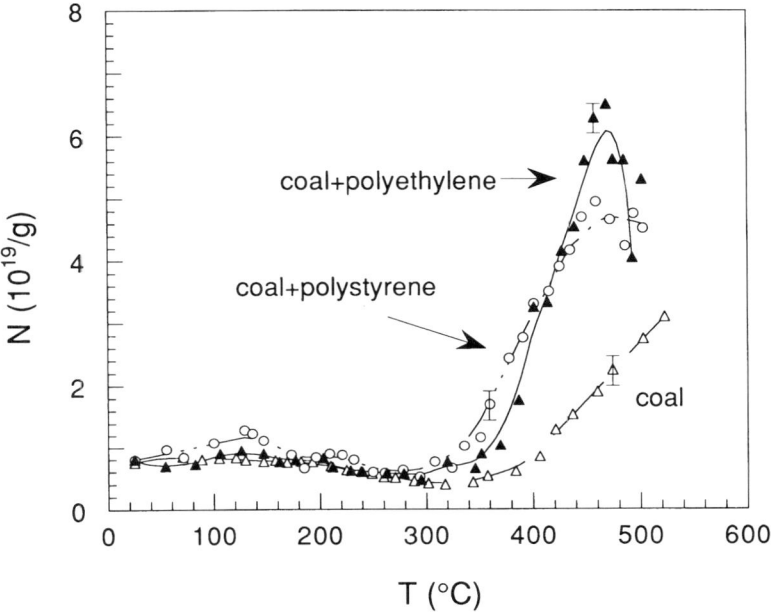

Figure 5 Variation of free radical density N with temperature for coal and for coal mixed with polyethylene and polystyrene for measurements under flowing H_2.

the coal alone. For samples sealed in the Pyrex tube (Figure 6), the variation of free radical densities is similar to that observed in hydrogen flow, still showing more cracking in coprocessing coal with these polymers. In separate experiments, polyethylene and polystyrene were taken alone and heated up to 500°C both in a sealed tube and in flowing hydrogen and no free radicals were detected by ESR. Note that depolymerization of polystyrene has been observed using ESR spectroscopy [16]. However, the half-time of depolymerization is about 102 minutes (time required to get one half of the maximum radical concentration formed) at 400°C, and the free radical densities are about two orders of magnitude smaller than the coal free radical densities [16]. The measurements reported here were typically done after 10 minutes at each temperature, and at these small time scales the increase in N cannot be attributed to depolymerization of polystyrene alone. This is probably due to the increased cracking of coal and/or polyethylene and polystyrene promoted by coprocessing.

To further understand the role of tire rubber in coprocessing, ESR experiments were conducted using the modified sample tube shown in the inset of Figure 1. In this experiment, the coal was heated in situ in H_2 flow up to 480°C at a heating rate of about 20°C/m and free radical densities were monitored as a function of time at 480°C. By 50 minutes equilibrium was established and there was no change in the free radical density with time (Figure 7). At about

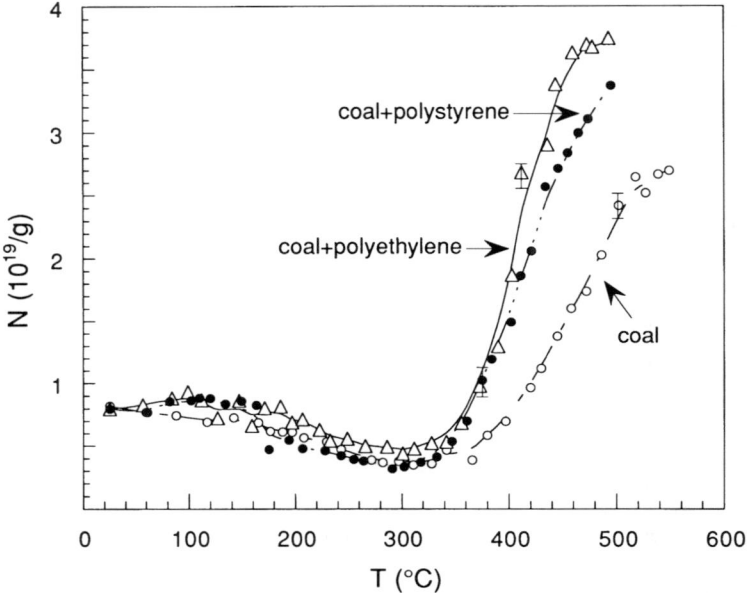

Figure 6 Variation of N with temperature for coal and coal mixed with polystyrene and polyethylene for samples sealed in a Pyrex tube.

53 minutes a tire rubber sample (Goodyear Vector) (2:1 by weight to coal) was added in situ via the side tube and the free radicals were monitored as a function of time for an additional 50 minutes. Figure 7 clearly shows that the addition of the tire rubber resulted in quenching of the free radicals (probably due to hydrogen transfer) and the free radical densities continued to decrease with time.

In order to determine mass changes, a thermogravimetry (TG) experiment was also done by heating coal to 480°C at a heating rate of 20°C/m and observing mass changes with time; a tire rubber sample was added at about 53 minutes and the change in mass was monitored as a function of time (Figure 7). There was no significant change in mass with time after the insertion, so the change in the free radical density with time shown in Figure 7 was not due to a need for mass correction.

Shin et al. [17] reviewed the effect of hydrogen on the free radicals and argued that, in addition to the generally accepted view of the capping of the free radicals by hydrogen in the hydrogenation process, hydrogenolysis can occur in the initial stages of the reaction (hydrogenolysis should increase N, whereas hydrogenation should lower N). The results presented in Figures 2 and 3 show an initial increase in N at lower temperatures due to the coal–tire rubber interaction, and this may be due to hydrogenolysis. The rapid decrease in the free radical density at the higher temperatures in Figures 2 and 3 is most likely due

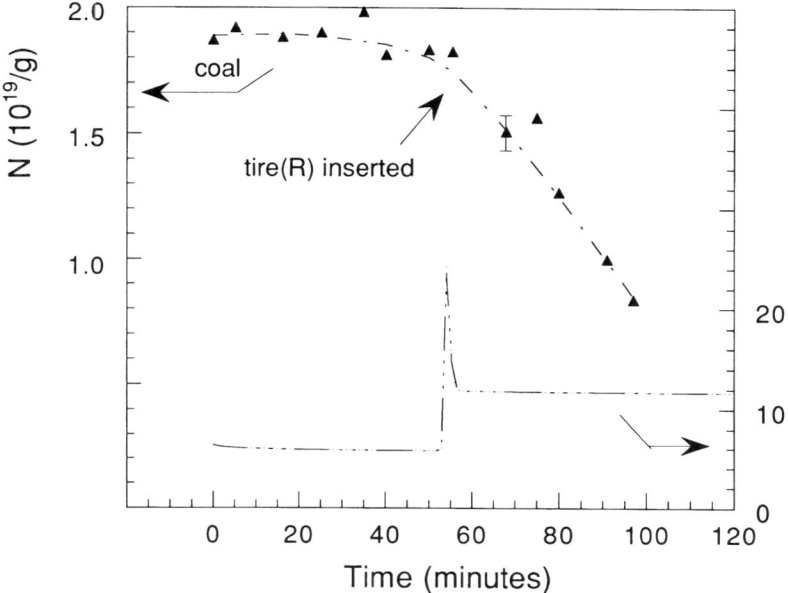

Figure 7 Variation of N with time for measurements at 480°C. The tire rubber sample was added to coal in situ at about 53 minutes. Variation of mass with time from TG measurements at 480°C is also shown.

to hydrogenation facilitated by the transfer of hydrogen from tire rubber polymers to coal fragments. The liquefaction experiments of Farcasiu and Smith [12] carried out at 425°C indicate increased conversion of coal in coprocessing with waste tire rubbers. Work by Liu et al. [13] on coprocessing of a waste tire rubber with Blind Canyon coal also shows that tire rubber acts like a donor solvent, similar to tetralin, and increases the total conversion of coal. Our results also indicate that in the presence of tire rubbers, the coal free radical densities decrease, probably as a result of hydrogen transfer facilitated by tire rubbers. Polyethylene and polystyrene mixed with coal, on the other hand, promote rapid hydrocracking of coal and polymers.

At a recent conference of the American Chemical Society, a number of papers on coliquefaction of coals with waste tire rubbers and polymers were presented [18–20]. They all indicate that waste tire rubbers and polymers improve the liquefaction yields of coal, in agreement with the results presented here.

In summary, the results presented here on the free radical density of Blind Canyon coal during coprocessing with tire rubber and polymers suggest that waste tire rubber lowers the free radical densities, probably as a result of hydrogen transfer. On the other hand, polyethylene and polystyrene promote a rapid increase in free radical densities above 350°C. Both of these findings are positive indicators of some synergistic effects during coprocessing of coal with waste tire rubbers and polymers.

ACKNOWLEDGMENTS

This research was supported in part by the U.S. Department of Energy through CFFLS under contracts DE-FC22-90PC90029 and DE-FC22-93PC93053.

REFERENCES

1. Farcasiu, M., and C. M. Smith. Method for co-processing waste rubber and carbonaceous materials. U.S. Patent 5,061,363, October 29, 1991.
2. Farcasiu, M. Another use for old tires. *Chemtech,* January: 22–24, 1993.
3. Retcofsky, H. L., J. M. Stark, and R. A. Friedel. Electron spin resonance in American coals. *Anal. Chem.* 40:1699–1704, 1968.
4. Silbernagel, B. G., L. A. Gebhard, G. R. Dyrkacz, and C. A. A. Bloomquist. Electron spin resonance of isolated coal macerals: preliminary survey. *Fuel* 65:558–565, 1986.
5. Seehra, M. S., B. Ghosh, and S. E. Mullins. Evidence for different temperature stages in coal pyrolysis from in situ e.s.r. spectroscopy. *Fuel* 65:1315–1316, 1986.
6. Seehra, M. S., and B. Ghosh. Free radicals, kinetics and phase changes in the pyrolysis of eight American coals. *J. Anal. Appl. Pyrolysis* 13:209–220, 1988.
7. Fowler, T. G., K. D. Bartle, and R. Kandiyoti. Low temperature processes in a bituminous coal studied by in situ electron spin resonance spectroscopy. *Fuel* 66:1407–1412, 1987.
8. Neavel, R. C. Liquefaction of coal in hydrogen-donor and non-donor vehicles. *Fuel* 55:237–242, 1976.
9. Petrakis, L. and D. W. Grandy. *Free radicals in coals and synthetic fuels.* New York: Elsevier Science Publishers, 1983.
10. Ibrahim, M. M., and M. S. Seehra. Depolymerization of coals promoted by zinc halides near 100°C. *Energy & Fuels* 5:74–78, 1991.
11. Ibrahim, M. M., and M. S. Seehra. An apparatus for in-situ high temperature/high pressure ESR spectroscopy and its applications in coal conversion studies. *ACS Fuel Div. Preprints* 37:1131–1140, 1992.
12. Farcasiu, M., and C. Smith. Coprocessing of coal and waste rubber. *ACS Fuel Div. Preprints* 37:472–479, 1992.
13. Liu, Z., J. W. Zondlo, and D. B. Dadyburjor. Tire liquefaction and its effect on coal liquefaction. *Energy & Fuels* 8:607–612, 1994.
14. Seehra, M. S. New method for measuring the static magnetic susceptibility by paramagnetic resonance. *Rev. Sci. Instrum.* 39:1044–1047, 1968.
15. Retcofsky, H. R., M. R. Hough, M. M. Maguire, and R. B. Clarkson. Some cautionary notes on ESR and ENDOR measurements in coal research. *Appl. Spec.* 36:187–189, 1982.
16. Carniti, P., A. Gervasini, P. L. Beltrame, and G. Audisio. Evidence of formation of radicals in the polystyrene thermo degradation. *J. Polymer Sci. A* 27:3865–3873, 1989.
17. Shin, S.-C., R. M. Baldwin, and R. L. Miller. Effect of radical quenching on hydrogen activity: Model compound and coal hydroliquefaction studies. *Energy & Fuels* 3:71–76, 1989.
18. Anderson, L. L., and W. Tuntawiroon. Coliquefaction of coal and polymers to liquid fuels. *ACS Fuel Div. Preprints* 38:816–822, 1993.

19. Ibrahim, M. M., and M. S. Seehra. Thermogravimetric and free radical evidence for improved liquefaction of coal with waste tires. *ACS Fuel Div. Preprints* 38:841–847, 1993.
20. Liu, K., E. Jakab, W. H. McClennen, and H. L. C. Meuzelaar. Microscale simulation of high pressure thermal and catalytic conversion processes in coal and waste polymers with on-line GC/MS. *ACS Fuel Div. Preprints* 38:823–830, 1993.

PART THREE

BIOSOLIDS, BIOMASS, AND MUNICIPAL SOLID WASTES

CHAPTER
EIGHT
FUNDAMENTALS OF HYDROTHERMAL PRETREATMENT OF BIOSOLIDS (SLUDGE): FEEDSTOCKS FOR CLEAN ENERGY

M. Rashid Khan, C. Gorsuch, R. Zang, and S. DeCanio

INTRODUCTION

Texaco Inc. has investigated techniques for preparing dewatered sewage, industrial, and biological sludges as feedstocks for gasification/oxidation processes as part of its waste gasification program [1–5]. One of the most promising pretreatment methods has proved to be the hydrothermal treatment developed by Texaco R&D. A number of patents have been issued [2–4]. The objective of this chapter is to present the fundamental aspects of hydrothermal pretreatment of sludge. Based on extensive bench-scale and pilot-scale runs, a demonstration plant was built to process 6 dry tons of centrifuged sludge cake per day. The results of the inaugural run are presented.

The Texaco Gasification Process produces synthesis gas, principally carbon monoxide and hydrogen, from carbonaceous feedstocks by partial oxidation. One hundred commercial plants have been licensed to use this technology over the past 30 years, and the synthesis gas generated has been used for end products ranging from hydrogen, ammonia, and alcohol to electric power. The feedstocks have included natural gas, distillate oils, petroleum coke, and coal.

The EPA Standards for the Use or Disposal of Sewage Sludge (40 CFR Part 503) cover the following options: incineration, land application, and surface disposal. Most large treatment plants would prefer incineration or land application. However, the first option could be limited to plants with existing incineration facilities and the next two, the so-called beneficial use options, could be limited by land availability and assurances of pathogen reduction. Solids handling is a significant operating cost for municipal wastewater treatment plants. As an example, greater Chicago spent over half of its 1990 operating and maintenance budget on solids processing and disposal [6]. These costs will increase at all plants because new regulations and public concerns are limiting old options. Ocean dumping is banned, landfills are closing, and the public is demanding that sewage sludge be put to beneficial reuse [7, 8].

Obviously, there is a market for new sludge management technologies. Recently, wastes have been considered for gasification because they are a good source of hydrocarbons and cost less than traditional feedstocks. Gasification research and development is conducted by Texaco at its Montebello Research Laboratory in Los Angeles, California. Montebello was established in 1945 and had three 15- to 25-ton-per-day pilot units. In 1988, under a Hazardous Waste Reduction Grant from the State of California Department of Health Services, a demonstration at Montebello proved that low-Btu hazardous waste could be gasified safely with coal and with little impact on thermal efficiency [1–5]. The successful testing of hazardous wastes showed that gasification would be an environmentally acceptable method of disposal for other high-volume, low-hazard wastes such as sewage sludge.

In order to maintain the highest efficiency in gasification, there is a need to limit the amount of water carried in the feed. The high viscosity of partially dried sludge limits its suitability as a feedstock, yet too much water in a less viscous slurry decreases its heat content. It was observed that even after complete drying, sewage sludge had a tendency to reabsorb water and did not form a good slurry for gasification. This phenomenon is known to occur with low-rank coals. Several steam and hot water processes were developed to thermally dewater low-rank coal. The first of these was the Fleissner process, patented in 1927 [9]. When coal is heated, various oxygen functional groups are decomposed, releasing carbon dioxide and water from the coal structure. The tendency to reabsorb moisture is reduced so that the slurring characteristics are similar to those of higher rank coals [10].

EXPERIMENTAL
Bench-Scale Results and Discussion

Bench-scale and pilot tests were begun at the Texaco Research & Development Center at Beacon, New York, to investigate various techniques for improving the physical, chemical and transport properties of sludge [8, 11]. An indicator viscosity for all feedstocks was established as 1000 centipoise to allow comparison of different pretreatment methods.

FUNDAMENTALS OF HYDROTHERMAL PRETREATMENT OF BIOSOLIDS (SLUDGE) 139

The Texaco Research & Development Center at Beacon conducted extensive autoclave and other testings including pilot-plant runs to demonstrate that hydrothermal pretreatment of sludge can significantly improve its physical and chemical properties. Hydrothermal treatment of sludge is somewhat similar to that of low-grade coals; however, there are significant differences between the pretreatment of coal and that of sludge. The viscosity of the product can be greatly reduced depending on the conditions used during pretreatment.

Figure 1 shows the influence of heat treatment temperature on the oxygen content of the sludge. The greater the pretreatment temperature, the lower the sludge oxygen content. However, the overall solids loading for a pumpable slurry is increased as a function of pretreatment temperature. The treated sludge was mixed with a bituminous coal at a ratio of 3:7 (sludge to coal) on a dry basis. Figure 2 shows the similar trends for a sludge received from Passaic, New Jersey. The sludge was mixed with coal after the sludge was hydrothermally pretreated. The ratio of mixture in this case (Figure 2) was 1:2 (sludge to coal) on a dry basis.

Figure 3 shows the influence of sludge pretreatment on the various functional groups of (dried or hydrothermally treated) sludges. The treated sludge contains significantly fewer oxygenated functional groups than the feed (dried) sludge. Some of the effects of pretreatment on relative concentrations of various functional groups as measured by Fourier transform infrared spectroscopy (FTIR) are listed below:

A significant reduction in the phenolic OH functional groups [from 7.5 to 3.75 wt % dry mineral matter free basis (dmmf)].
A significant increase in the ether oxygen content (from 2.8 to 2.25 wt %).

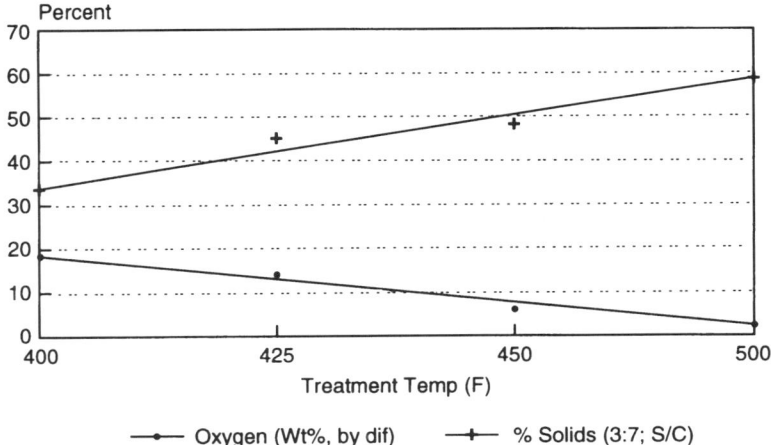

Figure 1 Relationship between heat treatment temperature and slurry solids loading as a function of sludge oxygen content.

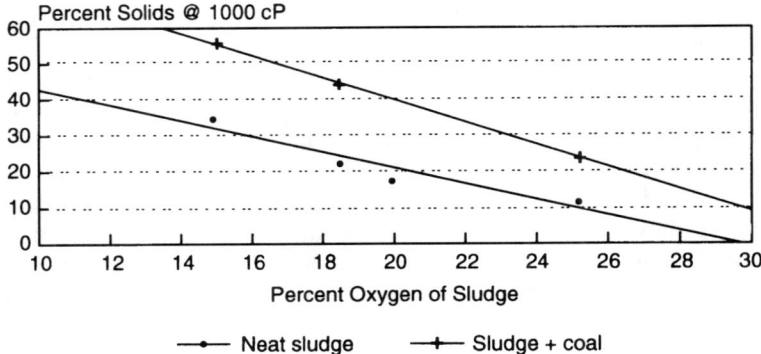

Figure 2 Relationship between slurry solids loading as a function of sludge oxygen content. Slurry viscosities at room temperature. Municipal sludge to coal ratio 1:2. Sludge sample from Passaic Valley, NJ.

A significant reduction in the carbonyl functional groups.
A small decrease in the hydrogen content in the hydroxyl groups (0.47 to 0.23 wt %).
An increase in aromatic hydrogen (from 0.46 to 1.32 wt %).
A small increase in the aliphatic carbon content of the sludge (from 31.7 to 34.4 wt %).

Whereas the hydrothermal treatment reduces the oxygen functional groups, flash drying of the same sludge (performed in the air-oxygen atmosphere) increases the oxygen functional groups. Similar results were also noted for an

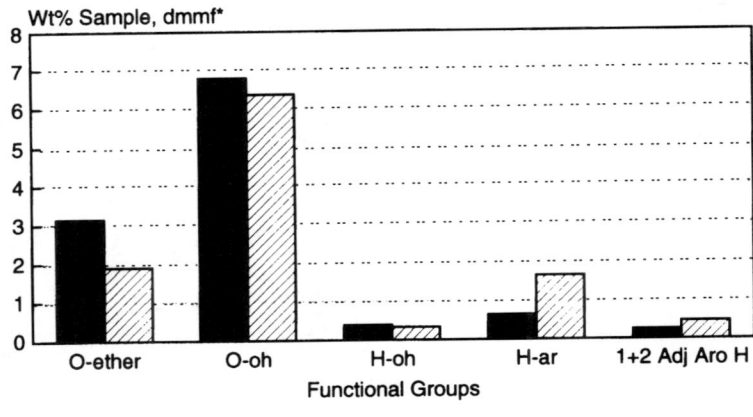

Figure 3 FTIR analysis of a municipal sludge. IR calibrated using model compds.

industrial sludge (Figure 4). Hydrothermal treatment of sludge therefore markedly reduces the oxygen functional groups present in the sludge. A reduction in oxygen content renders the sludge more hydrophobic and thus more amenable to slurry making with a higher solids loading (compared with untreated sludge). The influence of hydrothermal treatment on the structure and composition of sludge is depicted graphically in Figure 5.

Table 1 summarizes the interesting changes that occur in the sludge structure or morphology as a function of heat treatment temperature based on our observations.

A typical "coalification" curve as applied to sludge as a function of sludge pretreatment temperature is presented in Figure 6. The sludge O/C ($\times 100$) and atomic H/C ratios of various sludges along with various other fossil fuel feedstocks are plotted in this figure. It is interesting to note that as the pretreatment temperature of sludge is increased, a pathway is followed by sludge that is similar to the pathway followed in the geological maturation for type II kerogen. Various fossil fuels including lignite and subbituminous coals undergo similar changes during geological evolution.

Process Demonstration Unit Results

Based on these results, it was decided that construction of a larger process demonstration unit (PDU) was justified. An industrial services company was retained to provide detailed design and construction services for a trailer-mounted PDU with a capacity of 900 kg/h or 6 dry tons per day at 25% solids. The process is shown schematically in Figure 7. The effect of PDU Hydrother-

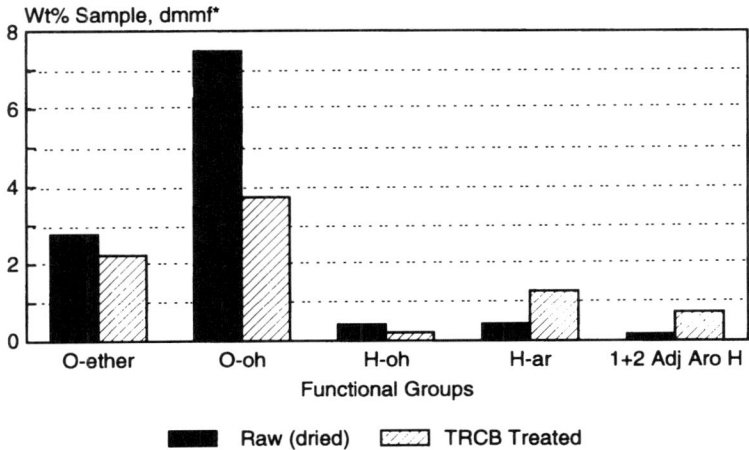

Figure 4 FTIR analysis of an industrial sludge. IR calibrated using model compds.

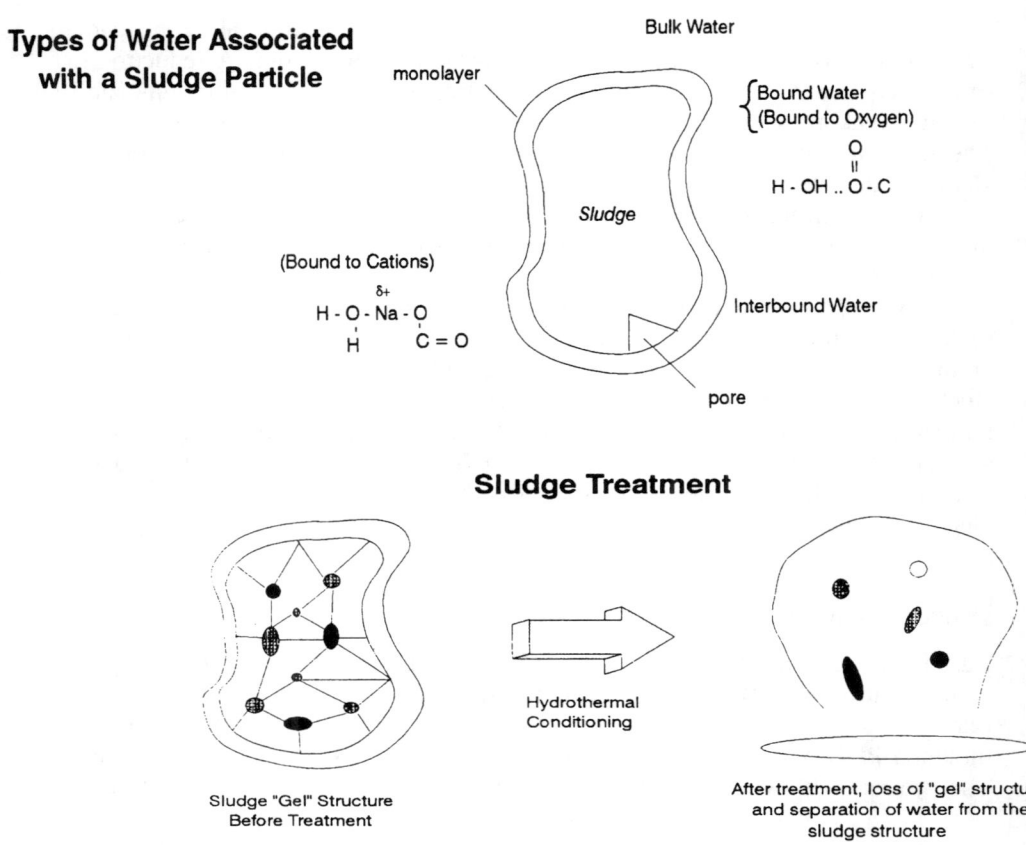

Figure 5 Hydrothermal treatment of sludge.

mal treatment on the viscosity of the slurry is shown in Figure 8. The dramatic effect is qualitatively similar to our pilot test results (not shown) and demonstrates that the pilot-scale and bench-scale results can be achieved in a commercial unit.

CONCLUSIONS

The Hydrothermal Pretreatment Process, developed on the basis of fundamental investigations, converts municipal sewage and industrial sludges into attractive feedstocks for gasification/combustion and offers municipalities an environmentally acceptable alternative to other disposal options.

Table 1 Genealogy of sludge pretreatment: events at sludge treatment temperatures

Sample	Observations
As received	Material clumpy, pasty, no free liquid, medium brown color, can centrifuge clear, colorless liquid
275°F	Sample sweaty with musty odor, still pasty like raw sludge
300°F	Sample more sweaty than at 275°F, clear water beaded on surface, but no real water separation, still pasty and clumpy with distinctive musty odor (typical of sludge heated to disinfect)
325°F	Small amount of separated liquid that is light brown in color, transparent, sludge smells more burned, color darkens, consistency more fluid
350°F	Similar to 325°F sample, but more decantable liquid, centrifuged liquid is very dark
400°F	A lot of decantable liquid, very obvious consistency change, small particles in decant that were not present at lower temperature, burned smell much stronger
425°F	Very similar to 400°F run, but more fluid
450°F	Increased amount of liquid decantable from the product, product very fluid, slushy, no pastiness left, dark and smooth, burned
500°F	Darkest, most fluid product, easily pours out of vessel, approximately half of the product is decantable liquid, smells burned, coffee-like

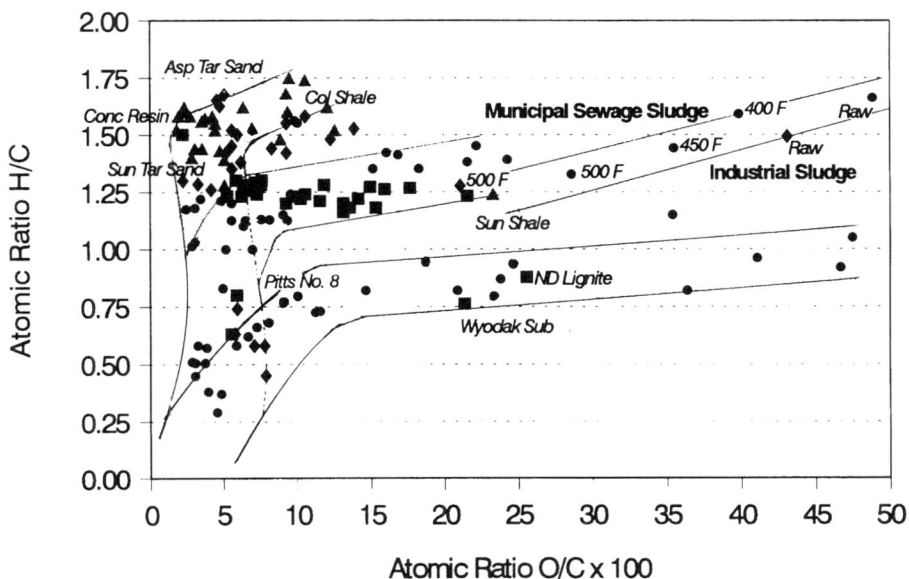

Figure 6 "Coalification" curve: atomic O/C ratio versus atomic H/C ratio.

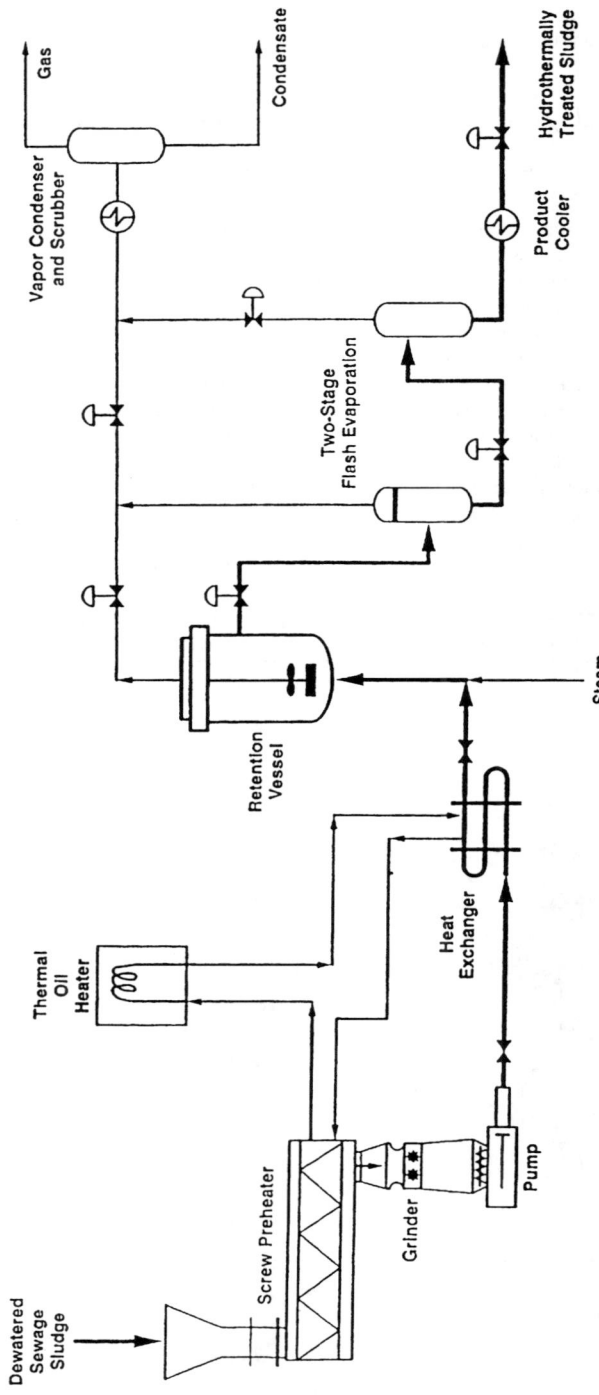

Figure 7 Texaco hydrothermal treatment.

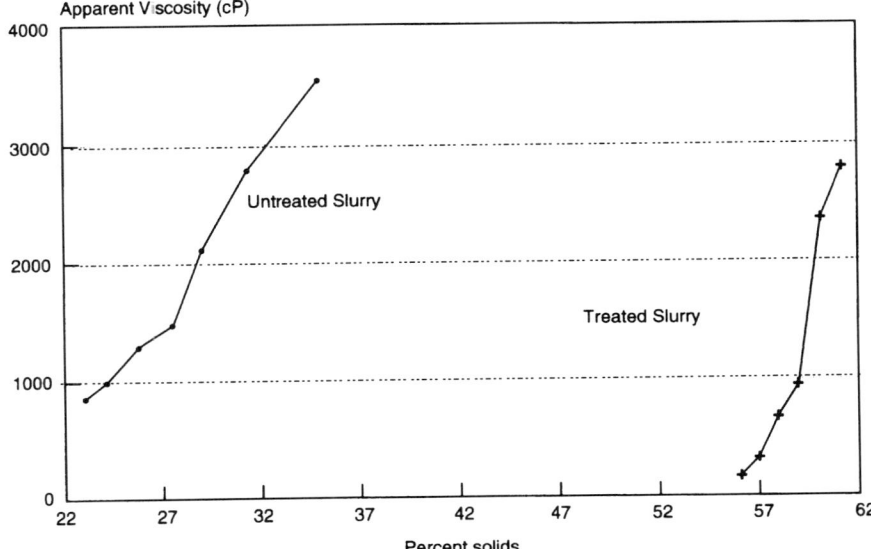

Figure 8 Texaco hydrothermal PDU results sludge/coal 1:2; slurry rheology.

REFERENCES

1. Davis, L. A., J. E. Miranda, and M. Amrheim. Pilot Plant Demonstration Gasification of Petroleum Production, RCRA Exempt, Low Btu, Hazardous Wastes. Final Report Contract No. 88-T0339, June 1990.
2. Khan, M. R. U.S. Patent 5,188,740; Zang, R., and M. R. Khan, U.S. Patent 5,188,741, issued February 23, 1993.
3. Khan, M. R., and R. McKeon. European Patent 93300425.1, 1993.
4. Zang, R., and M. R. Khan. European Patent 93300800.5, 1993.
5. Khan, M. R., R. Zang, and C. Albert. European Patent 92310995.3, 1992.
6. Ganelli, B. A., Improved Centrifugal Dewatering. AC91-042-004, Water Environment Federation, Alexandria, VA, October 1991.
7. Khan, M. R., M. A. McMahon, and S. J. DeCanio. Estimating the heating value of sewage sludge: A new correlation. In *Clean energy from waste and coal,* ed. M. R. Khan, Chapter 12. ACS Symposium Series No. 515. Washington, DC: American Chemical Society, 1992.
8. McMahon, M. A., and M. R. Khan. Preparing pumpable mixtures of sewage sludge and coal for gasification. In *Clean energy from waste and coal,* ed. M. R. Khan, Chapter 13. ACS Symposium Series No. 515. Washington, DC: American Chemical Society, 1992.
9. Fleissner, H. Drying of coal. U.S. Patents 1,632,829, July 21, 1927, and 1,678,078, July 31, 1928.
10. Khan, M. R., and T. Potas. Hot-water upgrading of low-grade coal: Fundamentals and slurry rheology. Preprints, Fuel Chemistry Division of ACS, American Chemical Society Annual Meeting, Atlanta, 1991.
11. Zang, R. B., and M. R. Khan. Gasification of sewage sludges. Presented to the New York Water Pollution Control Association Annual Meeting, January 1991.

CHAPTER
NINE

LAND APPLICATIONS OF BIOSOLIDS

M. Rashid Khan and K. Mitall

INTRODUCTION AND BACKGROUND

The term "biosolids" is used to promote sewage sludge as a valuable product. Compared with "sewage sludge," the term biosolids connotes the presence of a positive resource. Land application, the process of spreading biosolids on agricultural soil as a fertilizer, implies the utilization of sludge as a resource rather than the disposal of a waste. But disposal of sewage sludge can pose a serious problem. In spite of the fact that a large amount of sludge produced in the United States is land applied, the land application of sludge raises a number of serious concerns.

Despite the support of land application by the U.S. Environmental Protection Agency (EPA) and others, the general public has considerable reservations. A survey of Pennsylvania residents showed that only 34% of the population supports land application, while 45% are opposed [1]. According to the survey, sources of concern include the composition of sewage sludge and the odor coming from treated fields. The survey also showed that those most likely to support sludge application were males, middle-aged citizens, rural residents, and those with a higher education or higher income level. Men, in fact, are twice as likely as women to support land applications, and rural residents are nearly 50% more likely to support them than city dwellers [1].

The high content of nitrogen, phosphorus, and other nutrients in sludge makes it a cheaper fertilizer option for farmers. In addition to the crop benefits, land application improves soil quality. Application to clay soil increases the aeration, permeability, and water infiltration of the soil, in addition to decreasing the amount of runoff [2]. Water retention increases when sludge is applied to sandy soils [3].

A major limitation of application of biosolids is that it can occur only at certain times of the year and growing season. Biosolids cannot be applied during the winter or in the early spring, when the ground is still frozen. They also cannot be applied when crops have broken the surface. As a result, the only feasible times for application are at the beginning and the end of the growing season.

Some sludge must be stored for later use. However, if it stands in a stagnant and wet environment for too long, reaccumulation of pathogens is possible, so alternative options are needed. Some wastewater treatment facilities have opted to compost their sludge, either alone, with municipal solid waste (MSW), or with yard wastes. These options are discussed later.

Another option, currently being researched at the University of Maryland's Agricultural Experiment Station, is to apply excess sludge to forest lands [4]. With this option, land application can occur year-round without any danger of contaminating food supplies or other crops. Few regulations exist regarding the quality of sludge that can be applied to forest lands. The long-term impact of this application is not well known.

A serious limitation of land application is the amount of available land that is readily accessible to major metropolitan areas where sewage sludge is produced.

Factors that determine the amount and quality of sludge that can be applied to land are soil nutrient needs and sludge composition (including organic and inorganic constituents of sludge). These issues are discussed in the following sections.

NUTRIENT CONTRIBUTION

The amount of sludge that can be applied to a particular field is determined by the yearly nutrient needs (nitrogen, phosphorus, and potassium) of the field [5]. Although sludge contains large quantities of both nitrogen and phosphorus, it contains only small amounts of potassium. A farmer using biosolids as opposed to commercial fertilizers would need to add potassium but would still be saving money. For example, biosolid application costs about $7.50 an acre plus about

$10 an acre for added potassium, a total of less than $20 an acre. Commercial fertilizers would cost roughly twice that amount [2].

The concentration of nitrogen in the sludge must be strictly monitored to reduce the amount of nitrates leaching into the soil. Leaching will occur if the amount of applied nitrogen exceeds the nitrogen requirements of the crop [6]. Excess nitrogen in the form of ammonia can be nitrified to form nitrates [6], which are harmful to the plant, the soil, and the groundwater. Research at the University of Maryland [4] has also indicated that more nitrogen leaches into groundwater during drought conditions, when plants absorb less nitrogen. To reduce nitrate contamination of the soil and groundwater, nitrogen concentrations must return to baseline levels before reapplication of biosolids.

SOIL AND SLUDGE PARAMETERS

In addition to total nitrogen, ammonia, and nitrate loadings, concentrations of phosphorus, potassium, metals, and other parameters must be monitored in both the sludge and soil. For example, salt concentrations must be kept below 10 dry tons/acre. Levels above that will affect seed germination and early growth of the crop.

Pathogen levels in sludges must be limited to prevent the spread of disease (Table 1). All sludges that will be applied to fields growing food crops must be stabilized by aerobic digestion, anaerobic digestion, lagoon storage, or a similar process [6]. Sludges that are composted must be raised to a temperature of 50° to 60°C, the temperature at which coliform bacteria, *Salmonella,* and most viruses are considered to be destroyed [3]. In addition, sludge should not be applied to a field if food crops are to be grown that season.

Table 1 Indicator organisms and pathogens monitored under 40 CFR Part 503[a]

Organism to be Monitored	Allowable Level in Sludge
Fecal coliform	1000 most probable number (MPN) per gram (class A) of total solids (dry weight)
Salmonella sp. bacteria (in lieu of fecal coliform)	3 MPN per 4 g total solids (dry weight)
Enteric viruses	Less than one plaque-forming unit per 4 g total solids (dry weight)
Viable helminth ova	Less than one viable helminth ovum per 4 g of total solids (dry weight)
Fecal coliform	Less than 2×10^6 MPN or less than 2×10^6 colony-forming units per gram of total solids (dry weight) (expressed as geometric mean of the results of seven individual samples)

[a]Modified to limit discussion of class A alternative only.

Table 2 Pollutant concentrations

Pollutant	Monthly average concentrations (mg/kg)
Arsenic	41
Cadmium	39
Chromium	1200
Copper	1500
Lead	300
Mercury	17
Molybdenum	18
Nickel	420
Selenium	36
Zinc	2800

Metals

Although nutrient loadings are used to determine the amount of yearly applications of sludge, the total metal loadings determine the amount of long-term applications. The primary problems caused by high metal concentrations in the soil are plant toxicity and accumulation in plant tissues [5]. Residential areas do not usually contribute to high metal loadings in sludge, but local industries may provide significant amounts of chromium, mercury, lead, copper, zinc, cadmium, and other metals of concern. The levels of these metals in sludge must be strictly monitored (Table 2).

From a human health point of view, cadmium levels are of primary concern because of the easy uptake of cadmium into plant tissues. For this reason, both annual and long-term limits have been set for sludge applications of cadmium [6], as well as other metals (Tables 3 and 4). Because cadmium tends to accu-

Table 3 Cumulative pollutant loading rates

Pollutant	Cumulative pollutant loading rate (kg/ha)
Arsenic	41
Cadmium	39
Chromium	3000
Copper	1500
Lead	300
Mercury	17
Molybdenum	18
Nickel	420
Selenium	100
Zinc	2800

Table 4 Annual pollutant loading rates

Pollutant	Annual pollutant loading rate (kg/ha)
Arsenic	2.0
Cadmium	1.9
Chromium	150
Copper	75
Lead	15
Mercury	0.85
Molybdenum	0.90
Nickel	21
Selenium	5.0
Zinc	140

mulate in the leafy parts of plants, application should be avoided on fields growing plants like lettuce, spinach, and oats [6].

Other metals can have deleterious effects when applied to crop soils. Lead is also easily taken up into plants and can be ingested by grazing livestock. In addition, the presence of zinc, copper, or nickel in soils can reduce crop yields [6].

pH

The pH of the soils to which sludge is applied must be kept above 6.5 [5]. The presence of lime residues in some sludges can help to maintain this pH. If sludge application reduces the pH, lime must be applied to raise the pH back to acceptable levels.

The pH of the soil has a direct effect on the uptake of metals. A decreased pH will increase metal solubility in the soil, increasing the update and toxicity to plants. A nearly neutral pH will immobilize many metals, preventing uptake from the soil [5].

Cation Exchange Capacity

Another factor affecting metal applications to the soil is the cation exchange capacity (CEC) of the soil. The CEC measures the number of cation exchange sites in a particular soil, expressed by the amount of clay and organic materials in the soil. The CEC directly reflects the soil's ability to retain metals. A soil with a higher CEC will retain larger amounts of metals, preventing their uptake into plants. Long-term loadings of metals depend on the CEC (Table 5) of the soil [5].

Table 5 Maximum metal loading levels (lb/acre) according to CEC

Metal	Soil CEC (mg/100 g)		
	<5	5–15	>15
Pb	500	1000	2000
Zn	250	500	1000
Cu	125	250	500
Ni	50	100	200
Cd	5	10	20

Organics

Loading limits have also been established for the amount of organics applied to soils. For example, applied sludge cannot contain more than 10 mg/kg of polychlorinated biphenyls (PCBs) [6]. Organic compounds are metabolically stable and not easily broken down once applied to the soil. When they have entered the root zone of the soil, they move quickly to contaminate local groundwater [7]. In addition, organics present in chlorinated sludge form carcinogenic chlorinated compounds such as trihalomethanes. Chlorinated sludge also oxidizes any bromine present to its active form. Brominated organics, which are considered to be more carcinogenic than chlorinated ones, will also be formed [1].

Odor

One of the primary public arguments against land application of sludge is odor control. Most odor is a result of gases produced by metabolic reactions of microorganisms living in the sludge. However, most of these organisms can be destroyed along with pathogens in heating processes. Most sludge-treated fields have an "earthy" aroma [2] rather than the smell of a sewage treatment plant. Odor from composts results from high rainfall and poor mixing of the compost material [3]. Keeping the compost dry and properly mixing the material is the first step in reducing compost odor.

EPA REGULATIONS

On February 19, 1993, the U.S. EPA promulgated Rule 40 CFR Part 503 regulating the use and disposal of sewage sludge [8]. This rule included regulations for land-applied sludge. Part 503 set limits on heavy metals, pathogens, hydrocarbons, and other substances that may be present in sludge (Tables 2–4). The deadline for compliance with these new regulations was February 19, 1994. If construction of new facilities was needed, compliance could be delayed until February 19, 1995 [7].

The original proposal of Part 503 was reviewed by two peer groups and any other interested parties during a commentary period of 180 days [8]. One of the peer groups, sponsored by the United States Department of Agriculture (USDA) Cooperative State Research Service Regional Research Group (W-170), examined all aspects of the proposed rule except those pertaining to incineration. Other reviews were received from several environmental groups, municipal wastewater treatment authorities, and consultants [8].

Most commentators felt that the proposed regulations were overly strict. They felt the rule strongly discouraged land application of sludge, perhaps to the point of ending the practice altogether [8]. The strongest controversy was over molybdenum and cadmium levels [9]. Commentators felt the molybdenum limits were too high to be economically maintained. Other commentators stated that cadmium limits were not high enough to maintain health standards. These commentators submitted data on both issues to scientifically justify changing the regulations [8]. After further evaluation, the EPA did, in fact, make changes in the final rule.

OPINIONS AND SUPPORT FOR LAND APPLICATIONS

Federal agencies other than the EPA have had mixed opinions with regard to land applications. The Food and Drug Administration (FDA) has given the EPA its support and expressed an interest in helping to revise and promulgate the new regulations. However, the Forest Service has some doubts about forest land applications. Most of these reservations are due to staff shortages in terms of monitoring applications and soil quality. The Bureau of Land Management (BLM) expressed similar reservations but supports land applications of sludge overall [8].

The USDA also supports land application but does not feel the cadmium loading limits are sufficient. Because cadmium is so readily taken up into plants, the USDA argues that plant uptake should be the limiting factor. The suggestion is to lower cadmium limits from 39 kg/ha to 21 kg/ha [10]. If the EPA does not lower this limit, the USDA plans to issue their own advisory. Technically, this advisory will not have any legal status; it will merely be a guideline for farmers who are considering sludge application as a fertilizer option.

PRACTICAL LAND APPLICATIONS

In 1972 Israel established the National Sewage Reuse Project and began a process utilizing about two thirds of that country's total sewage. In comparison, the United States currently utilizes only about 2.4% of total sewage produced [11]. The project uses treated sewage effluents to irrigate crop fields.

The method of sewage treatment depends on the location of the treatment plant. Smaller municipalities base their treatment on an oxidation pond, and cities use a larger mechanical system. Treated sewage is stored in a reservoir until it can be used for irrigation during the dry season, which occurs from May to October [11].

This type of irrigation provides both water and essential nutrients for the soil and crop. As the water need determines the irrigation rate, the amounts of nutrients and contaminants applied are not controlled [11]. In addition, there is high variability in the applied effluent, which cannot be continuously monitored. The main goal of the project was to use the maximum amount of wastewater, so some compromises in water quality had to be made [11]. On a more positive side, since there is not much industry in Israel, the heavy-metal content in the effluent is low [11].

Another concern of the Israeli irrigation system is the potential spread of pathogens present in the effluents. This is most strongly affected by the method of irrigation. A drip irrigation system uses pipes spread along the ground. Contamination could occur only if the crop touched the ground. The probability of contamination is even lower if the pipes are buried in the ground or covered by protective sheeting [11].

If a sprinkler irrigation system is used, spraying of the effluents can cause the formation of bioaerosols. Bioaerosols are inhalable microorganisms such as *Aspergillus fumigatus*, endotoxins, beta-1,3-glucans, and myotoxins [4], all of which are capable of causing illness. These pathogens can be carried up to 750 m downwind [11]. Inhalation of bioaerosols can cause flu-like symptoms such as watery eyes, runny noses, or more serious reactions [12]. Initial research has indicated that these symptoms are self-correcting as the individual builds up an immune response [12].

The Delaware Solid Waste Authority (DSWA) has operated a compost system that utilizes both sewage sludge and MSW since 1984. The sewage treatment plant at Wilmington generates about 250 tons of sludge each day, all of which is sent to the composting facility at Pigeon Point. Pigeon Point sends only about a fifth of the sludge received to landfills and composts the remainder with equal amounts of MSW. About 20% of the MSW and sludge received is ultimately composted and sold to local landscapers, golf courses, and private citizens [13].

The compost mixture of MSW and sludge is aerobically digested for 7 days. A small amount is then removed for use as a daily cover at a local landfill. The remainder is aged for 30 days and screened for contaminants. Because several industries discharge into the Wilmington treatment plant, the resulting compost is high in lead content. The compost is considered class II and cannot be used on any food crops [13].

Bristol, Rhode Island, opened a composting facility in January 1993 that composts yard wastes with sewage sludge. The facility uses all of the sludge, about 12 tons/day, that is treated at the town's wastewater treatment facility.

Although the compost is generally used as a top cover for landfills or public works projects, it meets Part 503 regulations for heavy-metal concentrations [14].

Rhode Island's regulations for heavy metals are much more stringent than the Part 503 regulations. Although the compost meets federal regulations, it does not quite meet the Rhode Island state limits for heavy metals. Lead seems to be the most limiting parameter. Unlike the DSWA's sludge, however, high lead concentrations come primarily from leaves in the yard wastes rather than from industrial sewage [14].

The hydrothermal sludge pretreatment process (Chapter 8) developed by Texaco Inc. would allow one to prepare a pumpable slurry that could be pumped to a remote field and sprayed directly onto the land. One of the advantages of this treatment is that elevated temperature utilized in this process would destroy all the biological materials present in the biosludge. However, the inorganics present in the sludge would still remain in the treated materials.

CONCLUDING REMARKS

With an ever-decreasing number of landfill sites, disposal of wastes has become a serious problem. Sewage sludge has historically fit into this category as a waste. On the other hand, natural fertilizers like animal manure and composted yard wastes have been considered beneficial to crop soils. Why hasn't human waste been treated the same way?

The problem is that human waste is not all that is treated in municipal sewage treatment plants. Whether from industrial waste or home cleaning products, "unnatural" substances are going down the drain that should not be reintroduced to the environment. These substances limit the extent to which biosolids can be applied to land for beneficial use. Tenenbaum [2] summarized the central issue when he stated: "If you don't want pollutants distributed on the fields that grow your food, keep them out of your wastewater."

REFERENCES

1. Study says PA residents oppose sewage sludge land application. *Sludge Newsletter* 18(8):69–70, 1993.
2. Tenenbaum, D. Sludge. *Garbage* 4:32–37, October/November 1992.
3. Loehr, R. C. Environmental impacts of sludge disposal. In *Sludge and its ultimate disposal*, ed. J. A. Borchardt, W. J. Redman, G. E. Jones, and R. T. Sprague, 1179–1196. Ann Arbor, MI: Ann Arbor Science Publishers, 1981.
4. U-M studying forest application of sludge. *Sludge Newsletter* 18(9):76–77, 1993.
5. Jacobs, L. W. Agricultural application of sewage sludge. In *Sludge and its ultimate disposal*, ed. J. A. Borchardt, W. J. Redman, G. E. Jones, and R. T. Sprague, 109–126. Ann Arbor, MI: Ann Arbor Science Publishers, 1981.

6. Sommers, L. E., and D. W. Nelson. Monitoring the response of soils and crops to sludge applications. In *Sludge and its ultimate disposal,* ed. J. A. Borchardt, W. J. Redman, G. E. Jones, and R. T. Sprague, 217–239. Ann Arbor, MI: Ann Arbor Science Publishers, 1981.
7. U.S. Environmental Protection Agency. Standards for the use and disposal of sewage sludge. *Federal Register* 58(32):9095–9099 and 9248–9415, 1993.
8. Walker, J. M. Proposed technical sludge regulation update. *BioCycle,* November:46–48, 1989.
9. Focus of Part 503 regs turns to implementation. *Sludge Newsletter* 18(5):40, 1993.
10. USDA likely to issue own advisory on cadmium if EPA does not lower limit in Part 503 rule. *Sludge Newsletter* 18(14):114, 1993.
11. Avnimelech, Y. Irrigation with sewage effluents: The Israeli experience. *Environmental Science and Technology* 27(7):1278–1281, 1993.
12. Compost bioaerosol study will cite need for risk assessment research. *Sludge Newsletter* 18(24):194, 1993.
13. Composting is key to Delaware sludge management technique. *Sludge Newsletter* 18(8):71, 1993.
14. Bristol, R. I. Composts sludge and yard wastes. *Sludge Newsletter* 18(9):79–80, 1993.

TEN

ETHANOL FROM BIOMASS BY GASIFICATION/FERMENTATION

E. C. Clausen and J. L. Gaddy

INTRODUCTION

The United States currently imports about 20% of its total energy requirements of about 70 quads annually. As petroleum reserves decline and prices rise, the United States must develop alternative energy resources.

As shown in Table 1, the nation has about 1.5 billion tons of biomass residues and wastes that could be used as an energy source [1]. These residues could furnish 10 quads, or about 15% of our requirements, if converted into energy at 50% efficiency. In addition, if energy crops were grown on idle arable rangeland and forestland (about 200 million acres), another 25 quads could be produced (2). Therefore, the United States could supply half of its energy needs from renewable biomass and wastes.

Lignocellulosic matter may be used as a solid fuel and burned directly to produce energy. However, efficiencies are low and handling problems are serious. Consequently, biomass must be converted into gaseous or liquid forms of energy to be utilized in conventional energy processes. The major components of cellulosic biomass are hemicellulose, cellulose, and lignin. The compositions of the biomass resources vary; however, most materials contain 15–25% hemicellulose, 30–45% cellulose, and 5–20% lignin. The carbohydrates may be hy-

Table 1 Annual biomass residues in the United States

Residue source	Annual amount (MM tons)
Forestry	1000
Agricultural	300
Municipal	200
Total	1500

drolyzed to sugars and fermented to ethanol, or they may be converted into methane by anaerobic digestion. These technologies are under development and may become economical in the future.

Biomass may also be gasified to yield a low-Btu gas consisting of H_2, CO, CO_2, and N_2. Technology for pyrolysis or gasification of biomass has been under intensive development during the past two decades [3, 4]. Large-scale demonstration facilities have been tested [5] and small-scale commercial facilities are now in operation. A major advantage of gasification of biomass is that all the carbohydrate and lignin are converted into energy forms, whereas the lignin is not converted by hydrolysis or digestion.

The problems with the application of biomass gasification have not been technical but economic. The product from gasification is a heat source, which is very cheap today. Even nominal capital and operating costs for gasification cannot be justified with such low income. If the components of synthesis gas were converted into a higher value fuel product, biomass gasification could become a viable alternative energy technology.

Biological Production of Ethanol from Biomass Synthesis Gas

It was discovered that the anaerobic bacterium *Clostridium ljungdahlii* reacts CO and water to produce ethanol as follows [6–10]:

$$6CO + 3H_2O \longrightarrow C_2H_5OH + 4CO_2 \qquad (1)$$

The same organisms also convert CO_2 and H_2 to ethanol:

$$2CO_2 + 6H_2 \longrightarrow C_2H_5OH + 3H_2O \qquad (2)$$

Consequently, all the components of synthesis gas can be converted into ethanol. Biomass may be pyrolyzed by the approximate (and oversimplified) stoichiometry

$$(CH_2O)_n + 0.1nO_2 \xrightarrow{\Delta} 0.8nCO + 0.2nCO_2 + nH_2 \qquad (3)$$

Energy to drive this reaction is supplied by combustion, and CO_2 and H_2O are also present in the synthesis gas. Nitrogen will also be present if air is used instead of oxygen. Equation (3) may be combined with Eqs. (1) and (2) to yield

$$3(CH_2O)_n + 0.3nO_2 \longrightarrow 0.9nC_2H_5OH + 1.2nCO_2 + 0.3nH_2O \qquad (4)$$

Since nearly all the biomass (including the lignin but not the ash) can be converted into gas by Eq. (3), yields of ethanol of about 50% of the total biomass are possible. This compares to yields of only about 30% for enzymatic or acid hydrolysis/fermentation processes.

Synthesis gas compositions from biomass are variable, depending on the raw material, temperature, and process used. Typical compositions show a CO/H_2 ratio of about 1 with CO and H_2 compositions of about 35% (11–13). Yields of gas of about 90% are common. Gas composition can also be tailored for production of fuels [14].

Fuel ethanol production from grain is about 1 billion gallons per year and is expected to increase steadily as usage of oxygenated fuels is mandated in many metropolitan areas. Prices have been as high as $1.85 per gallon but have stabilized at about $1.20 per gallon (wholesale gasoline price plus tax credit of $0.54 per gallon) in recent years. The potential market for fuel ethanol is 10 billion gallons annually, in a 10% blend with gasoline. Higher percentages are, of course, possible and pure ethanol is marketed in Brazil. Therefore, the market for this product is quite large and the price is sufficient to support commercialization.

The biological process for producing ethanol from synthesis gas would be quite simple. The process would consist of an exchanger to cool the hot gases, a biological reactor, and a separator for ethanol purification. The cool gases would be passed through a liquid-phase reactor in which a culture of the desired microorganism is maintained. The microorganism would carry out the reaction to produce ethanol, converting only the CO, H_2, or CO_2 present. The reactor would be operated at mild temperature (95°F) and atmospheric pressure. Higher pressure may be desirable to enhance gas mass transfer and can be used where the synthesis gas may be at elevated pressure. Ethanol would be removed in the aqueous phase from the reactor and recovered by extraction and distillation. Separation of ethanol from water is done with standard commercial technology.

The advantages of biological reactions over chemical reactions include operation at atmospheric conditions, which generally provides a more energy-efficient process. Also, microorganisms give high yields (>95%), with only small amounts of raw materials used for growth and maintenance. Microorganisms are also quite specific in producing a single product, with only small quantities of by-products formed and requiring relatively simple recovery processes. Resistance to toxicity from substances that degrade other catalysts can often be developed in biological systems. Finally, the biocatalyst is continually regenerated in the biological process, which allows more dependable and consistent operation than catalytic processes.

Ethanol Yields/Energy Savings

A comparison of the yields of ethanol from biomass by several biological and nonbiological conversion techniques is summarized in Table 2. Four alternatives are shown:

1. Conversion of biomass to sugars by acid hydrolysis, followed by fermentation of the six-carbon sugars to produce ethanol.
2. Conversion of biomass to sugars by enzymatic hydrolysis, followed by fermentation of the six-carbon sugars to produce ethanol.
3. Conversion of biomass to syngas by gasification, followed by catalytic conversion to alcohols, including ethanol.
4. Conversion of biomass to syngas by gasification, followed by fermentation of CO, CO_2 and H_2 to produce ethanol using Eqs. (1) and (2).

These comparisons are for biomass material containing 50% cellulose, 25% hemicellulose (one half pentoses), and 20% lignin. As noted in the table, the route to ethanol through syngas and employing fermentation is by far the most efficient route.

The processes utilizing either acid hydrolysis or enzymatic hydrolysis to produce sugars suffer from conversion inefficiencies because the lignin fraction of biomass cannot be converted to fuel and the pentose sugars formed from hemicellulose cannot be fermented to ethanol with proven technology. The catalytic process using syngas requires a CO/H_2 ratio of 0.5, and a large portion (one third) of the CO must be used to produce additional H_2 in the synthesis gas by water-gas shift, substantially lowering the yield (15, 16). Also, a mixture of alcohols is produced, lowering the yield and complicating product recovery. The biocatalytic route converts most of the carbon and hydrogen in the biomass

Table 2 Comparison of yields of ethanol from biomass by various conversion technologies

Conversion process	Yield (gal ethanol/ton biomass)
Biomass $\xrightarrow{acid}$ sugars $\xrightarrow{ferm.}$ ethanol	80
Biomass $\xrightarrow{enzyme}$ sugars $\xrightarrow{ferm.}$ ethanol	85
Biomass $\longrightarrow$ syngas $\xrightarrow{cat}$ ethanol	77
Biomass $\longrightarrow$ syngas $\xrightarrow{biocat}$ ethanol	139

into ethanol, except that needed for gasification. Furthermore, either CO and H_2O or H_2 and CO_2 are used to produce ethanol, and yields are maximized with this process.

Based on the yield of ethanol by gasification/fermentation in Table 2, the application of this technology to save 10 trillion Btu annually would utilize only 1.5 million tons of biomass wastes. This computation is based on conversion of 50% of the synthesis gas in the reactor, with the remaining 50% used for the energy for ethanol purification and other process needs. If 10% of the residues in Table 1 are used, the technology would save 1 quad annually. Many biomass waste streams have a negative value. Therefore, there is substantial economic incentive to apply this technology to save significant amounts of energy.

Purpose

The purpose of this chapter is to present information on a unique bioprocess for the conversion of biomass to ethanol by gasification/fermentation. Following gasification of the biomass, CO, CO_2, and H_2 in synthesis gas are converted to ethanol using *C. ljungdahlii*. Data are presented for this fermentation, including results for both batch and continuous reactors. In addition, the effects of the sulfur gases H_2S and COS on growth, substrate uptake, and product formation are presented and discussed.

MICROBIOLOGY OF ETHANOL PRODUCTION

The bioconversion of the gases CO_2, CO, and H_2 by anaerobic bacteria has been known for many years and has been demonstrated in our laboratories to have commercial potential [7, 17–20]. Several bacterial species are capable of converting these gases into acetate, which is an intermediate in many biological pathways. However, only one bacterium has been shown to produce ethanol from the components of synthesis gas.

In 1985, Barik et al. [21] isolated a bacterium from animal waste that was capable of converting CO, CO_2, and H_2 into ethanol by the equations:

$$6CO + 3H_2O \longrightarrow C_2H_5OH + 4CO_2 \qquad (1)$$

$$2CO_2 + 6H_2 \longrightarrow C_2H_5OH + 3H_2O \qquad (2)$$

$$4CO + 2H_2O \longrightarrow CH_3COOH + 2CO_2 \qquad (5)$$

$$2CO_2 + 4H_2 \longrightarrow CH_3COOH + 2H_2O \qquad (6)$$

This strain was found to be a new bacterial species and was named *Clostridium ljungdahlii* [22]. Under normal conditions, the "wild strain" produced approximately 20 moles of acetate per mole of ethanol [9]. However, by manipulating the culture and employing low pH and minimal nutrients, the culture has been found to be capable of producing only ethanol, with minimal amounts of acetate [8].

A kinetic analysis was performed on the data to determine kinematic parameters for growth and CO uptake. The following models were obtained with $P_{CO}^L \leq 1.1$ atm:

$$\mu = \mu_m = 0.04 \text{ h}^{-1} \tag{7}$$

and

$$q = q_m = 42.7 \text{ mmol CO/g cell/h} \tag{8}$$

If the specific uptake rate of CO is converted to a carbon mass basis, a value of 0.22 g C/g cell/h is obtained for q_m, which is comparable to the rate of glucose uptake by *Saccharomyces cerevisiae* with a q_m of 0.27 g C/g cell/h [23]. This rate indicates that *C. ljungdahlii* has reaction rates equivalent to those of other organisms that are used for commercial fermentations.

Culture Manipulation

As shown in Eqs. (1), (2), (5), and (6), both ethanol and acetate are produced from the fermentation of CO, CO_2, and H_2 by *C. ljungdahlii*. Early results with the culture showed that acetate was the predominant product, with ethanol/acetate ratios of 0.1 or lower typically found in the wild strain. Many researchers have studied solvent/acid formation in clostridial cultures and reported that factors including medium manipulation, decreased pH, and addition of reducing agents have brought about solvent formation in favor of acid production.

Research with *C. ljungdahlii* has shown that lowering the pH to 4–4.5 coupled with a nutrient-limited medium brings about a drastic shift in product formation in favor of ethanol [24]. Figure 1 shows ethanol and acetate production from synthesis gas using *C. ljungdahlii* in the continuous stirred tan R reactor (CSTR) at reduced pH and with a specially designed nutrient-limited medium. As noted, ethanol concentrations exceeding 20 g/L with corresponding acetate concentrations of only 2–3 g/L are obtained. The cell concentration in these studies was approximately 1.5 g/L. Thus, an increase in ethanol concentration coupled with high product ratios was obtained with a lower reaction pH in combination with a specially designed nutrient medium.

Bioreactor Design

The choice of a suitable reactor for gas-liquid reaction or absorption is often a question of matching the reaction kinetics with the capabilities of the proposed reactor. In the case of biological systems, special care must be taken to ensure the viability of the biocatalyst at the operating conditions. The specific interfacial

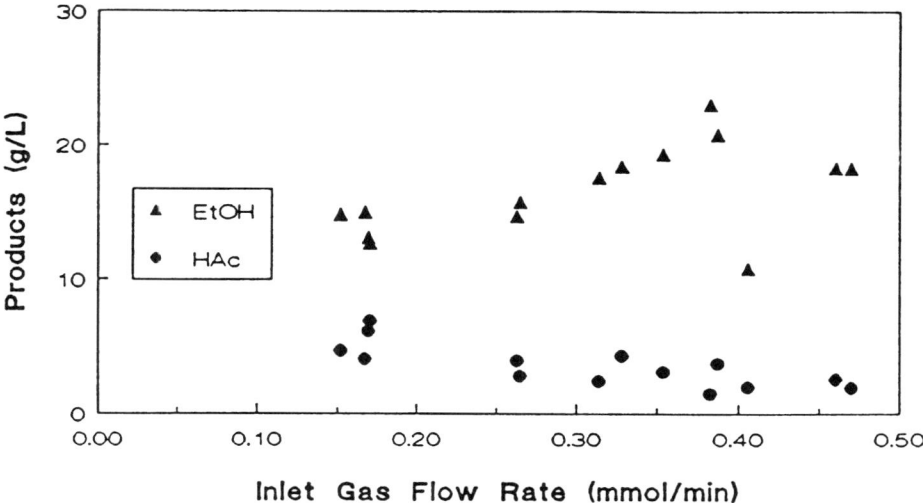

Figure 1 Product concentrations from growth of *C. ljungdahlii* in designed medium in the CSTR.

area, liquid holdup, and mass transfer coefficients are the most significant characteristics of a reactor, and special schemes have been devised to maximize mass transfer. Mechanically agitated reactors, bubble columns, packed columns, plate columns, spray columns, and gas-lift reactors are examples of various kinds of contacting systems employed in these type of processes.

The rate of disappearance of CO from the gas phase can be related to the partial pressures in the gas and the liquid phase and the cell concentration, X, by the equation:

$$-\frac{1}{V_L}\frac{dN_{CO}^G}{dt} = \frac{K_L a}{H}[P_{CO}^G - P_{CO}^L] = \frac{Xq_m P_{CO}^L}{K_P' + P_{CO}^L + [P_{CO}^L]^2/W'} \quad (9)$$

Under mass transfer limiting conditions, P_{CO}^L approaches zero, so the rate of disappearance of CO is proportional to the gas-phase CO concentration. As the cell concentration, X, reaches a value at which mass transfer is controlling, the concentration of carbon monoxide in the liquid becomes zero and the reaction rate is controlled by the rate of transport of the substrate into the liquid phase. Thus, both high cell concentrations and fast gas transport are necessary to minimize reactor size. High cell concentrations may be obtained in the reactor by employing a cell recycle system in which the cells are separated from the effluent and returned to the reactor. Fast mass transfer can be achieved by employing increased pressure or solvents to increase CO solubility.

Use of Cell Recycle in the CSTR. A cell recycle apparatus was used in conjunction with a standard CSTR as a method for increasing the cell concentration inside the reactor. This is particularly important because total product formation with *C. ljungdahlii* has been shown to be proportional to the cell concentration inside the reactor.

Figures 2 and 3 show cell concentration and product concentration profiles for the CSTR with cell recycle. As noted, the cell concentration increased (with agitation rate and gas retention time increases) from approximately 800 mg/L to over 4000 mg/L. The maximum in the previous CSTR study without cell recycle was 1500 mg/L. The CO conversion was consistently around the 90% level after 150 hours of operation. The corresponding H_2 conversion, on the other hand, averaged 70% up to a time of 500 hours. At this time, the H_2 conversion fell, probably because of accumulation of CO in the liquid phase. The ethanol concentration ranged from 6 g/L at the beginning of the study to 48 g/L with 560 hours of operation. The corresponding acetate concentrations at these times were 5 and 3 g/L, respectively. The ratio of ethanol to acetate ranged from 1.2 to 16 g/g. Thus, very high ethanol concentrations are possible and acetate production is nearly eliminated with high cell concentrations.

High-Pressure Operation. As mentioned previously, elevated pressure may be used to increase the rate of mass transfer of CO into the liquid phase. A high-pressure system (both CSTR and trickle-bed reactor) has been constructed in the University of Arkansas laboratories. The system has a maximum operating pressure of 5000 psig and can be used for both photosynthetic and nonphotosynthetic

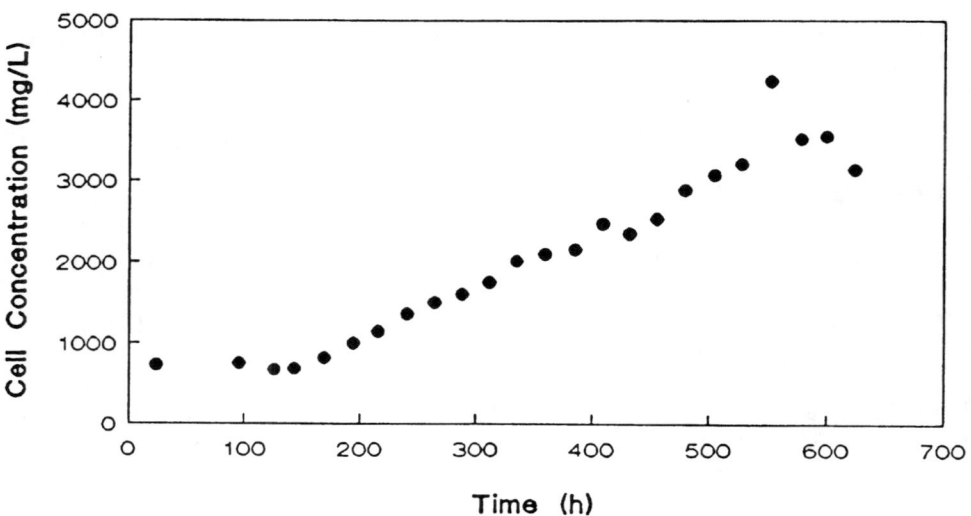

Figure 2 Cell concentration measurements for *C. ljungdahlii* in the CSTR with cell recycle.

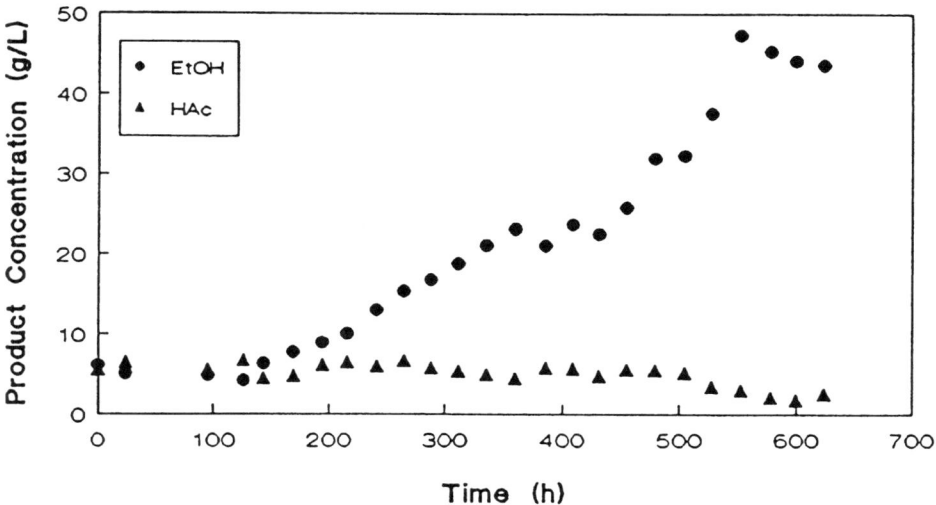

Figure 3 Product concentration measurements for *C. ljungdahlii* in the CSTR with cell recycle.

bacterial systems. The system is presently being used to gradually acclimate *C. ljungdahlii* to increased pressure.

Sulfur Gas Tolerance of *C. ljungdahlii*

Typical synthesis gases may contain 1–2% sulfur, primarily present as H_2S. Most CO-utilizing bacteria have been found to be quite tolerant of sulfur gases, with negligible decreases in the rate of CO uptake and cell growth found in the presence of H_2S concentrations of 5% or less.

The effects of H_2S on growth and substrate uptake by *C. ljungdahlii* are shown in Figures 4 and 5, respectively. As noted in Figure 4, growth was not significantly slowed at H_2S concentrations as high as 5.2%. Similar results are noted in Figure 5 for substrate uptake in the presence of H_2S. The presence of H_2S slowed the rate of substrate uptake only slightly at 5.2% H_2S concentration and had negligible effects on substrate uptake at 2.7% H_2S.

CONCLUSIONS

The anaerobic bacterium *C. ljungdahlii* has been shown to be effective in converting CO, CO_2, and H_2 to ethanol. Rates of carbon uptake by *C. ljungdahlii* comparable to the rate of carbon uptake by the yeast *S. cerevisiae* have been

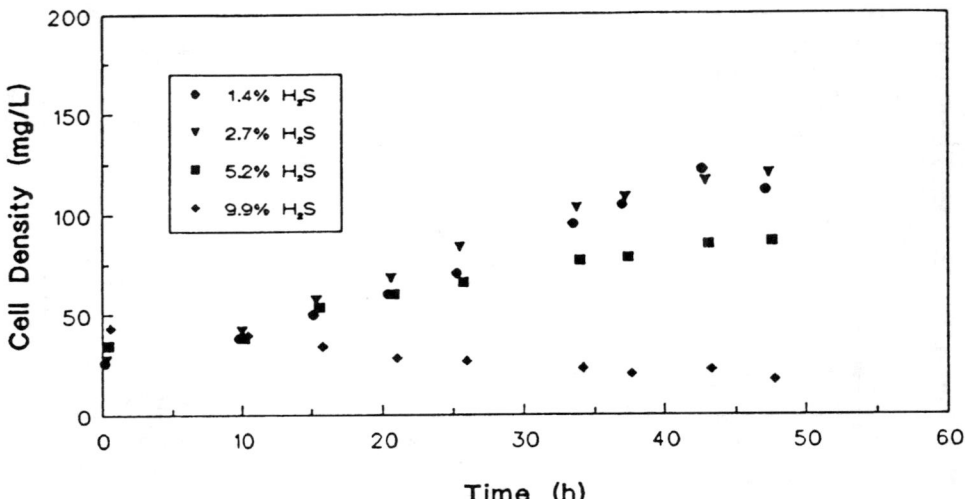

Figure 4 Effects of H_2S on the growth of *C. ljungdahlii* in batch culture.

obtained. A CSTR cell recycle system has been shown to be effective in permitting the cell concentrations necessary for high concentrations of ethanol. An ethanol concentration of 47 g/L with a corresponding acetate concentration of 3 g/L has been attained. *C. ljungdahlii* has been shown to be tolerant of H_2S or COS in concentrations exceeding typical levels in synthesis gas.

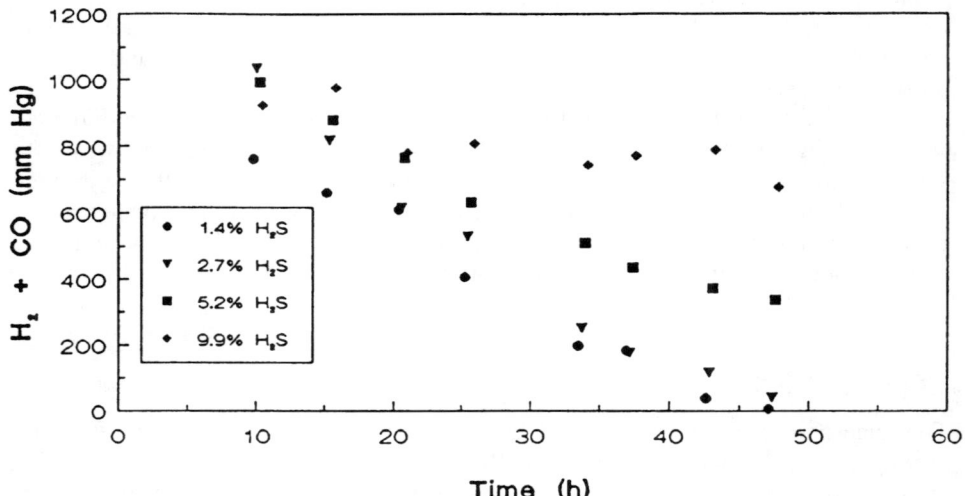

Figure 5 Effects of H_2S on CO and H_2 uptake by *C. ljungdahlii* in batch culture.

REFERENCES

1. Sitton, O. C., G. L. Foutch, N. L. Book, and J. L. Gaddy. Ethanol from Agricultural Residues, *Chem. Eng. Prog.,* 75(12):52, 1979.
2. Clausen, E. C., O. C. Sitton, and J. L. Gaddy. Converting Crops into Methane, *Chem. Eng. Prog.,* 73(1):71, 1977.
3. Steinberg, M. *Proc. Energy from Biomass and Wastes,* IGT, 1988.
4. Stasson, H. E. M. and H. N. Stiles. *Proc. Energy from Biomass and Wastes,* IGT, 1988.
5. Fisher, T. F., M. L. Kasbohm, and J. R. Rivero. The PUROX system, *Proc. of Nat'l. Waste Proc. Conf.,* Boston, May 1976.
6. Klasson, K. T., M. D. Ackerson, E. C. Clausen, and J. L. Gaddy. Bioconversion of Synthesis Gas into Liquid and Gaseous Fuels, *Proceedings Int'l Symp. on Biol. Proc. of Coal,* EPRI and DOE, Orlando, May 1990a.
7. Klasson, K. T., B. B. Elmore, J. L. Vega, E. C. Clausen, and J. L. Gaddy. Biological Production of Liquid Fuels from Synthetic Gas, *Applied Biochem. and Biotech.,* 24/25, 857, 1990b.
8. Klasson, K. T., M. D. Ackerson, E. C. Clausen and J. L. Gaddy. Bioliquefaction of Coal Synthesis Gas, presented at the San Fran. ACS Mtg April 1992.
9. Barik, S., S. Prieto, S. B. Harrison, E. C. Clausen and J. L. Gaddy. Biological Production of Alcohols from Coal Through Indirect Liquefaction. *Appl. Biochem. Biotechnol.* 18:363, 1988.
10. Barik, S., R. E. Corder, E. R. Johnson, E. C. Clausen, and J. L. Gaddy. Coal Gas Conversion by Anaerobic Bacteria, Abstr., Amer. Soc. Microbiol., Annual Meeting, Washington, D.C. Paper No. 024, p. 265, 1986a.
11. Borgwardt, R. H., M. Steinberg, E. W. Grohse and Y. Tang, *Energy from Biomass and Wastes,* IGT, March 1991.
12. Alden, H., E. Bjorhman, and L. Waldheim, *Energy from Biomass and Wastes,* IGT, March 1991.
13. Kaurkela, E., P. Stahling, and M. Nieminen, *Energy from Biomass Wastes,* IGT, March 1991.
14. Collaninno, J. and Mansour, Proc. *Energy from Biomass and Wastes,* IGT, 1988.
15. Elsaway, A. and D. Gray, *Proceedings Indirect Liquefaction Review Meeting,* PETC, November 1988.
16. Dombeck, D. B., *Proc. Indirect Liquefaction Review Mtg,* PETC, November 1988.
17. Barik, S., R. E. Corder, E. C. Clausen, and J. L. Gaddy. Biological Conversion of Coal Synthesis Gas to Methane, presented at the American Institute of Chemical Engineers Summer National Meeting, Seattle, August 1985a.
18. Barik, S., G. C. Magruder, R. E. Corder, E. C. Clausen, and J. L. Gaddy. Biological Conversion of Coal Gas to Chemicals and Methane, presented at the 86th American Society of Microbiology National Meeting, Washington, D.C., March 1986b.
19. Clausen, E. C. and J. L. Gaddy. Biological Conversion of Coal Synthesis Gas. Annual Report, US DOE—Morgantown Energy Technology Center, May 1985.
20. Gaddy, J. L. and E. C. Clausen. Biological Conversion of Coal Synthesis Gas. *Proceedings of the EPRI Workshop on Biological Coal Conversion,* 1987.
21. Barik, S., R. E. Corder, J. L. Vega, E. C. Clausen and J. L. Gaddy. Biological Production of Liquid Fuels and Chemicals from Coal Synthesis Gas. *Proceedings of the Coal-Liquid Fuels Technology Symposium,* 1985b.
22. Vega, J. L., S. Prieto, B. B. Elmore, E. C. Clausen and J. L. Gaddy. The Biological Production of Ethanol from Synthesis Gas, *Appl. Biochem. and Biotech.* 20/21: 781, 1989.
23. Vega, J. L. M.S. Thesis, University of Arkansas, 1985.
24. Phillips, J. R., K. T. Klasson, E. C. Clausen, and J. L. Gaddy. Biological Production of Ethanol from Coal Synthesis Gas: Medium Development Studies. Presented at the 14th Symposium on Biotechnology for Fuels and Chemicals, Gatlinburg, TN, May 1992.

CHAPTER
ELEVEN

GENETIC ENGINEERING FOR PRODUCTIVITY IN THE FERMENTATION OF XYLOSE TO ETHANOL

Jie Xu, Anu Das, Jarrod Erbe, L. M. Hall, and Kenneth B. Taylor

INTRODUCTION

With the increased awareness of the limitations on the world supplies of petroleum came an increased interest in gasoline extenders and additives. The oxygenated additives *tert*-butyl ether, methanol, and ethanol were found to have the additional advantages of octane enhancement and lower rates of air pollution. Although the oxygenated additives are associated with lower rates of air pollution by most contributing compounds, there is concern about higher emissions of volatile organics by the overall fuel [1, 2]. Nevertheless, these concerns are not so intractable as to force the abandonment of an otherwise fruitful approach. The use of ethanol on a national scale is associated with further advantages [3, 4]: security of fuel sources, a more favorable foreign trade balance, and renewability of the raw materials for production. Although ethanol can be produced either from a petroleum component or by fermentation, the latter source is the only one associated with all of these advantages. The most attractive raw materials for ethanol fermentation are cane sugar, corn sugar, and sugar derived from wood and crop residues (lignocellulosic materials). In this country cane sugar is too expensive to be a source of fuel ethanol and corn ($98–$104 per ton) [4] can be competitive only in special circumstances, such as with government subsidies. However, lignocellulosic material ($20–$70 per ton) could be competitive if the technical limitations associated with processing and fermentation could be overcome. Although the approximate price of corn-derived ethanol is currently $1.20 per gallon, it is estimated that the cost of fuel ethanol from lignocellulosic materials should be $0.60–$0.80 per gallon [5].

The Process

The most productive and cost-effective processing of lignocellulosic material currently consists of mild acid hydrolysis to release a fraction containing primarily xylose (15–40% of dry weight) followed by strong acid treatment to release the glucose (30–60%), [6–8]. The sugars are then fermented, usually separately, to ethanol. However, there are two steps in particular that limit the productivity and increase the cost of this process: the hydrolysis of cellulose and the fermentation of xylose. The most promising solution to the former problem is enzymatic hydrolysis, but currently the enzymes remain too expensive and the hydrolysis is incomplete. The problem with xylose fermentation results from a poor overall yield, which compromises the productivity and increases the cost. It is estimated that efficient conversion of xylose would reduce the cost of ethanol by about $0.40 [9].

Organisms

A number of yeasts and bacteria convert xylose to ethanol and other products [10–12], but most of the attention has been given to three yeast strains: *Pachysolen tannophilus, Candida shehatae,* and *Pichia stipitis*. Yeast has the advantages of growing in a simple, inexpensive medium; growing at a low pH to minimize contamination; and being the subject of much past experience. Although both bacteria and yeast generally use transketolase and transaldolase to convert D-xylulose to hexose (as the phosphate esters) and convert the hexose to ethanol by the glycolytic enzymes, they differ with respect to the methods used to convert D-xylose to D-xylulose. Most bacteria carry out the conversion with a single enzyme, xylose isomerase, whereas most yeast reduce the xylose to xylitol and subsequently reoxidize it to xylulose.

The principal limitation to the ethanol yield in the yeast fermentation is the accumulation of xylitol in molar amounts equal to or exceeding that of ethanol [13]. Furthermore, the accumulated xylitol is not further utilized. Fermentation under microaerobic conditions reduces the amount of xylitol [14, 15] accumulated, but the oxygen also promotes further oxidation of ethanol.

Pathways

Besides the accumulation of xylitol, the evidence for the pathway converting xylose to xylulose consists of the identification and isolation of both xylose reductase [16–19] and xylitol dehydrogenase [20, 21] (Table 1), as well as the demonstration that xylose induces both of them [22, 23]. The gene for each of the two enzymes from *P. stipitis* has been identified and sequenced [24, 25], and part of the amino acid sequence of xylose reductase from *P. tannophilus* is known [22]. Workers in several laboratories have described two forms of xylose reductase, one of which is NADH specific and the other of which uses NADH or NADPH [26], but other results suggest that there is a single protein with dual specificity [27]. In any case, the enzyme that requires NADPH predominates in *P. tannophilus*.

The evidence for the pathway for the pentose-to-hexose conversion consists of the demonstration of transaldolase and transketolase in *P. tannophilus*.

However, Ratledge and co-workers [28, 29] demonstrated the enzyme pentulose-5-phosphate phosphoketolase in a number of yeast strains including some *Candida*, some *Pichia*, and *P. tannophilus*. Although the activity was lower in these three strains than in some others, the results bring up the possibility of an alternative pathway that may be associated with a higher theoretical molar yield, 2.0 instead of 1.67.

Xylitol Metabolism

Whole, resting cells of *P. tannophilus* converted xylose to ethanol (yield, 1.11 mol/mol) but did not convert xylitol [30]. However, a cell-free extract, prepared by using a French pressure cell and low-speed centrifugation, produced ethanol from both xylose (yield, 1.1 mol/mol) and xylitol (yield, 1.64 mol/mol). Therefore, the cell-free extract converts xylitol to ethanol with the theoretical yield for the transaldolase-transketolase pathway. In order to test the hypothesis that the permeability of the cell membrane to xylitol limits its metabolism, experiments were done with whole cells in the presence of nystatin, amphotericin B,

Table 1 Summary of the enzymes of xylose metabolism

	Enzyme	
	Xylose reductase	Xylitol dehydrogenase
Protein identified	*P. stipitis*	*P. stipitis*
	P. tannophilus	*P. tannophilus*
Protein purified	*P. stipitis*	*P. tannophilus*
	P. tannophilus	
Protein sequenced	*P. stipitis*	
	P. tannophilus	
Gene identified	*P. stipitis*	*P. stipitis*
Gene sequenced	*P. stipitis*	*P. stipitis*
Gene expressed in	*S. cerevisiae*	*S. cerevisiae*
	E. coli	

and filipin, all agents that increase the cell membrane permeability of yeast and fungi [31, 32]. The rate of xylitol conversion to ethanol by resting cells of *P. tannophilus* increased 40-fold in the presence of nystatin (100 μg/mL), and the other two agents, amphotericin B and filipin, produced similar results [33]. Nystatin had little or no effect on the metabolism of either xylitol or xylose in cell-free extracts. Therefore, the rate of metabolism of extracellular xylitol by this yeast is limited by its transport rate across the cellular membrane. Acetone is apparently effective in the stimulation of xylitol metabolism because it too increases the permeability of the cell membrane. This action of acetone on cell membranes is well known in a variety of organisms.

Requirements

Although the aerobic conversion of xylitol to ethanol by the cell-free extract was quantitative, the anaerobic rate was immeasurably small [30]. In addition, the supernatant, soluble fraction of the cell-free extract, after centrifugation at 100,000 × g for 1 hour, failed to metabolize xylitol aerobically in the absence of the pellet, membrane fraction, which was shown to have oxygen uptake activity in the presence of succinate. Therefore, we have demonstrated that oxygen is required for the metabolism of xylitol. The oxygen requirement is predicted from the fact that xylitol is at an overall lower oxidation state than ethanol plus carbon dioxide.

According to the transaldolase-transketolase pathway, both ADP and NAD should be required in catalytic amounts for xylitol metabolism. The ADP is required for enzymes that generate ATP and for those that require ATP. The NAD is converted to NADH in the oxidation of xylitol to xylulose and of glyceraldehyde phosphate to diphosphoglycerate. Part of the NAD is regenerated in the reduction of acetaldehyde to ethanol and the rest in the reduction of molecular oxygen. Confirmation of these requirements was provided by incubation with xylitol of a fraction of the cell-free extract isolated by gel filtration to remove the low-molecular-weight compounds. Both ADP (100 μM) and NAD (50 μM) were necessary for the conversion of xylitol to ethanol [30].

The requirement for both NAD and oxygen is difficult, if not impossible, to explain if significant metabolic flux is through the phosphoketolase pathway as described above. The generation of ethanol from pentose by the latter pathway requires 2 moles of reduced nucleotide per mole of ethanol to reduce the acetylphosphate. This putative requirement for reduced nucleotide is inconsistent with the demonstrated requirement for oxygen, presumably to oxidize reduced nucleotide. Therefore, the experimental results support the hypothesis that little metabolic flux utilizes the phosphoketolase pathway.

Process Improvements

The ethanol yield has been improved by the addition to the medium of fatty acids [34], polyethylene glycol [35], azide [36], or acetone [37]. In addition, mutants have been reported [38–40]. However, none of these strategies seems to have become widely used in practice. Xylose isomerase has been utilized in conjunction with the fermentation to facilitate the conversion of xylose to xylulose [41], but neither the equilibrium of the reaction nor the pH optimum of the enzyme is favorable [42].

Genetic alterations have been attempted by cell fusions of *Saccharomyces cerevisiae* with *P. stipitis* and with *C. shehatae* [43], but the desired properties were not stable. Several workers have described strains of *S. cerevisiae* [44] and *Schizosaccharomyces pombe* [45] into which the xylose isomerase gene has been cloned, but the productivity was low, presumably because of low levels of transport systems, xylulokinase, transketolase, and transaldolase. In the former organism much of the xylose isomerase activity was insoluble and inactive, but the soluble portion was active. The latter organism produced ethanol with 80% of the theoretical yield but made significant quantities of glycerol and xylitol in the presence of xylose, because it normally has xylose reductase activity but not xylitol dehydrogenase. Others have expressed the gene from *P. stipitis* in *S. cerevisiae* [24], but, predictably, the transformant made xylitol but no ethanol even though some strains have both xylose reductase and xylitol dehydrogenase [46].

In this laboratory the approach has been to investigate the biochemistry of xylose utilization in a xylose-utilizing yeast until a logical and compelling strategy for genetic engineering could be developed. We report the results of some of these investigations below.

MATERIALS AND METHODS

Materials

Xylose isomerase (SPEZYME SGI glucose isomerase) was a gift from Finnish Sugar Co. Tri-sil reagent was purchased from Pierce (Rockford, IL) and the rest of reagents were from Sigma Chemical Co. (St. Louis, MO).

Strains and Growth Conditions

Pachysolen tannophilus (NRRL Y-2460) was from the National Center for Agricultural Utilization Research. Wild-type *Escherichia coli* K-12 was from American Type Culture Collection (Rockville, MD).

The growth medium (per liter) for *P. tannophilus* consisted of 45 g of either xylose or glucose, 3 g of yeast extract, 5 g of peptone, 7.5 g of NH_4Cl, 10 g of KH_2PO_4, and 0.244 g of $MgSO_4$ [38]. The initial pH before sterilization was adjusted to 5.2 with KOH. The liquid culture of 800 mL was incubated in a 2-L Erlenmeyer flask at 31°C.

The growth medium (per liter) for *E. coli* K-12 consisted of 5 g of sodium succinate, 18.9 g of Na_2HPO_4, 6.39 g of KH_2PO_4, 0.2 g of $MgSO_4 \cdot 7H_2O$ and

2 g of $(NH_4)_2SO_4$ [33]. The initial pH before sterilization was adjusted to 7.6 with NaOH. The liquid culture of 500 mL was incubated in a 2-L Erlenmeyer flask at 37°C.

The flasks were inoculated and then incubated in a rotary shaker operated at 200 rpm and an amplitude of 3 cm. The growth was monitored by measurement of the turbidity at 650 nm, after appropriate dilution was made so that the reading did not exceed 1.0.

Crude Extract Preparation

P. tannophilus cells at late log phase were harvested by centrifugation (1500 × g, 10 min), washed twice with one volume of phosphate buffer (73 mM KH_2PO_4, 1.0 mM $MgSO_4$, pH 6.6, 4°C), and resuspended to the final cell density of 40 g/L dry cell weight. The cells were disrupted by a single passage through a French press at 20,000 lb/in² (4°C). After the intact cells and the cell debris were removed by centrifugation, 1500 × g for 20 min, the resulting supernatant solution was the crude cell extract.

High-Speed Supernatant Fraction

The high-speed supernatant fraction was obtained by centrifugation of the crude cell extract at 100,000 × g for an hour to remove membrane organelles.

High-Molecular-Weight Fraction

The high-molecular-weight (HMW) fraction (molecular weight >1000) was obtained by gel filtration of either the crude cell extract or the high-speed supernatant fraction on a PD-10 Column (Sephadex G-25M, Pharmacia, Uppsala, Sweden). After 2.5 mL of sample was loaded on the column, the HMW fraction was eluted with 3.5 mL of phosphate buffer.

Charcoal Treatment

Activated charcoal was added to the HMW fraction (200 mg charcoal/mL liquid), and the mixture was gently mixed on a vortex mixer for several minutes before the charcoal was removed with syringe filter (0.22 μm). This treatment should remove residual NAD^+ completely [9, 36].

Xylose Isomerase

The commercial xylose isomerase was diluted 1.7-fold with the phosphate buffer. Unless otherwise noted, it was treated with PD-10 column three times and charcoal once to remove the low-molecular-weight species.

Transhydrogenase

The partially purified transhydrogenase was prepared as described by Sweetman [33] with some modification. *E. coli* K-12 cells at exponential phase were harvested and washed as described above and resuspended to the final cell density of 0.5 g/mL wet cell weight. The cells were disrupted by a single passage through a French press at 20,000 lb/in^2 (4°C). After the intact cells and the cell debris were removed by centrifugation, 30,000 × g for 20 min, the resulting supernatant solution was centrifuged again at 140,000 × g for 3 hours. The resulting dark brown pellet was resuspended with Tris-HCl buffer (40 mM Tris-HCl, 150 mM KCl and 10 mM MgCl$_2$, pH 7.4) to the final concentration of approximately 50 mg protein/mL.

Analytical Instrument Conditions

Ethanol, xylose, and xylitol were analyzed with a Hewlett Packard 7590A gas chromatograph equipped with a flame ionization detector and a Hewlett Packard HP3396A integrator. The flow rate of the carrier gas, helium, was 20 mL/min. For ethanol determination, the injector and the column (10% C1500 on 80/100 chromosorb, WHP, Alltech) were maintained at 90°C. For xylose and xylitol analysis, as the trimethylsilyl derivatives [32], the injector and the column (3% OV-17 on 100/120 Supelcoport from Supelco, Bellefonte, PA) were maintained at 160°C.

Ethanol, Xylose, and Xylitol Analysis

For ethanol determination, a sample (100 μL) of liquid phase was mixed with 100 μL of 0.4% (v/v) *n*-butanol (internal standard) before being analyzed by gas chromatography. For the determination of xylose and xylitol, a sample (50 μL) of the liquid phase was lyophilized with 106 μg of D-sorbitol (internal standard) and the trimethylsilane (TMS) derivatives were prepared by adding Tri-sil reagent according to the method provided by the manufacturer.

The ratio of the area of the peak of each chemical species to that of an internal standard was calculated. The amount of each chemical species was determined by interpolation of the ratio in a calibration curve. The amounts of the products formed and substrate consumed were calculated as the net change of each chemical species during the incubation.

Transhydrogenase Activity

The reaction mixture (1.0 mL) to determine transhydrogenase activity contained phosphate buffer (73 mM KH$_2$PO$_4$, 1.0 mM MgSO$_4$, pH 6.6), NADH (0.3 mM), acetylpyridine analogue of NAD$_+$ (0.6 mM), and *E. coli* transhydrogenase preparation (0.6 mg protein/mL) and the absorbance was measured at 375 nm.

Protein Concentration

The protein concentrations were determined with a Bio-Rad protein assay kit (Bio-Rad Laboratories, Richmond, CA) with bovine serum albumin as a standard.

Incubation for Ethanol Formation

The reaction was conducted at 31°C under atmospheric pressure in a 50-mL centrifuge tube with a screw cap, and it was placed in a shaker with rotary shaking at 200 rpm and an amplitude of 3 cm. The incubation time was 2 hours unless noted otherwise.

RESULTS

In order to simplify the investigation of the requirements for xylose metabolism, the putative pathway was separated into two parts by the inclusion of xylose isomerase (Spezyme GI, Finnsugar Biochemicals, Inc.) in the incubations with the fractions from the cell-free extract. Preliminary experiments showed that ethanol formation from xylose by the crude cell extract was stimulated ninefold by added xylose isomerase.

In the presence of xylose isomerase the activity of the high-speed supernatant fraction, subjected to gel filtration (fraction HMW), was restored by the addition of both ADP and NAD^+ (Table 2). Although NAD^+ stimulated ethanol formation in this preparation (fraction A), its complete removal required treatment three times with gel filtration and once with charcoal (fraction B, Table 2). In contrast to our previous results with xylitol [10], xylose is metabolized in the absence of oxygen, and xylose is metabolized by soluble fractions of the

Table 2 Effects of Sephadex column and charcoal treatment of the high-speed supernatant fraction

Fraction	Additional components		EtOH (mg/mL)
HMW	—	—	<0.01
HMW	—	NAD^+	<0.01
Fraction A[a]	ADP	—	0.12
Fraction A[a]	ADP	NAD^+	0.64
Fraction B[b]	ADP	—	<0.01
Fraction B[b]	ADP	NAD^+	0.62

[a]Fraction A, the HMW fraction of the high-speed supernatant fraction, has undergone two consecutive gel filtration treatments.
[b]Fraction B; fraction A was treated with charcoal.

cell-free extract from which the membranes and organelles have been removed by centrifugation.

Although the addition of NAD^+ and ADP restored the production of ethanol from xylose by the HMW fraction in the presence of xylose isomerase, their addition failed to restore ethanol formation from xylose by the same fraction in the absence of xylose isomerase. In addition, inclusion of $NADP^+$, NADPH, or ATP in addition to NAD^+ and ADP, as well as inclusion of a variety of other coenzymes and metal ions in the presence of NAD^+, $NADP^+$, and ADP, failed to restore activity. Because of the fact that added NADPH was rapidly consumed by the HMW fraction in the absence of xylose and the fact that, at a higher concentration (>100 μM), both NADPH and $NADP^+$ inhibited ethanol formation from xylose by the unfractionated cell extract, three NADPH generation systems were tested to maintain a constant supply of the reduced coenzyme at low concentration: partially purified transhydrogenase from *E. coli* K-12, NADP-linked isocitrate dehydrogenase from porcine heart, and NADP-linked alcohol dehydrogenase from *Thermoanaerobium brockii*. Although transhydrogenase activity was ineffective in the restoration of ethanol production, both the NADP-linked isocitrate dehydrogenase and the alcohol dehydrogenase stimulated the formation of ethanol from xylose by about 60- and 80-fold respectively (Table 3).

A number of factors contribute to xylitol accumulation. Certainly, the fact that xylose reductase requires NADP whereas xylitol dehydrogenase requires NAD (nucleotide imbalance) is one of the factors. Another is the fact that extracellular xylitol is not normally utilized by whole cells.

In experiments containing purified cell-free fraction, xylose isomerase, NAD, and ADP, xylitol was produced during xylose metabolism (Table 4). Furthermore, the gas chromatographic assay for ethanol contained an additional peak that had the same retention time as acetaldehyde. Therefore, some of the

Table 3 Restoration of ethanol formation in the presence of an NADPH generation system

Reaction mixture	Substrate	EtOH made (mg/mL)
None	Xylose	<0.01
IC^a + $ICDH^a$	None	0.057
IC + ICDH	Xylose	0.710
$cPen-OH^a$ + ADH^a	None	0.016
cPen-OH + ADH	Xylose	0.925
TH^b + NADPH	Xylose	<0.01

[a]IC, sodium isocitrate; ICDH, $NADP^+$-linked isocitrate dehydrogenase; cPen-OH, cyclopentanol; ADH, $NADP^+$-linked alcohol dehydrogenase; TH, transhydrogenase preparation.

Table 4 Ethanol and xylitol production

	Xylose used (μmol)	Xylitol made (μmol)	EtOH made (μmol)	Ratio EtOH/ xylitol
Whole cell				5.75[a]
Crude extract				3.63[a]
High-speed supernate				0.63[a]
Fraction B[b] + XI[b] + ADP + NAD$^+$	15.4	7.9	11.0	1.39
Fraction B + XI + ADP + NAD$^+$ + NADP$^+$	29.1	19.9	9.7	0.49

[a]From Reference 30.
[b]Same as in Table 2; XI, xylose isomerase.

NADH generated by glyceraldehyde-phosphate dehydrogenase is reoxidized in the generation of xylitol, presumably from xylulose by xylitol dehydrogenase. The addition of NADP$^+$ had little effect on the amount of ethanol made from xylose, but it increased the xylitol accumulation by 2.5-fold, increased the xylose consumption by 2-fold, and eliminated the acetaldehyde peak (Table 4). Therefore, most of the xylitol is generated by xylose reductase in these extracts containing all of the necessary cofactors and presumably in the whole cells as well.

Xylose is reduced to xylitol with electrons (NADPH) from the pentose-hexose cycle, but the poor coupling of the rate with the rate of xylitol dehydrogenase causes xylitol to accumulate faster than it is utilized. The intracellular xylitol leaks out of the cell and becomes diluted in the medium, from which it is not reutilized because the transport back into the cell is limiting. The intracellular xylitol can be metabolized if oxygen is present. Thus oxygen is required for optimum xylose metabolism.

The logical strategy for process improvement according to the foregoing results would be the transformation and expression of the gene for xylose isomerase in one of the xylose-fermenting yeasts. However, it would also be necessary to inhibit both the xylose reductase and xylitol dehydrogenase, in order to prevent xylitol accumulation. The transformants should be able to ferment xylose to ethanol anaerobically.

REFERENCES

1. Calvert, J. G., J. B. Heywood, R. F. Sawyer, and J. H. Seinfeld. Achieving acceptable air quality: Some reflections on controlling vehicle emissions. *Science* 261:37–45, 1993.
2. Anderson, E. V. Ethanol's role in reformulated gasoline stirs controversy. *Chem. Eng. News* 70: 7–13, 1992.
3. Lynd, L. R. Large-scale fuel ethanol from lignocellulose. *Appl. Biochem. Biotechnol.* 24/25: 695–719, 1990.
4. Wyman, C. E., and N. D. Hinman. Ethanol, fundamentals of production from renewable feedstocks and use as a transportation fuel. *Appl. Biochem. Biotechnol.* 24/25:735–753, 1990.

5. Lynd, L. R., J. H. Cushman, R. J. Nicholas, and C. E. Wyman. Fuel ethanol from cellulosic biomass. *Science* 259:1318–1323, 1991.
6. Whistler, R. L., J. Bachrach, and D. R. Bowman. Preparation and properties of corn cob hemicellulose. *Arch. Biochem.* 19:25–33, 1948.
7. Weihe, H. D., and M. J. Phillips. Hemicellulose from wheat straw. *Agric. Res.* 60:781–786, 1940.
8. Rosenberg, S. L. Fermentation of pentose sugars to ethanol and other neutral products by microorganisms. *Enzyme Microb. Technol.* 2:185–193, 1980.
9. Hinman, N. D., J. D. Wright, W. Hoagland, and C. E. Wyman. Xylose fermentation, an economic analysis. *Appl. Biochem. Biotechnol.* 20/21:391–401, 1989.
10. Skoog, K., and B. Hahn-Hagerdal. Xylose fermentation. *Enzyme Microb. Technol.* 10:66–80, 1988.
11. Toivola, A., D. Yarrow, E. van den Bosch, J. van Duken, and W. A. Scheffers. Alcoholic fermentation of D-xylose by yeasts. *Appl. Environ. Microbiol.* 47:1221–1223, 1984.
12. Nigram, J. N., R. S. Ireland, A. Margaritis, and M. A. Lachance. Isolation and screening of yeasts that ferment D-xylose directly to ethanol. *Appl. Environ. Microbiol.* 50:1486–1489, 1985.
13. Taylor, K. B., M. J. Beck, D. H. Huang, and T. T. Sakai. The fermentation of xylose: studies by carbon-13 nuclear magnetic resonance spectroscopy. *J. Ind. Microbiol.* 6:29–41, 1990.
14. du Preez, J. C., B. A. Prior, and A. M. T. Monteiro. The effect of aeration on xylose fermentation by *Candida shehatae* and *Pachysolen tannophilus*. *Appl. Microbiol. Biotechnol.* 19:261–266, 1984.
15. Debus, D., H. Methner, D. Schulze, and H. Dellweg. Fermentation of xylose with the yeast *Pachysolen tannophilus*. *Appl. Microbiol. Biotechnol.* 17:287–291, 1983.
16. Bolen, P. L., J. A. Bietz, and R. W. Detroy. Aldose reductase in the yeast *Pachysolen tannophilus*: Purification, characterization, and N-terminal sequence. *Biotechnol. Bioeng. Symp.* 15:129–148, 1985.
17. Rizzi, M., P. Erlemann, B. T. N. Anh, and H. Dellwig. Xylose fermentation by yeasts. 4. Purification and kinetic studies of the xylose reductase from *Pichia stipitis*. *Appl. Microbiol. Biotechnol.* 29:148–154, 1988.
18. Ditzelmuller, G., C. P. Kubicek, W. Woeher, and M. Roeher. Xylose metabolism in *Pachysolen tannophilus:* Purification and properties of xylose reductase. *Can. J. Microbiol.* 30:1330–1336, 1984.
19. Verduyn, C., R. Van Kleef, J. Frank, H. Schreuder, J. P. Van Dijken, and W. A. Scheffers. Properties of the NAD(P)H-dependent xylose reductase from the xylose-fermenting yeast *Pichia stiptitis*. *Biochem. J.* 226:669–677, 1985.
20. Ditzelmuller, G., C. P. Kubicek, W. Wohrer, and M. Rohr. Xylitol dehydrogenase from *Pachysolen tannophilus*. *FEMS Microbiol. Lett.* 25:195–198, 1984.
21. Morimoto, S., M. Matsuo, K. Azuma, and A. J. Sinskey. Purification and properties of D-xylulose reductase from *Pachysolen tannophilus*. *J. Ferment. Technol.* 64:219–225, 1986.
22. Bolen, P. L., and R. W. Detroy. Induction of NADPH-linked D-xylose reductase and NAD-linked xylitol dehydrogenase activities in *Pachysolen tannophilus* by D-xylose, L-arabinose, or D-galactose. *Biotechnol. Bioeng.* 27:302–307, 1985.
23. Verduyn, C., J. Frank, J. P. van Dijken, and W. A. Sheffers. Multiple forms of xylose reductase in *Pachysolen tannophilus* CBS4044. *FEMS Microbiol. Lett.* 30:313–317, 1985.
24. Takuma, S., N. Nakashima, M. Tantirungkij, S. Kinoshita, H. Okada, T. Seki, and T. Yoshida. Isolation of xylose reductase gene of *Pichia stipitis* and its expression in *Saccharomyces cerevisiae*. *Appl. Biochem. Biotechnol.* 28:327–340, 1991.
25. Kotter, P., R. Amore, C. P. Hollenberg, andn M. Ciriacy. Isolation and characterization of the *Pichia stipitis* xylitol dehydrogenae gene, *XYL2,* and construction of a xylose-utilizing *Saccharomyces cerevisiae* transformant. *Current Genet.* 18:493–500, 1990.
26. Verduyn, C., J. Frank, J. P. van Dijken, and W. A. Scheffers. Multiple forms of xylose reductase in *Pachysolen tannophilus* CBS4044. *FEMS Microbiol. Lett.* 30:313–317, 1985.

27. Bolen, P., K. A. Roth, and S. N. Freer. Affinity purifications of aldose reductase and xylitol dehydrogenase from the xylose-fermenting yeast *Pachysolen tannophilus*. *Appl. Environ. Microbiol.* 52:660–664, 1986.
28. Evans, C. T., and C. Ratledge. Induction of xylulose-5-phosphate phosphoketolase in a variety of yeasts grown on D-xylose: The key to efficient xylose metabolism. *Arch. Microbiol.* 139:48–52, 1984.
29. Ratledge, C., and J. E. Holdsworth. Properties of a pentulose-5-phosphate phosphoketolase from yeasts grown on xylose. *Appl. Microbiol. Biotechnol.* 22:217–221, 1985.
30. Xu, J., and K. B. Taylor. Characterization of ethanol production from xylose and xylitol by a cell-free *Pachysolen tannophilus* system. *Appl. Environ. Microbiol.* 59:231–235, 1993.
31. Bolard, J. How do the polyene macrolide antibiotics affect the cellular membrane properties? *Biochim. Biophys. Acta* 864:226–234, 1986.
32. Sutton, D. D., P. M. Arnow, and J. O. Lampen. Effect of high concentrations of nystatin upon glycolysis and cellular permeability in yeast. *Proc. Soc. Exp. Biol. Med.* 108:170–175, 1961.
33. Sweetman, A. J., and Friffeths, D. E. *Biochem J.* 121:117–124, 1971.
34. Dekker, R. F. H. Lipid-enhanced ethanol production from xylose by *Pachysolen tannophilus*. *Biotechnol. Bioeng.* 6:605, 1986.
35. Hahn-Hagerdal, B., B. Joensson, and E. Lohmeier-Vogel. Shifting product formation from xylitol to ethanol in pentose fermentations using *Candida tropicalis* by adding polyethylene glycol (PEG). *Appl. Microbiol. Biotechnol.* 21:1173–1175, 1985.
36. Hahn-Hagerdahl, B., T. W. Chapman, and T. W. Jeffries. Mutants of *Pachysolen tannophilus* for ethanol production from xylose. *Appl. Microbiol. Biotechnol.* 24:287–292, 1986.
37. Alexander, N. Acetone stimulation of ethanol production from D-xylose by *Pachysolen tannophilus*. *Appl. Microbiol. Biotechnol.* 25:203–207, 1986.
38. Jeffries, T. W. Mutants of *Pachysolen tannophilus* showing enhanced rates of growth and ethanol formation from D-xylose. *Enzyme Microb. Technol.* 6:254–258, 1984.
39. Lochke, A. H., and T. W. Jeffries. Levels of the enzymes of the pentose phosphate pathway in *Pachysolen tannophilus* Y-2460 and selected mutants. *Enzyme Microb. Technol.* 8:353–359, 1986.
40. Lee, H., A. P. James, D. M. Zahab, G. Mahmourides, R. Maleszka, and H. Schneider. Mutants of *Pachysolen tannophilus* with improved production of ethanol from D-xylose. *Appl. Environ. Microbiol.* 51:252–258, 1986.
41. Gong, C. S., L. F. Chen, M. C. Flickinger, L. C. Chiang, and G. T. Tsao. Production of ethanol from D-xylose by using D-xylose isomerase and yeasts. *Appl. Environ. Microbiol.* 41:430–436, 1980.
42. Olivier, S. P., and P. J. du Toit. Sugar cane bagasse as a possible source of fermentable carbohydrates. II. Optimization of the xylose isomerase reaction for isomerization of xylose as well as sugar cane bagasse hydrolyzate to xylulose in laboratory-scale units. *Biotechnol. Bioeng.* 28:984–699, 1986.
43. Gupthar, A. S. Segregation of altered parental properties in fusions between *Saccharomyces cerevisiae* and the D-xylose fermenting yeasts *Candida shehatae* and *Pichia stipitis*. *Can. J. Microbiol.* 38:1233–1237, 1992.
44. Sarthy, A. V., B. L. McConaughy, Z. Lobo, J. A. Sundstrom, C. E. Furling, and B. D. Hall. Expression of the *Escherichia coli* xylose isomerase gene in *Saccharomyces cerevisiae*. *Appl. Environ. Microbiol.* 53:1996–2000, 1987.
45. Chan, E. C., P. P. Ueng, and L. Chen. D-Xylose fermentation to ethanol by *Schizosaccharomyces pombe* cloned with xylose isomerase gene. *Biotechnol. Lett.* 8:231, 1986.
46. Batt, C. A., S. Carvallo, D. D. Easson, Jr., M. Akedo, and A. J. Sinskey. Direct evidence for a xylose metabolic pathway in *Saccharomyces cerevisiae*. *Biotechnol. Bioeng.* 28:549–553, 1986.

CHAPTER
TWELVE

GAS RECOVERY AND UTILIZATION FROM MUNICIPAL SOLID WASTE LANDFILLS

Todd A. Potas

INTRODUCTION

The combination of landfill gas recovery and utilization can be a viable energy resource. Landfill gas, which is composed mostly of methane, carbon dioxide, and minor amounts of nonmethane organic compounds (NMOCs), is generated by anaerobic degradation of the municipal solid waste (MSW) in place in MSW landfills. Although landfills are required to place a clay cap over the MSW to contain the gas generated, as well as to minimize environmental exposure to any other biological and chemical hazards, the cap can be breached by many mechanisms and pathways, including gas migration through and around the cap, leachate migration, rainwater runoff or erosion, and heavy equipment activities. Therefore, MSW landfill gas will eventually migrate out of landfills and be emitted to the atmosphere. Air emissions of NMOCs and methane from landfills are of concern, as there were over 6000 active MSW landfills nationwide as of 1987 with the potential to emit annually over 283,000 tons of NMOCs and over 12 million tons of methane [1]. In addition, there are over 32,000 closed solid waste disposal facilities in the United States, many of which received MSW during past operations [2].

Gas recovery systems installed at landfills have proved to capture high-quality gas (greater than 500 Btu/ft^3). New source performance standards (NSPS) for new and existing MSW landfills have thus been proposed by the U.S. Environmental Protection Agency that outline minimum control requirements for gas recovery and subsequent combustion or utilization of the methane and NMOCs in the landfill gas. These rules are proposed as 40 CFR Parts 51, 52, and 60 [1]. The EPA Office of Management and Budget completed a finalized draft of the NSPS on August 24, 1994, which appeared in the *Federal Register* as a final rule in April 1995. The updated compliance thresholds and default values are included in this paper.

Landfills required to comply with this proposed rule will be any new or operating landfills with the potential to emit more than 55.1 tons of NMOCs per year (50 megagrams per year) and closed landfills that have accepted waste since November 8, 1987, or that have not been formally closed under the provisions of the Resource Conservation and Recovery Act (RCRA). Assuming an average gas generation potential of 170 m^3/Mg (6003 ft^3/ton) and an average NMOC concentration of 4000 ppm (as hexane) in the landfill gas, any landfill that has a design capacity of over 2,800,000 tons of MSW per year (2,500,000 Mg/year) and has accepted waste since November 8, 1987, would be required to comply with the proposed rule [1]. All MSW landfills that have accepted waste since the cutoff date will need to file a design capacity report showing that the landfill is under the design capacity and emission thresholds. Testing the actual NMOC concentrations and gas flow rate from the landfill to show that the landfill gas emissions are actually under the threshold is possible, and the procedures to show this are also outlined in the NSPS. An MSW landfill of this size will generate an average of approximately 300,000 ft^3 of landfill gas per day or, assuming 500 Btu/ft^3 in the gas, 6.3 MMBtu/h, for 20–30 years, if the landfill accepts no more waste.

Landfills with emissions over the waste capacity and emission thresholds will be required to show that their emissions are below the 55.1 tons/year threshold of NMOCs using a three-tiered source testing procedure or implement the minimum gas collection and combustion requirements of the pending NSPS, which is referred to as "best demonstrated technology" (BDT) to control the NMOC emissions. BDT is the combination of an active gas collection system and a combustion device with 98% NMOC combustion efficiency or achieves an outlet concentration of 20 parts per million, volume basis (ppmv). Figure 1 shows a diagram of a model landfill with an active gas collection system, an energy recovery unit, and two types of combustion flares (open and enclosed) [3]. The diagram also shows other necessary ancillary equipment for gas recovery and utilization, such as a slurry cutoff wall, a leachate collection system, and perimeter vents. A landfill with the equipment illustrated would be expected to comply with the proposed NSPS, provided that the gas wells are sufficiently spaced for an adequate radius of influence. The radius of influence is the area around each well in which gas recovery can occur.

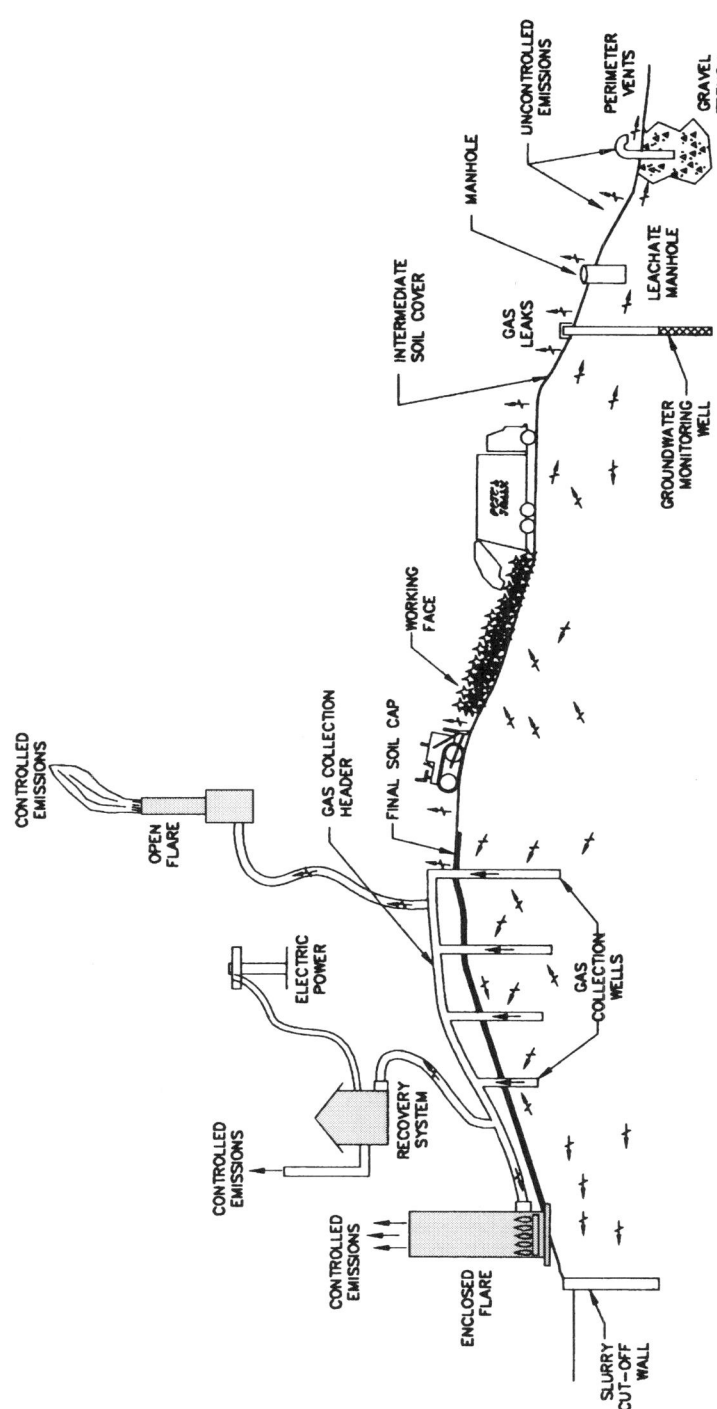

Figure 1 Diagram of an MSW landfill with gas collection and recovery systems.

DISCUSSION

The minimum requirement for complying with the proposed rule in terms of simplicity of operation is an active gas collection system and an open flare. Candle flame–type, open flares are relatively inexpensive to install and operate. However, it is not possible to sample air emissions accurately from open flares. In addition, open flares have no defined residence time at temperature, create noise pollution, and are aesthetically unpleasant because of the sight of the candle flame. Open flares are also difficult to control and can be subject to flameout during high winds. Therefore, it is difficult to predict the air emissions from open flares over an extended period of time. It is likely that to guarantee compliance with the pending NSPS and applicable air permitting requirements, state agencies will probably not accept open flares in air permit applications, although the open flare is a dramatic emission control improvement over venting landfill gas directly to the atmosphere.

Alternatively, enclosed flares (1) provide for stack emissions measurement, (2) have a defined residence time, and (3) keep the flame hidden. This is due to the defined combustion chamber created by the stack of an enclosed flare. They also provide better mixing and temperature control in the combustion chamber, assuring more complete combustion. Therefore, the enclosed flare can be expected to perform with more dependability in keeping air emissions to a minimum than an open flare.

A deficiency of both types of flares is that they do not make use of the landfill gas as a fuel (energy) source. Two major types of stand-alone, heat engine electrical generation systems are capable of utilizing landfill gas for electrical power generation. The two types are reciprocating internal combustion engines and gas turbines. There are several advantages and disadvantages to each utilization option. In addition, it is often possible to sell the landfill gas to local industrial customers for boiler fuel or other combustion uses.

Reciprocating Engines

Reciprocating engines are popular because of their flexibility, low capital cost, low gas pressure, and small size [base units start at 0.8 MW (1200 HP)]. It is easy to add and subtract units with variations in the landfill gas generation or the size of the collection system, and they operate with low maintenance costs. In addition, when these engines are operated at a low operating pressure (12–30 psi), there is less gas compression condensate to handle and treat than if they are operating at a high operating pressure (60–160 psi). The disadvantage of reciprocating engines is their relatively low combustion efficiency compared with turbines; 80% for internal combustion engines compared with 99+% for gas turbines. This results in low overall efficiencies for power generation and relatively higher air emissions for the same amount of gas combusted.

Gas Turbines

Gas turbines have the advantage of being highly efficient systems for electrical power generation, which results in relatively low air emissions. However, these two advantages come at the expense of several disadvantages; high capital and operating cost, limited flexibility, larger base capacity and higher maintenance costs than reciprocating engines, and the creation of large amounts of gas condensate. The gas compression condensate from the high-pressure delivery turbine feed systems must often be treated as a hazardous waste. In addition, the turbines need a dependable gas source to operate efficiently. Both types of systems, turbines and reciprocating engines, are most efficiently operated when a flare is used as a backup and/or as a supplemental combustion source to respond to fluctuations in landfill gas generation.

Air Emissions Comparison

The air emissions generated from the landfill operations are obviously of highest concern when the landfill gas is vented directly to the atmosphere. The addition of an active gas collection system and combustion control system of the types mentioned above relieves this concern dramatically. Once installed, the emissions from an active gas collection system itself are nearly nonexistent due to the negative pressure maintained in the system and the relatively nonporous landfill cap. Collection efficiencies, in terms of potential gas generated, are estimated as 75–90% [1]. Only infrequent cases of gas well maintenance or isolated landfill gas pockets breaching the cap would probably cause any significant amount of emissions. Minor amounts of volatile organic compound (VOC) emissions are generated from condensate water collected from gas collection system pumps, exposed manholes and vents, water storage tanks, and turbine or reciprocating engine compressors, as the water will contain a small percentage of soluble and insoluble VOCs.

A comparison of the NMOC emissions from the three combustion sources discussed—enclosed flares, reciprocating interal combustion engines, and gas turbines—is shown in Figure 2 on a logarithmic scale. Data for the graph were generated from averaged actual compliance stack testing for each type of unit tested [4]. Data for direct venting of the gas to the atmosphere are included for comparison. The emissions data shown as referenced for each type of combustion system are from Reference 3. Data have been included for both low-pressure and high-pressure reciprocating internal combustion engines. Emissions factors for open flares, assuming well-maintained operation and temperature control, are available in the US EPA AP-42, Compilation of Emission Factors, Section 2.4, Landfills, January, 1995 (5).

Emissions comparison for the three different types of combustion systems are shown for carbon monoxide and nitrogen oxides (NO_x) in Figure 3 and 4. In general, the emissions from the landfill gas turbines are the lowest per Btu

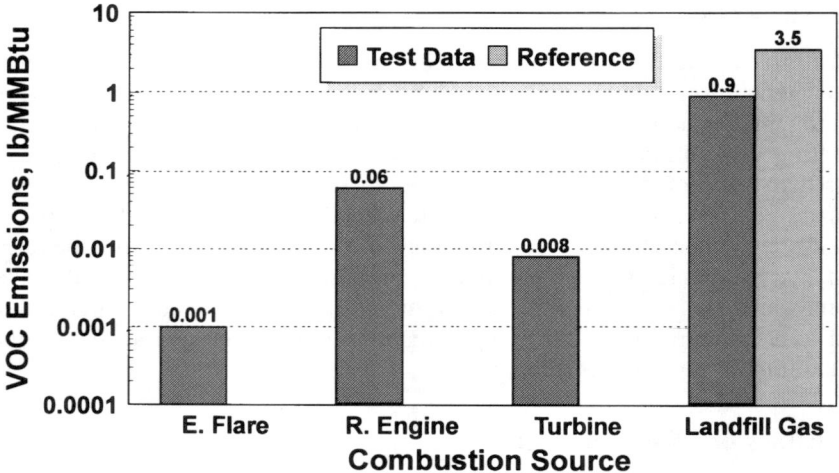

Figure 2 Volatile organic carbon emissions from landfill gas combustion sources.

of landfill gas fired. The averaged test data are very comparable to the reference data for most of the pollutants. NO_x emissions for the reciprocating engines were considerably higher for the averaged test data compared with the referenced data. There are several points to note when comparing the emissions data for the landfill gas combustion units discussed.

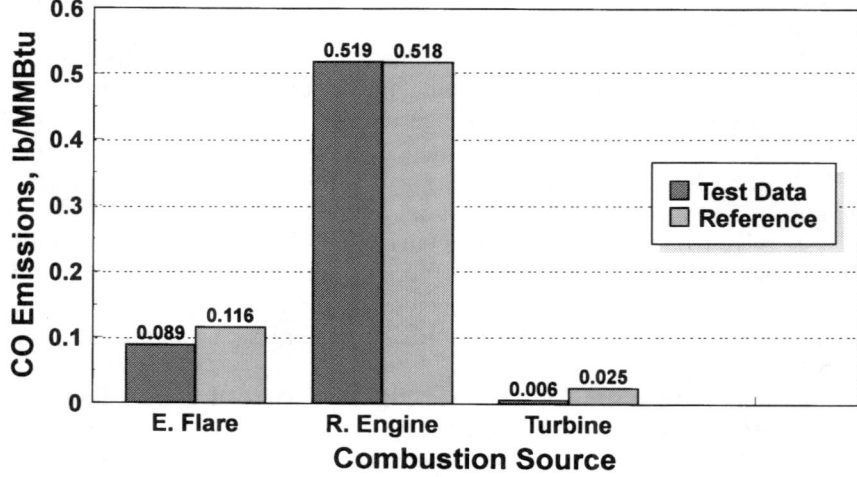

Figure 3 Carbon monoxide emissions from landfill gas combustion sources.

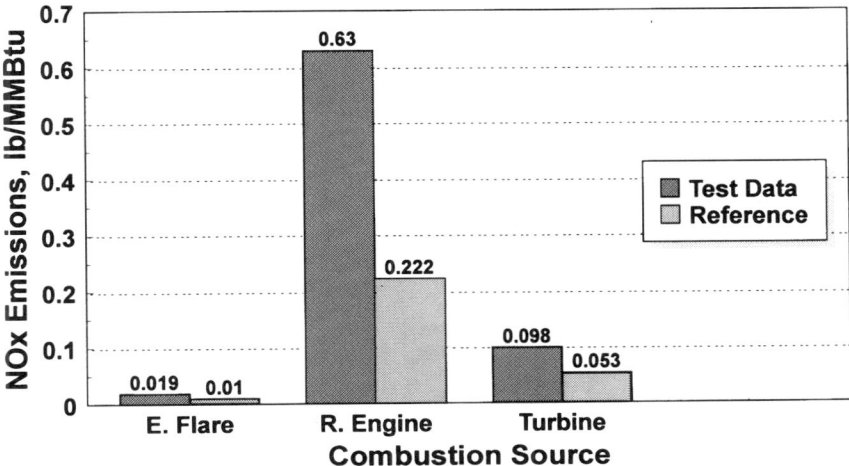

Figure 4 Nitrogen oxides emissions from landfill gas combustion sources.

1. The sulfur dioxide emissions for the turbines and the reciprocating engines were quite low, just as they are for natural gas combustion. This is expected as landfill gas is typically low in sulfur, usually below 200 ppm in total sulfur compounds. Sulfur dioxide was not measured in the test data shown for the enclosed flare. The reference information shows an average value of 0.006 lb/MMBtu for flares, reciprocating engines, or turbines. A comparison of sulfur dioxide emissions for other fuels compared with landfill gas is shown in Figure 5.
2. Carbon monoxide (CO) emissions were found to be relatively high for the reciprocating engines and the enclosed flares, while being low for the gas turbines. This can be an important consideration if the landfill is located in a metropolitan area that is nonattainment for CO. The turbine option is, therefore, a good possibility for many metropolitan areas of the United States.
3. NO_x emissions were higher for the reciprocating engines than for the gas turbines and the enclosed flares. The reference data showed the same trend, although the NO_x emissions for the actual reciprocating engine tests averaged over three times higher than for the reference data. Several factors could contribute to higher NO_x, including the facts that operating conditions for reciprocating engines are highly variable and the cylinder combustion temperatures of the engines are between 2500° and 3000°F.
4. The NSPS for new and existing MSW landfills has indicated that the destruction efficiency for the flare, the reciprocating engine, and the gas turbine is expected to be greater than 98% if the systems are operating properly and within their design capacities. Experience shows that it may prove difficult to achieve such efficiency in reciprocating engines in the field.

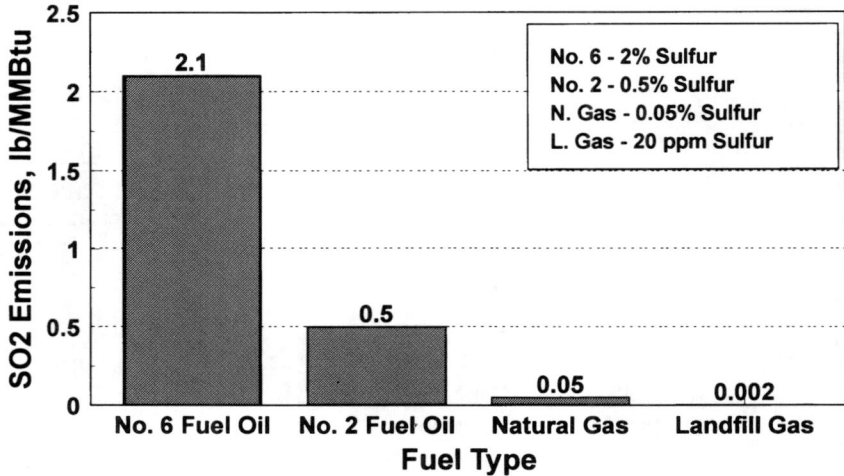

Figure 5 Typical sulfur dioxide emissions from various types of fuel combustion.

REFERENCES

1. Standards of performance for new stationary sources and guidelines for control of existing sources: Municipal solid waste landfills. *Federal Register* 56(104), May 30, 1991. Updated by EPA Office of Planning & Budget, August 24, 1994.
2. *Federal Register,* 53 FR 33324, August 30, 1988.
3. Buonicore, A. J., and W. T. Davis, eds. *Air pollution engineering manual,* Air and Waste Management Association, 863–873. New York: Van Nostrand Reinhold, 1992.
4. Accumulated project information, primarily from Waste Management Incorporated, Rust Environment and Infrastructure, Minneapolis, MN, 1991–1993.
5. US EPA AP-42, Compilation of Emission Factors, Section 2.4, Landfills, Research Triangle Park, NC, January 1995.

CHAPTER
THIRTEEN

EMISSIONS FROM BURNING FOREST FUELS: DEVELOPMENT OF A MODEL

Wei Min Hao, Lisa M. McKenzie, and Geoffrey N. Richards

INTRODUCTION

Burning biomass fuels emit a complex mixture of particulate matter and gases into the atmosphere. These emissions could provide an important feedstock of some compounds or a source of energy, but their potential harmful effects must also be considered. Many of the compounds may be toxic and/or carcinogenic, especially if exposure levels are high [1]. Some of the low-molecular-weight hydrocarbons and particulate matter are major air pollutants and/or greenhouse gases [2]. The conditions under which these compounds are produced and their concentrations must be known in order to fully and safely utilize biomass wastes and to protect the health and safety of people exposed to emissions of burning biomass.

The burning of biomass is widespread, with more than 6 million hectares burned each year in the United States alone, which accounts for only 2 to 3% of the global total of more than $6 \times 10^{+15}$ g [3]. Other regions of the world have even more complex air quality problems resulting primarily from the smoldering combustion of biomass. For example, concentrations of carbon monoxide

*Wei Min Hao is also at Intermountain Fire Sciences Laboratory, Intermountain Research Station, Forest Service, U.S. Department of Agriculture, P.O. Box 8089, Missoula, MT 59807.

of more than 50 ppm resulting primarily from biomass burning were measured in the urbanized areas of the Amazon Basin in Brazil [4]. Rapid changes of land-use practices in the tropics have significantly increased the use of fires for deforestation, shifting cultivation, domestic energy consumption, and clearing of agricultural residues during the past two decades.

Biomass fires are generally divided into two phases, flaming and smoldering combustion. These phases are distinguished by their combustion efficiencies. The combustion efficiency has been defined as the molar ratio of CO_2 emitted to the sum of CO_2 and CO emitted [5]. Flaming combustion occurs when the combustion efficiency is greater than 90%, and smoldering combustion occurs when the combustion efficiency is less than 90%.

Our work has focused on characterizing the condensable volatile fraction of smoke from the smoldering combustion of wood using bench-scale fires [6]. The emissions were condensed at $-45°C$, and the major condensable components of the smoldering combustion of ponderosa pine sapwood were identified and quantified using gas chromatography/mass spectrometry (GC/MS) [7]. The compounds identified and quantified are listed in Table 1. The quantifications were performed by integrating the molecular ion of a compound in the condensate and comparing it to the integration of molecular ions in authentic standards and the integration of the molecular ion of an internal standard. The major condensable compounds emitted from the smoldering combustion of sapwood were methanol (0.44%), acetic acid (0.43%), vinyl acetate (0.3%), pyruvic aldehyde (0.2%), and acetol (0.12%) [7]. The major components of the condensates were attributed to pyrolysis of the lignin and polysaccharide components of wood. We now present models of emissions of 2-furaldehyde, acetol, acetic acid, and guaiacol from 29 bench-scale fires of various components of ponderosa pine relative to the combustion efficiency of those fires. We also investigate correlations between components in the condensate.

EXPERIMENTAL

The experimental procedure has been described previously [7]. It is briefly outlined here. Ponderosa pine (*Pinus ponderosa*) needles, bark, and litter were ground in a Wiley mill to pass a 1-mm sieve. Ponderosa pine duff and humus were manually forced through a 0.022-inch wire sieve, and the rocks were removed. The litter, duff, and humus were collected from a relatively pristine (80 years undisturbed) ponderosa pine stand. Approximately 20 g of sample was burned via smoldering combustion in the bench-scale fire apparatus [7]. Three needle fires, three litter fires, three duff fires, and three humus fires were allowed to smolder for 40 minutes, with smoldering maintained by a radiant electric heater. Two bark fires were allowed to smolder in the same way for 45 minutes.

Table 1 Mean yields of components in smoke condensates of ponderosa pine wood

No.	Compounds	Yield (mg/kg)[a]	% RSD[b]
1	1-Hydroxy-2-propanone *(acetol)*	1200	33
2	2-Oxopropanoic acid, methyl ester	nq[c]	
3	3-Hydroxypropanal	nq	
4	2-Hydroxy-3-oxobutanal	nq	
5	3-Furaldehyde	nq	
6	2-Cyclopenten-1-one	170	27
7	2-Furaldehyde	790	15
8	3-Oxobutanoic acid, methyl ester	680	26
9	2-Furanmethanol	240	43
10	2-Cyclopenten-1,4-dione	nq	
11	gamma-Butyrolactone	150	10
12	2-(5H)-Furanone	190	15
13	2-Methyl-2-cyclopenten-1-one	nq	
14	2-Acetylfuran	48	16
15	Angelicalactone (not alpha)	nq	
16	5-Methyl-2-furaldehyde	130	17
18	Phenol	94	3
19	*o*-Hydroxybenzaldehyde	13	9
20	2-Methylphenol *(o-cresol)*	36	24
21	3 and/or 4-Methylphenol *(m/p-cresol)*	41	23
22	2-Methoxyphenol *(guaiacol)*	160	18
23	Benzoic acid	16	53
24	2-Methoxy-4-methylphenol *(4-methylguaiacol)*	230	25
25	2-Methoxy-4-ethylphenol *(4-ethylguaiacol)*	nq	
26	2-Methoxy-4-ethenylphenol *(4-vinylguaiacol)*	nq	
27	2-Methoxy-4-(1-prop-2-enyl)phenol *(eugenol)*	61	14
28	4-Hydroxy-3-methoxybenzaldehyde *(vanillin)*	67	17
29	2-Methoxy-4-(1-prop-1-enyl)phenol	nq	
30	2-Methoxy-4-(prop-1-en-3-one)phenol[d]	nq	
31	Vinyl acetate	3000	51
32	Pyruvic aldehyde	2000	13
33	Acetic acid	4400	29
34	Formic acid	970	24
35	Propanoic acid	250	22
36	Crotonic acid	68	10
37	Methanol	4300	37

[a]The mean yield of three separate fires as per kg fuel consumed.
[b]Relative standard deviation.
[c]Not quantified.
[d]Name used by Pouwels et al. [10].

These fires are defined as smolder fires. The radiant heater was removed, and a further three bark, three litter, and three duff fires were lit with matches and allowed to smolder until the smoldering ceased. These fires are defined as self-sustained smolder fires. Three fires of ponderosa pine wood sticks were started with matches and allowed to flame. These fires were extinguished when the flaming ceased. These are defined as flaming fires. The final products of the fires were a mixture of ash, char, and some unburned fuel (7–15 g), a creosote oil or tar on the side of the funnel (<2 g), condensed volatiles in the two cold traps (2 to 7 g, mostly water) and in the Teflon corrugated tube, and noncondensed gases, which is an integrated sample collected in a Tedlar bag.

The condensates were extracted into methylene chloride. The methylene chloride extracts and nonextracted condensate were analyzed by electron impact GC/MS at 70 eV on a Hewlett Packard (HP) 5890 GC interfaced to a 5970 series mass-selective detector. The gases collected in the Tedlar bag were analyzed on an HP 5890 series GC with a flame ionization detector.

RESULTS

The results for three smoldering fires of ponderosa pine sapwood that we reported previously [7] and the results for the 26 additional fires of ponderosa pine needles, bark, litter, duff, and humus fires have been statistically examined for correlations between the components of the condensates. The correlations between components of the condensates and combustion efficiencies have also been examined. The 29 fires include fires of six separate fuel components of ponderosa pine and three different fire types, smoldering, self-sustained smoldering, and flaming.

Correlations Between Components in the Condensates

The linear regression correlation coefficients (r) between the emissions in the condensates are presented in Table 2. Linear correlations between 2-furaldehyde, guaiacol, acetic acid, acetol, and some other components of the condensate were examined. 2-Furaldehyde and acetic acid are known to be products of the pyrolysis of hemicelluloses [8]. They were therefore chosen to be marker compounds for the pyrolysis of hemicelluloses. If a component is positively correlated with one of these compounds, it may also be a pyrolysis product of hemicellulose. For example, the emissions of propanoic acid were found to have a strong positive linear correlation ($r = 0.90$) with the emissions of acetic acid; therefore propanoic acid may also be a product of the pyrolysis of hemicelluloses. 2-Acetylfuran, vinyl acetate, methanol, 5-methyl-2-furaldehyde, and bu-

Table 2 Linear regression correlation coefficients (r) between various emissions

	2-Furaldehyde		Guaiacol
5-Methyl-2-furaldehyde	0.84	Methylguaiacol	0.85
Butyric acid	0.80	Eugenol	0.92
		Vanillin	0.71
	Acetic acid		Acetol
2-Acetylfuran	0.81	2-Furanmethanol	0.81
Propanoic acid	0.90	2-Butyrolactone	0.92
Vinyl acetate	0.67	2-($5H$)-Furanone	0.86
Methanol	0.77	Crotonic acid	0.82
		Pyruvic aldehyde	0.93
		2-Cyclopenten-1-one	0.73
		Formic acid	0.73
		3-Oxobutanoic acid, methyl ester	0.84

tyric acid may also be products of the pyrolysis of hemicelluloses, because they are positively correlated with either acetic acid or 2-furaldehyde.

Guaiacol is known to be produced in the pyrolysis of lignin [9, 10]. An emission in the condensate that has a positive linear correlation with the emission of guaiacol may also be a product of the pyrolysis of lignin. Methylguaiacol, eugenol, and vanillin may be lignin pyrolysis products, as they are positively correlated with guaiacol.

Acetol is known to be produced in the pyrolysis of polysaccharides (including hemicelluloses and celluloses) [11]. A positive linear correlation of a compound with acetol would suggest that the compound is also a pyrolysis product of polysaccharides. 2-Furanmethanol, 2-butyrolactone, 2-($5H$)-furanone, crotonic acid, pyruvic aldehyde, 2-cyclopenten-1-one, formic acid, and 3-oxobutanoic acid, methyl ester may all be polysaccharide pyrolysis products, as they are positively correlated with acetol.

2-Furaldehyde and Combustion Efficiency

The emission of 2-furaldehyde has a negative linear correlation with the combustion efficiency ($r = -0.86$). As the combustion efficiency decreases, the emission of 2-furaldehyde increases in accordance with the fact that 2-furaldehyde is a product of pyrolysis rather than combustion. The emissions of 2-furaldehyde in a fire of primarily of softwoods, such as ponderosa pine, can be predicted from the combustion efficiency with the following equation.

$$[\text{2-Furaldehyde}] = 5100 \text{ mg/kg} - 54 \text{ mg/kg} \times (\text{combustion efficiency})$$

Acetol and Combustion Efficiency

There was no apparent correlation between the emission of acetol and combustion efficiency over the 29 fires However, there were linear trends between the emission of acetol and combustion efficiency when the fires were separated by fuel component. The wood, needle, bark, and duff fires all exhibited negative linear correlations with the combustion efficiency. That is, as the combustion efficiency decreased the emissions of acetol increased.

The emission of acetol is likely to be dependent on the polysaccharide content of the fuel, because acetol is a pyrolysis product of polysaccharides. The wood of Douglas fir has three times the polysaccharide content of the needles and six times the polysaccharide of the bark [12]. The amount of polysaccharide in a tree also varies with location, height, and season [13]. In order to compare emission factors on a basis reflecting content of polysaccharide, normalization factors were used and are listed in Table 3. The estimated polysaccharide content of each fuel component was used to estimate the normalization factors. The normalized yield was then calculated by multiplying the actual yield by the appropriate polysaccharide normalization factor. The sapwood was assumed to contain 5 times the polysaccharide content of the needles, 9 times the polysaccharide content of the litter, 11 times the polysaccharide content of the bark, and 12 times the polysaccharide content of the duff and humus.

The normalized yield of acetol has a negative linear correlation with the combustion efficiency ($r = -0.78$). The emission of acetol can then be estimated using the following equation.

$$[\text{Acetol}] = \frac{14000 \text{ mg/kg} - 140 \text{ mg/kg (combustion efficiency)}}{\text{polysaccharide normalization factor}}$$

Acetic Acid and Combustion Efficiency

There was little correlation between the emission of acetic acid and the combustion over the 29 fires, but as in the case of acetol, there were negative linear

Table 3 Normalization factors

Fuel component of ponderosa pine	Polysaccharide normalization factor	Lignin normalization factor
Wood	1	1
Needles	5	3
Bark	11	3
Litter	9	1
Duff	12	1
Humus	12	1

correlations between acetic acid emissions and combustion efficiency when the fires were separated by fuel type. The same normalization factors were used to normalize the acetic acid emissions that were used to normalize the acetol emissions. The actual yield of acetic acid was multiplied by the polysaccharide normalization factor to give the normalized yield.

The normalized acetic acid yields increased exponentially as the combustion efficiency decreased ($r = -0.85$). Acetic acid emissions may be estimated using the following equation.

$$[\text{Acetic acid}] = \frac{(18 \times 10^7) \text{ mg/kg } e^{(-0.13 \times \text{combustion efficiency})}}{\text{polysaccharide normalization factor}}$$

Guaiacol and Combustion Efficiency

A correlation between the emission of guaiacol and the combustion efficiency was not apparent. Two linear correlations could be distinguished upon close inspection. One linear relationship includes the bark and needle fires; the other includes the wood, litter, duff, and humus fires. In both cases, the emission of guaiacol increased as the combustion efficiency decreased. This may be explained by considering the lignin content of the various fuel components of ponderosa pine.

Bark contains polyphenols other than lignin that are not present in sapwood. When the apparent lignin content of bark is determined by many commonly used methods, the nonlignin polyphenol content is included [14]. These polyphenols contain less of the guaiacol monomer than lignin, and this may account for the low guaiacol emissions in the bark fires. A similar situation may exist for needles. Lignin content is also known to vary with location within a tree, height, and season [13]. The guaiacol emissions may increase in the litter and duff layers due to decomposition of lignin and polyphenols in these layers. A normalization factor for the lignin content, as listed in Table 3, was assigned for each fuel component of the ponderosa pine. The bark and needles were assumed to have three times less lignin content than the other fuel components. A normalized yield was then calculated by multiplying the normalization factor by the actual yield in order to compare the emission factors on a basis reflecting the lignin content.

The normalized yield of guaiacol has a negative linear correlation with the combustion efficiency ($r = -0.85$). The emission of guaiacol can then be estimated using the following equation.

$$[\text{Guaiacol}] = \frac{1100 \text{ mg/kg} - 12 \text{ mg/kg (combustion efficiency)}}{\text{lignin normalization factor}}$$

DISCUSSION

The emissions of condensable volatiles increased with decreasing combustion efficiency in all of the emissions we measured. This was expected, because combustion efficiency is a measure of the completeness of combustion, and the condensable compounds are predominantly pyrolysis products. Thus, products of pyrolysis and of incomplete combustion increase as the combustion efficiency decreases. It has been shown that other products of incomplete combustion of biomass, such CH_4, CO, and nonmethane hydrocarbons, also increase with decreasing combustion efficiency [15, 16].

The results also indicate that combustion efficiency is not the only factor in determining gaseous emissions from fires. Other factors, such as the fuel's chemical composition, can influence the emissions produced during a fire, as was the case for acetol, guaiacol, and acetic acid.

The models presented here could be reasonably applied to a fire of predominantly softwoods, such as ponderosa pine. The CO_2 and CO emissions measured in the field, along with an estimate of a normalization factor based on the fuel chemistry of the predominant fuel being burned, could be used to estimate the emissions of acetol, acetic acid, and guaiacol. No normalization factor is required for the estimation of 2-furaldehyde. The estimates for 2-furaldehyde could then be used to estimate the emissions of 5-methyl-2-furaldehyde and butyric acid. The estimates for acetol could in turn be used to estimate the emissions of the compounds in the condensates correlated with it. The same approach could be used with the estimated emissions of acetic acid and guaiacol.

ACKNOWLEDGMENTS

The experimental assistance of S. P. Baker, M. C. Bos, and J. L. Pilon, helpful discussions with D. E. Ward, and financial support of the USDA Forest Service Intermountain Research Station (INT-91639-RJVA) and NASA's Global Change Research Program (4054-GC93-0134) are gratefully acknowledged.

REFERENCES

1. Ward, D. E., J. Peterson, and W. M. Hao. An inventory of particulate matter and air toxic emissions from prescribed fires in the USA for 1989. Presented at the 1993 Annual Meeting of the Air and Waste Management Association, Denver, 1993.
2. Crutzen, P. J., and M. O. Andreae. Biomass burning in the tropics: Impact on atmospheric chemistry and biogeochemical cycles. *Science* 250:1669–1678, 1990.
3. Hao, W. M., M. H. Liu, and P. J. Crutzen. Estimates of annual and regional releases of CO_2 and other trace gases to the atmosphere from fires in the tropics, based on the FAO statistics for the period 1975–80. In *Fire in the tropical biota*. Vol. 84. Ecological Studies. ed. J. G. Goldammer, 440–462. New York: Springer-Verlag, 1990.

4. Ward, D. E., R. A. Susott, A. P. Waggoner, P. V. Hobbs, and D. J. Nance. Emission factor measurements for two fires in British Columbia compared with results for Oregon and Washington. Presented at the 29th Annual Meeting of the Pacific Northwest International Section of the Air and Waste Management Association, Bellevue, WA, 1992.
5. Ward, D. E., and W. M. Hao. Projections of emissions from burning in studies of global climate and atmospheric chemistry. Presented at the 1991 Annual Meeting of the Air and Waste Management Association, Vancouver, British Columbia, Canada, 1991.
6. Edye, L. A., and G. N. Richards. Analysis of condensates from wood smoke: Components derived from polysaccharides and lignins. *Environ. Sci. Technol.* 25:1133–1137, 1991.
7. McKenzie, L. M., W. M. Hao, G. N. Richards, and D. E. Ward. Quantification of major components emitted from smoldering combustion of wood. *Atmos. Environ.* 28:3285–3292, 1994.
8. McGinnis, G. D. Pyrolysis of "xylan" and model compounds. Ph.D. thesis, University of Montana, 1970.
9. Faix, O., D. Meier, and I. Fortmann. Thermal degradation products of wood: Gas chromatographic separation and mass spectrometric characterization of monomeric lignin derived products. *Holz als Roh- und Werkstoff* 48:281–285, 1990.
10. Pouwels, A. D., A. Tom, G. B. Eijkel, and J. J. Boon. Characterisation of beech wood and its holocellulose and xylan fractions by pyrolysis–gas chromatography–mass spectrometry. *J. Anal. Appl. Pyrolysis* 11:417–436, 1987.
11. Faix, O., I. Fortmann, J. Bremer, and D. Meier. Thermal degradation products: A collection of electron-impact (EI) mass spectra of polysaccharide derived products. *Holz als Roh- und Werkstoff* 49:299–304, 1991.
12. Shafizadeh, F., and W. F. DeGroot. Combustion characteristics of cellulosic fuels. In *Thermal uses and properties of carbohydrates and lignins*, 3. New York Academic Press, 1976.
13. Keays, J. J., and G. M. Barton. Recent Advances in Foliage Utilization. Information Report VP-X-137, Department of the Environment, Canadian Forestry Service, Western Forest Products Laboratory, Vancouver, British Columbia, 1975.
14. Sjostrom, E. *Wood chemistry: Fundamentals and applications,* Chapter 6. New York: Academic Press, 1981.
15. Ward, D. E., R. A. Susott, J. B. Kauffman, R. E. Babbitt, B. Cummings, B. Dias, B. N. Hoben, Y. J. Kaufman, R. A. Rasmussen, and A. W. Setzer. Smoke and fire characteristics for cerrado and deforestation burns in Brazil: Base-B experiment. *J. Geophys. Res.* 97:14601–14619, 1992.
16. Hao, W. M., and D. E. Ward. Methane production from global biomass burning. *J. Geophys. Res. Atmos.* 98:20657–20661, 1993.

CHAPTER
FOURTEEN
COPROCESSING OF LIGNOCELLULOSIC WASTES AND COAL TO LIQUID FUELS

Palaniraja Sivakumar, Heon Jung, John W. Tierney, and Irving Wender

INTRODUCTION

Two hundred million tons of municipal solid wastes (MSW) are generated annually in the United States, about two thirds of which is currently deposited in landfills [1]. The number of operational landfills is projected to drop to 1800 by the year 2000, from 18,000 in 1960 [2]. At the same time, the volume of MSW generated will probably increase as a result of an increase in population and a change in living standards.

Products present in a typical MSW are shown in Table 1 [3]. It is composed of an organic fraction, an inorganic fraction, and moisture. The organic fraction is primarily lignocellulosic material, a potential source for energy recovery. About 85% of the moisture-free MSW is combustible or convertible to liquids [3].

Lignocellulosic wastes other than those in MSW include agricultural crop residues such as bagasse, corncobs, and straw; animal manure; and wood wastes from pulp mills, sawmills, and plywood mills. If all the organic municipal wastes generated in the United States were converted to synthetic crude oil, about 300 million barrels could be generated annually [4], which corresponds roughly to the amount of oil the United States imports in a month. Organic wastes could

Table 1 Total municipal solid waste generated in the United States in 1990

Material	Quantity (million tons)	Percentage by weight
Paper & paperboard	73.3	37.6
Yard waste	35.0	17.9
Metals	16.2	8.3
Plastics	16.2	8.3
Food waste	13.2	6.7
Glass	13.2	6.7
Other	28.6	14.6

Source: Reference 3.

make a significant contribution to the U.S. energy supply, reducing dependence on imported oil and helping to reduce our trade deficit. Alternatively, virgin biomass could be used for the production of fuel. Crops are now planted specifically for the production of fuel. Research on fuel crops requiring low input and high photosynthetic efficiency is the main goal of such projects.

Possible ways to convert the combustible portion of waste to energy are direct combustion or mass burn, burning of refuse-derived fuel (RDF), gasification, pyrolysis, biodegradation, and liquefaction. Direct combustion and burning of RDF can be used for the generation of energy. The heat of combustion of organic wastes varies from 3000 to 8000 Btu per pound of dry waste, values below the heating value of coal or fuel oil. Hazardous emissions from exhaust gases and high capital and operating costs have inhibited the widespread use of combustion. Biodegradation and fermentation have very slow reaction rates and need large reactors for waste conversion. Liquid products from the pyrolysis of biomass wastes contain olefins, acids, alcohols, aldehydes, ketones, ethers, and phenolic compounds. Gasification of wastes can produce synthesis gas (a mixture of CO and H_2), which can be converted to transportation fuels via the synthesis of methanol or the Fischer-Tropsch process. Liquefaction of wastes and coprocessing of wastes with coal or heavy oils are promising ways to convert organic wastes to transportation fuels, products needed in the United States. At present, 58% of U.S. petroleum is imported. Conversion of wastes to fuels can reduce our dependence on imported oil while reducing wastes consigned to landfills. Processes for the direct liquefaction of coal can also be used to convert high-molecular-weight material such as paper, lignocellulosic material, plastics, and rubber tires into lower-molecular-weight oils by treatment with hydrogen at elevated temperatures and pressures. The products can be processed to make transportation fuels.

Coal constitutes a major fraction of the U.S. proven reserves of fossil fuels. It can be liquefied in a hydrogen–hydrogen donor solvent environment in the presence of a catalyst [5–16] or in a carbon monoxide, water, alkali catalyst system (Costeam) [17]. It has been shown that lignocellulosic material can be liquefied in the same manner as coal but under less severe conditions [18–21]. Appell et al. [18] studied liquefaction of various lignocellulosic materials including newsprint, corncobs, and pine bark; they showed that conversions as high as 95% and oil yields as high as 42% could be obtained at 250°C using CO and water with sodium carbonate as the catalyst.

Coprocessing of coal and wastes could provide technical and economical advantages not achievable by processing either substrate alone. Collection and transportation of waste are expensive, and plants operating on waste alone are necessarily small and located near sources of waste. Supplementing with coal will make it easier to create an acceptably large feed supply to ensure steady operation, because supplies of waste are likely to be uneven. Coliquefaction experiments on coal and lignocellulosic wastes conducted by various investigators have shown improvement in the product quality compared with the liquefaction of coal alone [22–27]. It has been shown that coal residues from the direct liquefaction of coal can remove metals present in wastes [28]. Coprocessing of coal and heavy petroleum residues is being studied as a promising means of simultaneously liquefying coal and upgrading heavy oil [28]. There is a need for research to understand the various aspects of waste liquefaction and coliquefaction with coal to assess the viability of using this technology for waste disposal and for production of high-quality liquid fuels. The present work focuses on the coprocessing of lignocellulosic wastes with coal.

COAL LIQUEFACTION

Coal liquefaction involves the disintegration of the macromolecular network of coal to form smaller units with an increase in the H/C atomic ratio of the product. Liquefaction reactions are generally carried out at temperatures in the range 400–450°C. At these temperatures free radicals generated in the coal matrix are hydrogenated to form smaller molecules. Coal is usually liquefied in the presence of hydrogen, a hydrogen donor solvent, and a hydrogenation catalyst (termed route A). A recycle solvent acts as a hydrogen shuttling agent in the presence of hydrogenation/hydrogenolysis catalysts. Coal can also be liquefied in the presence of carbon monoxide, water, and an alkali catalyst (here termed route B or the Costeam reaction). Each route will be discussed separately.

Coal Liquefaction via the H_2/Donor Solvent System (Route A)

Either dispersed catalysts or supported catalysts can be used in coal liquefaction by this route. Many transition metals (Mo, Fe, Sn, Ni, W, and Ru) have been identified as potential dissolution catalysts [6]. A sulfide is usually the thermodynamically stable phase for the catalyst under coal liquefaction conditions [7].

The sulfide can be formed by reaction with sulfur inherent in the coal or with sulfur added in some form. Dispersed iron-based catalysts have been of great interest because of their low cost, moderate activity, and environmentally benign nature [6]. Pyrrhotite ($Fe_{1-x}S$), a nonstoichiometric sulfide of iron, has been postulated to be an active catalyst [8]. Ogawa et al. [9] reported the use of mineral pyrrhotite ore (<74 μm) as a coal liquefaction catalyst, observing a good correlation between the stoichiometry of iron sulfides and the partial pressure of H_2S in the reactor. High-activity pyrrhotite is formed at high partial pressures of H_2S; low-activity troilite (FeS) is found at low partial pressures. Oil-soluble precursors such as transition metal carbonyls of Fe and Mo and metal naphthenates are commonly used for coal liquefaction [10–12]. Certain anion-modified metal oxides, such as Fe_2O_3/SO_4, said to be superacids, are highly active catalysts for coal liquefaction in the presence of sulfur [12–14]. Pradhan et al. [16] have shown that bimetallic catalysts such as $Mo/Fe_2O_3/SO_4$ or $Mo/FeOOH/SO_4$ are good catalysts for coal liquefaction. A recent paper has focused attention on the role of catalysts and catalyst dispersion in coal liquefaction [29].

Coal Liquefaction via the CO/H_2O System (Route B)

The CO/H_2O system was investigated for the liquefaction of coal as early as 1921 by Fischer and Schrader [30]. Later work at the Pittsburgh Energy Research Center by Appell et al. [31] has shown that lignite and bituminous coals can be converted to high yields of benzene-soluble material using CO in alkali solution. Some possible advantages of route B in coal liquefaction are:

1. CO-rich synthesis gas produced from the gasification of coal or biomass can be used instead of the more expensive hydrogen needed in route A.
2. In route A, some of the solvent is incorporated irreversibly into the coal and the recycled solvent has to be augmented by a steady supply of fresh organic solvent. However, in route B, liquefaction can be accomplished without an added organic solvent.
3. Hydrogen generated in the water gas shift reaction is a valuable by-product and can be used in situ or later in the upgrading of fuels for transportation uses.
4. There is no metal catalyst in route B to be disposed of.

It has been proposed that the reactive intermediate in the CO/water/alkali system is the formate ion [31]. Using high-pressure nuclear magnetic resonance (NMR) and isotope studies, Horvath and Siskin [32] detected the formate ion in this system. From work with model compounds, Appell et al. [31] proposed the following mechanism:

$$M^+OH^- + CO \longrightarrow HCOO^-M^+ \qquad (1)$$

$$coal + HCOO^-M^+ \longrightarrow coal(H)^-M^+ + CO_2 \qquad (2)$$

$$coal(H)^-M^+ + H_2O \longrightarrow coal\text{-}H_2 + M^+OH^- \qquad (3)$$

$$HCOO^-M^+ + H_2O \longrightarrow M^+OH^- + CO_2 + H_2 \qquad (4)$$

Here the hydroxide ion (OH^-) reacts with CO to form the formate ion ($HCOO^-$), the postulated reactive intermediate, which transfers a hydride ion to coal to form CO_2 and "hydro coal" (coal H^-). The reaction of hydro coal with water completes the hydrogenation cycle by regenerating the hydroxide ion and hydrogenated coal. The water gas shift reaction ($CO + H_2O \rightarrow CO_2 + H_2$) proceeds in parallel with the coal hydrogenation reaction [18, 19]. An increase in water gas shift activity increases the conversion of coal. It was observed that this system was different from the conventional hydrogen donor system in the reduction of benzaldehyde to benzyl alcohol. Appell et al. [31] obtained a conversion of benzaldehyde to benzyl alcohol of 93% in route B, compared with only 6% in route A. In general, in route B carbonyl groups are reduced to alcohols in the presence of formate ion.

To study the mechanism of liquefaction via route B, isotope studies have been conducted by various investigators. Coal conversions in route A with tetralin as solvent showed a normal isotope effect, that is, a decrease in coal conversion using deuterated tetralin [33]. A normal isotope effect indicates hydrogen transfer as the rate-controlling step [34]. But when D_2O is used instead of H_2O in the Costeam reaction, coal conversion increases, indicating an inverse isotope effect [35]. It has been widely accepted that the formate is the reactive intermediate. The decomposition of formate ion to complete the water gas shift reaction shows a normal isotope effect, so that when D_2O is used instead of H_2O, the water gas shift reaction is retarded and the concentration of formate increases (reaction 4). The rate of the reaction involving transfer of hydrogen to coal is therefore increased when D_2O is used (reaction 2) [36]. A mechanism involving a formate intermediate is thus compatible with the observed inverse isotope effect.

Acetates of alkali metal catalysts have been used in the Costeam reactions, and it has been shown that activity decreases in the order $Cs > K > Na > Li$ [37]. This decrease in activity is expected because the water gas shift activity of the catalysts (and the formate intermediate) decreases in the same order. The concentration of the formate ion is dependent on the pH of the aqueous medium and a maximum conversion of Illinois No. 6 coal is obtained at pH values >12.6 [38]. Metal catalysts such as oxyanions (soluble in water) of the highest oxidation states of Mo, Cr, W, Mn (e.g., Na_2WO_4) are reported to be effective in improving coal conversion, benzene-soluble products, and H/C ratios [39].

It is not clear whether all or most of the hydrogen generated during the Costeam process is utilized in the liquefaction of coal. Cassidy et al. [40] have shown that hydrogenation catalysts such as Co, Ni, or Mo, which are active in H_2/tetralin systems, did not improve coal conversion in the CO/H_2O system without an alkali catalyst. Use of catalysts such as sodium aluminate and sodium silicate increased the conversion of Victorian brown coal; these workers observed that sodium and potassium carbonates did not improve the conversion of acid-washed brown coal, which contradicts findings by other investigators.

In studies by Cassidy et al. [37, 40], the coal was acid washed to remove the mineral matter. But reactions with beneficiated coal (treated with a series of gravity separations to remove mineral matter) did not show results different from those obtained with fresh coal [41]. It is possible that the acid treatment may have changed the structure or properties of the coal. In another work by Cassidy et al. [37], catalysts such as sodium aluminate and sodium silicate promoted by Co, Ni, or Cu in the $CO/H_2/H_2O$ system increased the conversion of brown coal to liquid products from 52% to 68%. The increase in conversion could be due to the additional water gas shift capability of the added promoters. Bianco and Girardi [42] studied the effect of molybdenum on the conversion of Illinois No. 6 bituminous coal with sodium carbonate and potassium hydroxide catalysts; they did not observe any increase in coal conversion and oil yields.

Studies using synthesis gas instead of CO have shown that addition of H_2S greatly enhances the conversion of lignite [43]. In the presence of H_2S, the water gas shift capacity increases (as shown below), which may help explain the observed increase in conversion:

$$H_2S + CO \longrightarrow COS + H_2$$

$$COS + H_2O \longrightarrow CO_2 + H_2S$$

COMPONENTS OF LIGNOCELLULOSIC WASTES

The components of the fibrous tissues in plants and trees are cellulose, hemicelluloses, and lignin. Inorganic compounds such as compounds of calcium, magnesium, potassium, manganese, and silicon are also present in significant amounts.

Cellulose

During photosynthesis, water and carbon dioxide combine to form glucose and other simple sugars, with oxygen as a by-product. Cellulose is synthesized from units of glucose. Each glucose unit is combined by a β-glucoside link to the C4 hydroxyl of another glucose unit as shown in Figure 1 [44].

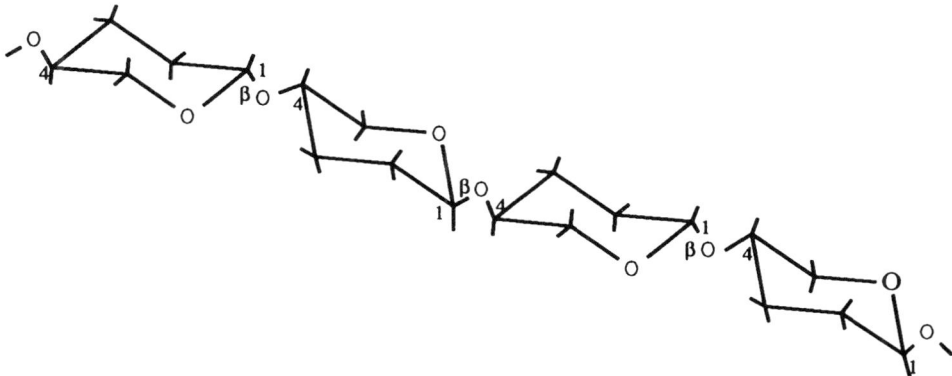

Figure 1 Linkages in a cellulose polymer.

The resulting polymer has as many as 30,000 glucose units. Cotton is 99% cellulose and is used in textiles and in the paper industry. Starch, another polysaccharide stored in the seeds, roots, and fibers of plants, is identical to cellulose except that it has an α arrangement of bonds around the corresponding 1–4 carbon atom in each monomer. Termites and ruminants can metabolize cellulose with the help of certain microorganisms in their intestinal tract. However, humans cannot digest cellulose because they lack enzymes needed to hydrolyze β-linkages.

Hemicelluloses

Glucose is not the only sugar produced during the process of photosynthesis. Six-carbon sugars such as galactose and mannose and five-carbon sugars such as xylose are also produced. These sugars form branched, low-molecular-weight polysaccharides called hemicelluloses.

Lignin

Lignin is the second most abundant natural organic product, next to cellulose, forming about one fourth of the woody tissue of plants. It acts as a cementing agent to bind the matrix of cellulosic fibers together into a rigid woody structure. The chemical structure and molecular weight of lignin vary, depending on the source (hardwoods, softwoods, straw, etc.) and the isolation procedures. It is practically impossible to isolate two lignin preparations with identical properties, even by the same procedure. Lignin is a noncrystalline, three-dimensional cross-linked polymer built up of various hydroxy/methoxyphenylpropane units. The phenylpropane units are linked by aliphatic and aromatic carbon bonds and ether linkages. A postulated structure of lignin is shown in Figure 2 [45].

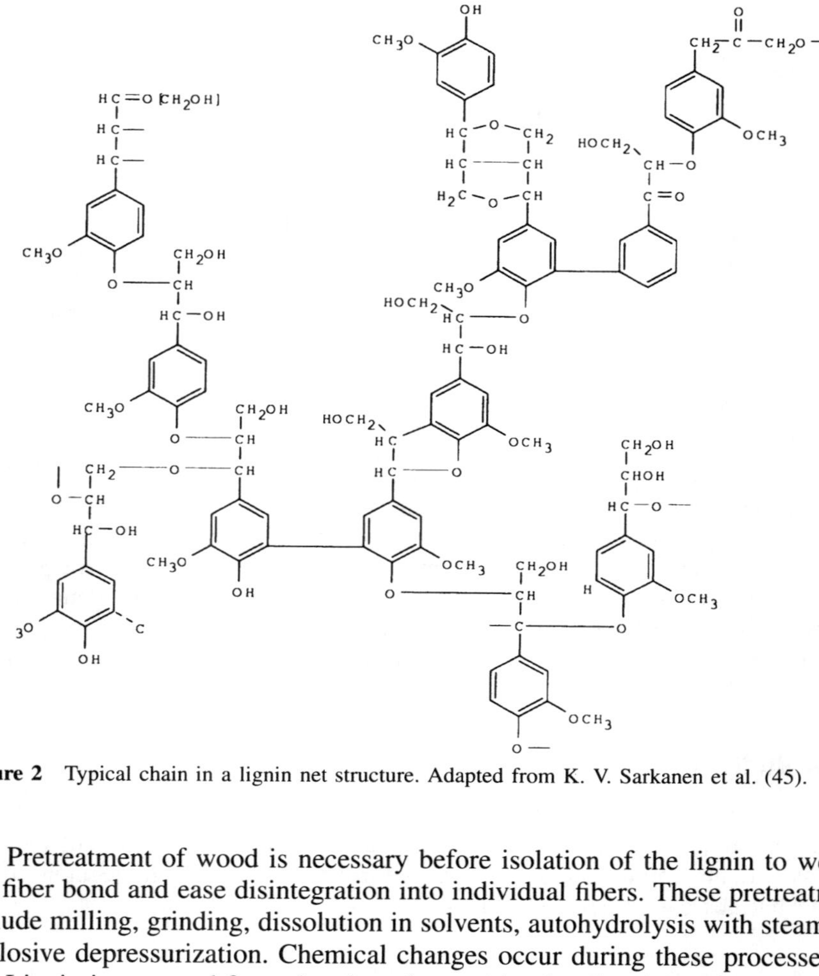

Figure 2 Typical chain in a lignin net structure. Adapted from K. V. Sarkanen et al. (45).

Pretreatment of wood is necessary before isolation of the lignin to weaken the fiber bond and ease disintegration into individual fibers. These pretreatments include milling, grinding, dissolution in solvents, autohydrolysis with steam, and explosive depressurization. Chemical changes occur during these processes.

Lignin is removed from the plant tissue either by dissolving the lignin from cellulose and other constituents or by removing cellulose by chemical treatment. In the fermentation industry, cellulose is hydrolyzed to glucose, which is converted to alcohols, leaving lignin as a residue.

There is a great deal of ambiguity in the use of the term "lignin." When referring to lignin in plant tissue, "lignin in situ" is preferred to "native lignin"; the lignin isolated by Brauns was termed native lignin [45]. Various chemical changes occur during the isolation procedures. It is important that investigators do not underestimate these changes. Lignin derivatives may be called lignins as long as it is clear that we are dealing with derivatives and not isolated lignin preparations. However, certain reactions change the phenylpropane characteristic structure of lignins, and it seems excessive to include these products as lignin derivatives.

In pulp making for the production of paper, lignin is removed using the sulfite process, the kraft and soda process, or the organosolv process. In the sulfite pulping process, a pulp is made by cooking wood chips under pressure in a solution of sulfurous acid and either calcium, magnesium, sodium, or ammonium bisulfite. Lignosulfonic acids and corresponding salts are formed in the spent liquor. The lignosulfonic acids and salts can be isolated by precipitation as insoluble basic sulfonates, a salting method, precipitation with aromatic bases, or by dialysis.

The kraft pulping process uses sodium hydroxide and sodium sulfide as the cooking liquor. The sodium hydroxide breaks down the lignin molecules and sodium sulfide provides elemental sulfur to lignin molecules, making them more soluble. The spent liquor is removed from the pulp and is burned to recover the cooking chemicals. The degraded lignin product, if isolated, is precipitated using an acid.

The organosolv process uses organic solvents such as ethanol to dissolve lignin from other components. This degraded lignin product has a lower molecular weight, is free of sulfur, and is used in the production of specialty chemicals.

It is important to note that most of so-called lignin liquefaction studies have been carried out using the by-product obtained from the kraft pulping process [22, 27]. These products are chemically modified products of lignin and are different from the lignin in the plant. Hence results of experiments on modified lignin products from the kraft process cannot be extended to the corresponding conversion of lignin in natural lignocellulosic material. It is really not appropriate to call these products lignins; however, for convenience, we shall refer to them as kraft lignins.

LIQUEFACTION OF LIGNOCELLULOSIC WASTES CONTAINING LIGNIN IN SITU

The synthesis of hydrocarbon fuels from lignocellulosic wastes is of major interest because they are renewable and abundant. Liquefaction of lignocellulosic wastes or biomass has been carried out via both routes A and B.

The H_2/Donor Solvent System (Route A)

Heinemann [46] has shown that sugarcane, wood, bamboo sticks, and cotton, which contain lignin, hemicelluloses, and cellulose of varying contents, can be converted to bitumen in the presence of sodium hydroxide at 650°C. The amount of the pyrobitumen (product resembling coal) obtained increases with the lignin content of the lignocellulosic material. Petroleum-like hydrocarbons can be synthesized when these bitumens are hydrogenated over molybdenum sulfide catalysts. The lighter hydrocarbon fraction increases and the heavier fractions decrease with decrease in the lignin content of the raw material. Gas formation also decreases with increase in the lignin content.

Wood has been used as the principal lignocellulosic feedstock for liquefaction processes. Meier et al. [47] used a palladium catalyst in the aqueous phase to hydrogenate spruce and birch wood, pine bark, and cellulose. The role of solvent in wood liquefaction is more critical than in coal liquefaction. The solvent not only acts as a hydrogen-donating agent but also should break down the physical morphology of the lignocellulosic network. If solubilization is not promoted, intramolecular elimination of oxygen takes place with heating to high temperatures, resulting in the formation of chars. For wood, creosote oil is a good solvent, allowing complete liquefaction of wood [48].

Gupta et al. [49] studied cellulose liquefaction via route A and suggested that, in the presence of hydrogen and paraffin oil, cellulose reacts over a nickel catalyst through a set of first-order unidirectional steps involving the formation of pyrobitumens. At low temperatures, the formation of pyrobitumens is favorable with a low oil yield. As the reaction temperature increases, the oil and gas yields increase. Because gas formation increases with the formation of oil, an optimum exists at which the oil yield is maximum.

Lignocellulosics have high H/C ratios but the oxygen content is also high, more than 25% depending on the source [50]. The heating value of the oil product varies with the oxygen content, and removal of oxygen is necessary to produce quality products. It is desired to remove oxygen during primary liquefaction without using molecular hydrogen, because removal of oxygen in the upgrading step requires expensive hydrogen. If oxygen is removed as CO_2, the result is a low oil yield due to loss of carbon; if all the oxygen is removed as water, the main product is char.

Oils from the liquefaction of lignocellulosic material have oxygen contents as high as 30% [51]. When tetralin is used as the solvent in cellulose liquefaction, the oxygen content of the oil can be reduced by decreasing the tetralin/cellulose ratio or the reaction time, but there is a corresponding decrease in conversion [51]. The observed decrease in oxygen content may be due to elimination of oxygen to produce char. The oxygen content can also be reduced by high reaction temperatures, but the oil yield is reduced.

The CO/H_2O/alkali System (Route B)

Route B has been used to convert cellulosic wastes and carbohydrates to oils. Most of the oxygen is removed as CO_2, whereas in route A H_2O is mainly formed. Lignocellulosic materials containing lignin in situ such as pine bark [18], corncobs [18], wood [18, 52, 53], and bagasse [21] have been studied in the Costeam route. A continuous unit (3 tons/day) for liquefaction of MSW, cellulose and newsprint, lignocellulosics, wood, bovine manure, and other organic wastes was designed and constructed by the Pittsburgh Energy Research Center of the U.S. Bureau of Mines [54]. Unfortunately, the Bureau property was situated in a first-class residential area. The mayor and other members of the town listened in silence to a speech dedicating the pilot unit to be used for the liquefaction of the wastes mentioned above. The mayor and council quickly voted that no wastes other than wood would be allowed on the Bureau property. For this reason, only wood was ever treated in the 3-ton-per-day plant.

The restriction to wood contributed to operational difficulties, limiting the concentration of the feed to the reactor to 8%. These limitations were later partially overcome, with the work at Lawrence Berkeley Laboratory, by adding an aqueous prehydrolysis step and processing in a colloid mill, which increased the feed concentration to 20–30% [55].

Liquefaction of sugarcane bagasse with CO, water, and sodium formate has resulted in 90% conversion with the maximum conversion at a feed/water ratio of 1:2.5 [21]. The extent of conversion, the oil yield, and the product quality improve with increase in the partial pressure of CO [53]. However as the pressure increases, more formate is thermally decomposed to CO_2, so that CO utilization efficiency decreases.

Synthesis gas is less costly than pure hydrogen. Use of a 1:1 H_2/CO synthesis gas ratio gives an oil yield that is 2 to 4% less than that obtained with pure CO [53]. Solvents such as anthracene and cresols have been shown to improve oil yields in the liquefaction of wood [53]. Solubilization of the lignocellulosic waste material in the solvent is important in this system. In the absence of water or in the presence of insufficient water, the main product from liquefaction of lignocellulosics is char [53]. With an effective catalyst and an optimum feed/water ratio, temperature, and CO partial pressure, high oil yields can be obtained [53, 56].

LIQUEFACTION OF CHEMICALLY MODIFIED LIGNIN

Every year about 20 million tons of kraft lignin and about 2 million tons of lignosulfonates are generated by the paper and pulp industry in the United States. Most of these modified lignin products are burned in recovery units in the mills. These modified lignin products can be converted to low-molecular-weight chemicals and could be used as a feedstock for the chemical industry. Studies on the liquefaction of lignocellulosic material (mostly via route A) have shown that kraft lignin can be converted to liquid fuels in the same way as coal [22, 23].

Schultz et al. [57] observed that, when chemically modified lignin obtained from sweet gum wood meal by treatment with concentrated hydrochloric acid, was hydrotreated using anthracene and tetralin as solvents at 375–425°C (route A), 11% of phenol was obtained. They also observed that the reaction to form phenolics had first-order irreversible kinetics, similar to the results for pure cellulose liquefaction.

In a study of the reactivity of brown coals, Stray et al. [58] hydrogenated modified lignin (extracted with NaOH and precipitated by chloroform) from hardwoods and softwoods at temperatures between 200 and 400°C in the presence of a copper(II) acetate catalyst (route A). The lignin products decomposed at temperatures below 300°C, while brown coal did not exhibit significant conversion to liquids. Below 300°C, lignins from hardwoods were more reactive than the lignins from softwoods, indicating that cleavage of ether linkages (more abundant in hardwoods than in softwoods) may be responsible for the low-temperature liquefaction of lignins.

High yields of low-molecular-weight oils are obtained by hydrotreating kraft lignin using water-soluble molybdenum catalysts at 430°C (route A). Addition of sodium hydroxide decreased the hexane-soluble product and increased the viscosity of the oil product.

Liquefaction of a newsprint (which contained 23.6 wt % modified lignin) by route B (Costeam) has shown that, with complete conversion, more than a 30% oil yield can be obtained [53].

COLIQUEFACTION OF LIGNOCELLULOSIC WASTES AND COAL

Most lignin-coal liquefaction studies have used the hydrogen/hydrogen-donor/ catalyst system; few investigators have used the Costeam system for this type of coliquefaction [5]. Kraft lignin has been used by most investigators in coprocessing experiments. Several investigators have reported that coliquefaction of coal and such lignins allows coal to be liquefied at lower temperatures [22–27]. Because kraft lignin contains phenolic groups, results of coal liquefaction using solvents containing phenolic groups can be helpful in understanding the pathways involved in the enhancement of coal liquefaction. Pott and Broche [59] reported that addition of cresol in coal liquefaction using tetralin resulted in a remarkable increase in coal conversion. Similar results were obtained by Orchin and Storch [60] using various hydroaromatic solvents containing a phenolic group. Kamiya et al. [61] studied the effect of the addition of phenolic compounds in the liquefaction of Japanese coals in the presence of 1-methylnaphthalene/tetralin (3:1 wt %) or tetralin as donor solvent. They observed that addition of phenol increased coal conversion with more than 80% recovery of the phenol. They suggested that the observed increase may be due to either scission of ether linkages present in coal by phenol or increased solubility of the coal fragments by hydrogen bonding.

Larsen et al. [62] observed that the use of phenol as a solvent, without any donor solvent, increased the conversion of Wyodak coal. They also found that extensive rearrangement of hydrogen in coal took place and a hydrogen-rich oil product and a hydrogen-poor residue were obtained. This finding is important because a hydrogen-rich oil product was obtained with minimum input of hydrogen. They suggested that when phenol is heated, phenoxy radicals ($C_6H_5O\cdot$) are formed. This radical abstracts hydrogen from coal and is regenerated to phenol. Phenol can thus be a hydrogen shuttling agent in coal liquefaction. But they also considered the possibility that the observed effect may be due to the increased solubility of coal in the solvent. Sharma et al. [63] reported that the cleavage of di-β-naphthyl ether occurs more rapidly in the presence of mixtures of polyethers in phenol than in the presence of phenol alone. Polyethers are possible decomposition products of lignin. In lignin-coal liquefaction, these products may help bring about the scission of the carbon-carbon bonds in coal.

Coughlin and Davoudzadeh [22, 23] reported that the enhancing effect of kraft lignin (not in situ or native lignin) in coal depolymerization is observed at temperatures below 350°C, but the conversion of coal at these temperatures is less than 40%. When lignin products are heated to temperatures of about 300°C, resonance-stabilized phenoxy radicals are formed [23]. The formation of these radicals may help explain the enhancement in coal liquefaction. In coal liquefaction, addition of guaiacol (o-methoxyphenol), a source of phenoxy radicals, yielded better results than those obtained when tetralin was used [23]. These phenoxy radicals, formed at temperatures of about 300°C, should be used in situ to avoid possible deactivation due to polymerization of these radicals with coal. No enhancing effect is observed when the reaction products from kraft lignin liquefaction at 300°C, obtained separately, are used in coal liquefaction. An explanation may be that sufficient phenoxy radicals are not available from the liquid product obtained from lignin liquefaction for enhancement in coal liquefaction or that the phenoxy radicals formed initially are incorporated into other molecules. But Lalvani et al. [24] observed a 30% enhancement in coal liquefaction at 375°C and 22–34 atm of hydrogen pressure with liquefaction products from kraft lignin products. This result is surprising, because Illinois coal and kraft lignin were used by the research groups. It is not valid to compare lignin-coal coliquefaction with chemically modified lignin products from different sources.

If phenol is added to enhance coal liquefaction, there are some deleterious effects in subsequent catalytic upgrading; hydrogen uptake in downstream processes is also increased. Hence it is better to recover phenol from the product for use as a recycling solvent [64]. If a large fraction of lignin products is used as solvent in coal liquefaction, more coal conversion is obtained, but the product contains more phenolic compounds.

The increased conversion of coal when coprocessed with kraft lignin is observed at 300°C but not at 350°C [23]. It has been suggested that extensive polymerization or radical recombination takes place at higher temperatures. Altieri and Coughlin [65] analyzed the products from coal liquefaction with kraft lignin at 400°C at 1900–2400 psig. They observed that the benzene-soluble fraction of the product increased with a greater portion of the benzene-soluble fraction being pentane soluble. Lalvani et al. [26] analyzed products from coal liquefaction with kraft lignin and found that the liquid product had a higher H/C ratio than the liquid product from coal alone. Also, the O/H ratio was lower than that obtained from this lignin alone. The product from coliquefaction contained low-molecular-weight compounds. ^{13}C NMR data showed an increase in C—O bonds in coal-lignin coliquefaction liquids, indicating incorporation of phenols into the coal molecules [26]; perhaps evidence for the observed repolymerization of radicals at higher temperatures. An optimum ratio of lignin to coal in feed exists for increased conversion; a large lignin product/coal ratio implies a large population of radicals and recombination of these radicals occurs. In one study by Coughlin and Altieri [27], using kraft lignin and Illinois No. 6 coal, this optimum ratio was found to be 0.7. At greater ratios, the conversion of coal leveled off.

Because newsprint constitutes a large portion of MSW, it is of prime importance in waste minimization. Akash et al. [25] have studied the feasibility of liquefying lignin products derived from hydrolysis of newsprint (which contains modified lignin) and coal in a hydrogen/tetralin system. They observed that the conversion of bituminous coal in the mixture increased from 50% to 90% as the temperature was increased from 300° to 325°C. They also observed, as did other investigators, that the enhancing effect decreased above 325°C.

EXPERIMENTAL

Wyodak subbituminous coal, obtained from the Argonne Coal Sample Bank, and newsprint, ordinary copy paper, cotton, and corncobs were used as lignocellulosic wastes. Table 2 lists the elemental and proximate analyses of samples. Route A experiments were carried out in a horizontally shaken 42-cm^3 microautoclave with an initial hydrogen pressure of 1000 psig. Because severe mass transfer limitations were observed in preliminary microautoclave experiments

Table 2 Ultimate and proximate analysis of the substrates

Sample	Ultimate analysis (wt % dry)					Proximate analysis (wt %)	
	C	H	N	S	O	Ash	Moisture
Corncobs	50.0	6.2	1.4	0.10	43.3	2.6	6.0
Newsprint	51.1	5.4	0.2	0.04	43.3	1.3	7.8
Copy paper	41.6	5.5	0.1	0.02	52.8	6.9	5.7
Cotton	44.4	6.2	—	—	49.4	0.0	6.6
Wyodak coal	51.4	4.2	0.7	0.32	11.5	6.3	28.1

using route B, a 300-cm³ stirred autoclave was used. A total of 8 g of reactants mixed with 24 g of water and 1.02 g of Na_2CO_3 was used with an initial pressure of 575 psig CO for most of the experiments. The reactor was heated to the reaction temperature in about 60 minutes and held there for 1 hour. Because of the large amount of water, pressures up to 4200 psig were recorded. At the end of the reaction, the autoclave was cooled to room temperature.

The nonvolatile water-insoluble products were contacted with tetrahydrofuran (THF) and then filtered through a cellulose thimble. The filtered materials (THF insolubles) were then extracted with boiling THF for 2 hours. After removing the THF in a rotary evaporator, an extraction with pentane was carried out. Pentane-soluble products are called oils, and THF-soluble but pentane-insoluble products are called asphaltenes. The water-soluble products, which were generally in the range of 1–3 wt %, were not recovered.

RESULTS AND DISCUSSION

Route A

Table 3 lists the liquefaction results for ordinary newsprint, copy paper, and Wyodak coal at 400°C using an $Mo(CO)_6$ catalyst with added sulfur. Under these conditions, the conversion of copy paper was almost complete (98.6%). A large yield of oil and gas was obtained with only about 10% high-molecular-weight asphaltenes. Newsprint with a lignin content of 23.6% also gave a high conversion at 400°C, but the asphaltene yield was higher than that from copy paper. Conversion of Wyodak coal was also high but with significant amounts of asphaltene formation. As shown in Table 3, the conversion of coal increased from 86.0% to 91.7% when a 2:1 coal/waste ratio by weight of copy paper was coprocessed. The increased conversion of coal resulted in an increase in asphaltenes, but the yield of oil and gas remained unchanged. Increasing the

Table 3 Liquefaction in the H_2/tetralin/$Mo(CO)_6$ system (route A) at 400°C for 1 hour[a]

Reactant	Conversion (%)	Asphaltenes (%)	Oil and gas (%)
Copy paper	98.6	10.2	88.4
Newsprint	95.5	18.0	77.5
Wyodak coal	86.0	39.9	46.1
Coal (67%) + copy paper (33%)	91.7[b]	46.0[b]	45.7[b]
Coal (50%) + copy paper (50%)	91.7[b]	44.0[b]	47.7[b]
Coal (50%) + newsprint (50%)	86.7[b]	49.1[b]	37.6[b]

[a]3 g of reactant and 12 g of tetralin with 5000 ppm Mo and 10,000 ppm S in a 42-mL microreactor. Initial pressure of H_2 was 6.9 MPa and at 400°C about 14.5 MPa.

[b]Conversion and yields are for coal, assuming that conversion and yield of waste remain unchanged.

amount of coprocessed copy paper to 50% (1:1 weight ratio) did not increase the coal conversion further. Coprocessing of newsprint and coal changed coal conversion only slightly, but it increased asphaltene yields and decreased oil and gas yields by about 10%.

It has been proposed that the thermal depolymerization of lignin forms phenoxy radicals, which then cause scission of carbon-carbon bonds in coal [23]. It has also been suggested that, at temperatures as high as 400°C, phenoxy radicals polymerize or recombine so that their enhancing effect is minimal [62]. The fact that the added newsprint, which contains 23.6% lignin, resulted in more asphaltenes from coal at 400°C may support the hypothesis of repolymerization of phenoxy radicals generated from lignin at higher temperatures.

Route B

Results from the liquefaction of various lignocellulosic wastes and Wyodak coal in $CO-H_2O$ environments with Na_2CO_3 catalysts at 400°C are summarized in Table 4. Conversions of all these wastes when reacted without coal were almost complete (>97%). The amount of oil (pentane-solubles) can be measured in $CO-H_2O$ runs, so oil yields and gas yields are listed separately in Table 4. Also listed in Table 4 are the acid-soluble lignin contents of the lignocellulosic wastes (moisture and ash-free basis) determined by the ASTM D-1106 procedure. As shown in Table 4, less gas and more oils and asphaltenes are produced from material that contains more lignin. During liquefaction, much of the oxygen is removed as CO_2 and H_2O. Liquefaction of lignin results in the production of more asphaltenes than are obtained from cellulose. More asphaltenes were expected from corncobs than from newsprint, because corncobs contain more lignin than newsprint. However, less asphaltenes were obtained from corncobs than from newsprint. Newsprint contains lignin that is modified during the pulping process. This lignin differs in structure from the lignin in corncobs. The difference in the type of lignin might be the reason for the lower yield of asphaltenes from corncobs.

Table 4 Liquefaction results for various lignocellulosic wastes in the $CO/H_2O/Na_2CO_3$ system (route B) at 400°C for 1 hour[a]

Material	Lignin (%)	Conversion (%)	Asphaltenes (%)	Oil (%)	Gas + water (%)
Cotton	0	97.4	4.5	16.1	76.8
Copy paper	0	97.4	3.0	16.8	77.6
Newsprint	23.6	98.8	13.6	17.0	68.2
Corncobs	29.7	100.0	10.1	22.8	67.1
Coal	—	90.0	28.6	31.3	30.1

[a] 8 g of reactant and 24 g of H_2O with 1.02 g of Na_2CO_3 in 300-mL autoclave. Initial pressure of CO was 4 MPa and at 400°C total pressure was about 29 MPa at 1500 rpm.

Table 5 Lignocellulosic waste–Wyodak coal coliquefaction results in the $CO/H_2O/Na_2CO_3$ system (route B) at 400°C for 1 hour[a]

Coprocessed Material	Lignin (%)	Conversion[b] (%)	Asphaltenes[b] (%)	Oil[b] (%)	Gas + water[b] (%)
Cotton	0	94.5	30.0	37.9	26.6
Copy paper	0	89.2	31.1	34.7	23.4
Newsprint	23.6	92.9	26.4	36.4	28.5
Corncobs	29.7	91.6	29.5	25.9	36.2
Coal	—	90.0	31.3	30.1	30.1

[a]4 g of coal, 4 g of coprocessed material, and 1.02 g of Na_2CO_3 in 300-mL autoclave. Initial pressure of CO was 4 MPa and at 400°C total pressure was about 29 MPa at 1500 rpm.
[b]Conversion and yields are for coal, assuming that conversion and yield of waste remain unchanged.

The conversion of newsprint via the Costeam reaction at 400°C was almost the same as that achieved with route A but the Costeam product contains fewer asphaltenes (Tables 3 and 4). Wyodak coal also produced fewer asphaltenes but gave higher conversion via Costeam than by route A (Tables 3 and 5).

The results of coprocessing runs when 50% of Wyodak coal was replaced with paper and other lignocellulosic wastes in the Costeam system at 400°C are listed in Table 5; conversion and yields listed in Table 5 are those for coal. Addition of these wastes to coal gave only a small increase in coal conversion. But the total liquid products (asphaltenes and oils) from coal increased except for coprocessing with corncobs. The increase in the liquid coal products from coal in coprocessing was manifested in increased oil yields. Asphaltene yields of coal from coprocessing with these wastes were almost the same as those from a coal-alone run. The observed enhancement of oil yield in coprocessing is greater with wastes that contain larger fractions of cellulose.

Table 6 shows results of gas analyses for typical route B runs at 400°C. Because our experiments were carried out in a batch reactor, the system pressures at room temperature after reaction were higher than the initial pressure of pure CO (580 psig) due to hydrogen produced by the water gas shift reaction

Table 6 Typical gas compositions after 1 hour in $CO/H_2O/Na_2CO_3$ system (route B) at 400°C (initial pressure of CO = 4 MPa)

Reactant	Final pressure (psig)	CO	H_2	CO_2	CH_4	CO conversion (%)	H_2 consumption (mmol/g)
Cotton	1002	5.5	47.1	47.5	0.11	90.6	5.2
Copy paper	998	6.2	46.1	47.7	0.09	89.4	5.6
Newsprint	945	6.8	43.5	49.8	0.16	89.0	10.1
Corncobs	957	6.8	42.9	50.3	0.17	89.3	11.1
Wyodak coal	912	7.2	44.6	48.2	0.16	88.7	10.2
Coal + cotton	934	6.2	45.1	48.8	0.14	90.1	9.7

(water is condensed at room temperature). From the gas composition data with the final pressure at room temperature, the total conversion of CO can be calculated and is listed in Table 6; about 89% of CO is converted at 400°C to form hydrogen, CO_2, and a small amount of light hydrocarbons. The total conversion of CO includes conversion to supply hydrogen to coal and to form hydrogen gas via the water gas shift reaction conversion. The difference between the total moles of CO converted and the moles of hydrogen produced give the moles of hydrogen consumed by the reactants. The moles of consumed hydrogen thus calculated are also listed in Table 6; more hydrogen is consumed for wastes with larger amounts of lignin (newsprint and corncobs) than for conversion of cellulosic wastes with more lignin. In the cotton-Wyodak coal coprocessing run, the consumption of hydrogen (9.7 mmol/g) was higher than the average amounts (7.7 mmol/g) of hydrogen consumption by coal alone (10.2 mmol/g) and cotton alone (5.2 mmol/g).

Significant amounts of valuable hydrogen are produced as a by-product from Costeam runs (Table 6). Hydrogen production is the result of the water gas shift reaction catalyzed by the alkali catalyst. This hydrogen could be separated and then used in a second upgrading stage. Further research should focus on using this hydrogen for hydrogenation during the coprocessing reaction.

ACKNOWLEDGMENTS

The authors gratefully acknowledge financial support from the U.S. Department of Energy under grant number DE-FC22-93PC93053. Elemental and proximate analyses by Consol, Inc. are also appreciated.

REFERENCES

1. Darcey, S. EPA finds growth in recycling and waste generation. *Management of World Wastes* 35(11):6, 1992.
2. Carra, J. S., and R. Cossu, eds. *International perspectives on municipal solid wastes and sanitary landfilling,* Municipal solid waste and sanitary landfilling in the United States of America, 990. New York: Academic Press, 1990.
3. Franklin Associates, Ltd. *Characterization of Municipal Solid Waste in the United States: 1992 Update.* U.S. EPA, Office of Solid Wastes and Emergency Responses, Washington, D.C., 1992.
4. Donald, L. K. Fuels from biomass. In *Encyclopedia of chemical technology.* 3rd ed. Vol. 11, 337–339. New York: Wiley, 1980.
5. Jung, H., J. W. Tierney, and I. Wender. Coprocessing of cellulosic waste and coal. *Prepr. Pap. Am. Chem Soc. Div. Fuel Chem.* 38:880–885, 1993.
6. Derbyshire, F. *Catalysis in direct coal liquefaction: New directions for research,* 10–40. London: IEA Coal Research Publication, 1988.
7. Weisser, O., and S. Landa. *Sulfide catalysts, their properties and applications,* 237–262. New York: Pergamon, 1973.

8. Sweeny, O., V. I. Stenberg, R. D. Hei, and P. A. Montano. Hydrocracking of diphenyl ether and diphenylmethane in the presence of iron sulphides and hydrogen sulphide. *Fuel* 66:532–535, 1987.
9. Ogawa, T., V. I. Stenberg, and P. A. Montano. Hydrocracking of diphenylmethane: Roles of H_2S, pyrrhotite and pyrite. *Fuel* 63:1661–1664, 1984.
10. Suzuki, T., H. Yamada, P. L. Sears, and Y. Watanabe. Hydrogenation and hydrogenolysis of coal model compounds by using finely dispersed catalysts. *Energy and Fuel* 3:707–713, 1989.
11. Watanabe, Y., O. Yamada, K. Fujita, Y. Takegami, and T. Suzuki. Coal liquefaction using iron complexes as catalysts. *Fuel* 63:752–755, 1984.
12. Garg, D., and E. Givens. Relative activity of transition metal catalysts in coal liquefaction. *Fuel Processing Technol.* 8:123–134, 1984.
13. Tanabe, K., H. Hattori, T. Yamaguchi, T. Iizuka, H. Matsuhashi, A. Kimura, and Y. Nagase. Function of metal oxide and complex oxide catalysts for hydrocracking of coal. *Fuel Processing Technology* 14:247–260, 1986.
14. Kotanigawa, T., S. Yokoyama, M. Yamamoto, and Y. Maekawa. Catalytic activities of sulphate and sulphide in sulphur-promoted iron oxide catalyst for coal liquefaction. *Fuel* 68:618–621, 1989.
15. Kotanigawa, T., H. Takahashi, S. Yokoyama, M. Yamamoto, and T. Maekawa. Mechanism for formation of sulphate in S-promoted iron oxide catalyst for coal liquefaction. *Fuel* 67:927–931, 1988.
16. Pradhan, V. R., D. E. Herrick, J. W. Tierney, and I. Wender. Finely dispersed iron, iron-molybdenum, and sulfated iron oxides as catalysts for coprocessing reactions. *Energy & Fuels* 5:712–720, 1991.
17. Appell, H. R., and I. Wender. The hydrogenation of coal with carbon monoxide and water. *Prepr. Am. Chem. Soc. Div. Fuel Chem.* 12(3):220–230, 1968.
18. Appell, H. R., Y. C. Fu, S. Friedman, P. M. Yavorsky, and I. Wender. *U.S. Bur. Mines Rep. Invest.* 7560, 1971; *Agric. Eng.* March:17–19, 1972.
19. Appell, H. R., I. Wender, and R. D. Miller. Conversion of urban refuse to oil. *U.S. Bureau of Mines Tech. Prog. Rep.* 25, 1970.
20. Cotton, J. D., J. C. Penrose, and P. R. Wells. Effect of added $Ru_3(CO)_{12}$ on the direct hydrogenation of cellulose. *Fuel* 51:873–874, 1982.
21. Schuchardt, U., and F. P. Matos. Liquefaction of sugarcane bagasse with formate and water. *Fuel* 61:106–110, 1982.
22. Coughlin, R. W., and F. Davoudzadeh. Lignin depolymerizes coal at 300°C. *Nature* 303:789–791, 1983.
23. Coughlin, R. W., and F. Davoudzadeh. Coliquefaction of lignin and bituminous coal. *Fuel* 65:95–106, 1986.
24. Lalvani, S. B., C. B. Muchmore, J. A. Koropchak, B. Akash, C. Chavez, and P. Rajagopal. Coal liquefaction in lignin-derived liquids under low-severity conditions. *Fuel* 70:1433–1428, 1991.
25. Akash, B. A., C. B. Muchmore, and S. B. Lalvani. Coliquefaction of coal and newsprint-derived lignin. *Fuel Processing Technology* 37:203–210, 1993.
26. Lalvani, S. B., C. B. Muchmore, J. Koropchak, B. Akash, P. Chivate, and C. Chavez. Lignin-augmented coal depolymerization under mild conditions. *Energy & Fuels* 5:347–353, 1991.
27. Coughlin, R. W., and P. Altieri. Enhanced coal liquefaction using lignin. In *Energy recovery from lignite, peat and lower rank coals,* ed. D. J. Trantolo and D. J. Wise, Chapter 6. Amsterdam: Elsevier Science Publishers, 1989.
28. Miller, T. J., S. V. Panvelkar, I. Wender, J. W. Tierney, and Y. T. Shah. Thermal non-catalytic coprocessing of Illinois No. 6 coal with Maya resid and Boscan crude. *Fuel Processing Technology* 23:23–38, 1989.
29. Weller, S. Catalysis and catalyst dispersion in coal liquefaction. *Energy & Fuels* 8:415–420, 1994.

30. Fischer, F., and H. Schrader. Über die Umwandlung der Kohle in Öl durch hydrierung, Part I: Über die hydrierung der Kohle und anderm fasten Brennstoffen Mittels natrium formiates, *Brenstoff-Chem.* 2:161–176, 1921.
31. Appell, H. R., R. D. Miller, and I. Wender. On the mechanism of lignite liquefaction with CO and H_2O. Presented before the Division of Fuel Chemistry, 163rd National Meeting, American Chemical Society, April 10–14, 1972.
32. Horvath, I. T., and M. Siskin. Direct evidence for formate ion formation during the reaction of coals with carbon monoxide and water. *Energy & Fuels* 5:933–934, 1991.
33. Franz, J. A. ^{13}C, ^2H, ^1H NMR and GPC study of structural evolution of a subbituminous coal during treatment with tetralin at 427°C. *Fuel* 58:405–412, 1979.
34. Wilberg, K. B. The deuterium isotope effect. *Chem. Rev.* 55:713–743, 1955.
35. Ross, D. S., G. P. Hum, T. C. Miin, T. K. Green, and R. Mansani. Isotope effects in supercritical water—kinetic studies of coal liquefaction. *Am. Chem. Soc. Div. Fuel Chem. Prepr.* 30:94–100, 1985.
36. Jackson, W. R., F. P. Larkins, and G. J. Stray. Reactions of coals with CO/D_2O in the presence of alkaline catalysts: Establishment of mineral matter in coal as the origin of kinetic isotope effects. *Fuel* 71:343–345, 1992.
37. Cassidy, P. J., W. R. Jackson, F. P. Larkins, M. B. Loney, and R. J. Sakurovs. Promoters for the liquefaction of wet Victorian brown coal in carbon monoxide. *Fuel Processing Technol.* 14:231–246, 1986.
38. Ross, D. S., J. E. Blessing, Q. C. Nguyen, and G. P. Hum. Conversion of bituminous coal in CO/H_2O systems. 2. pH dependence. *Fuel* 63:1206–1210, 1984.
39. Ross, D. S., Q. C. Nguyen, and G. P. Hum. Conversion of bituminous coal in CO/H_2O systems. 3. Soluble metal catalysts. *Fuel* 63:1211–1213, 1984.
40. Cassidy, P. J., W. R. Jackson, F. P. Larkins, R. J. Sakurovs, and J. F. Sutton. Hydrogenation of brown coal: The effect of added promoters and water on the liquefaction of Victorian brown coal using hydrogen, CO and synthesis gas. *Fuel* 65:374–379, 1986.
41. Ross, D. S. Coal conversion in carbon monoxide–water system. In *Coal science,* Vol. III, ed. M. L. Gorbaty, J. W. Larsen, and I. Wender. New York: Academic Press, 1984.
42. Bianco, A. D., and E. Girardi. Studies on coal reactivity by using the CO/H_2O/base system. *Fuel Processing Technol.* 23:205–213, 1989.
43. Sweeny, P. G., V. Gutenkunst, M. E. Martin-Schwan, and V. I. Stenberg. Effect of temperature programming on the liquefaction of Indianhead lignite in an inorganic (H_2S-H_2O) solvent system. *Am. Chem. Soc. Div. Fuel Chem. Prepr.* 32(1):637–642, 1987.
44. Roberts, J. D., and M. C. Caserio. *Basic principles of organic chemistry.* New York: Benjamin, 1964, pp. 402–403.
45. Sarkanen, K. V., and C. H. Ladwig. *Lignins: Occurrence, formation, structure and reactions.* New York: Wiley-Interscience, 1971.
46. Heinemann, H. Hydrocarbons from cellulosic wastes. *Petroleum Refiner* July:161–163, 1951.
47. Meier, D., D. R. Larimer, and O. Faix. Direct liquefaction of different lignocellulosics and their constituents: 1. Fractionation, elemental composition. *Fuel* 65:910–915, 1986.
48. Vanasse, C., E. Chornet, and R. P. Overend. Liquefaction of lignocellulosics in model solvents: Creosote oil and ethylene glycol. *Can. J. Chem. Eng.* 66:112–120, 1988.
49. Gupta, D. V., W. L. Kranich, and A. H. Weiss. Catalytic hydrogenation and hydrocracking of oxygenated compounds to liquid and gaseous fuels. *Ind. Eng. Chem. Process* 15:256–260, 1976.
50. van Heek, K. H., B. O. Strobel, and W. Wanzl. Coal utilization processes and their applications to waste recycling and biomass conversion. *Fuel* 73:1135–1143, 1994.
51. Vasilaakos, N. P., and D. M. Austgen. Hydrogen-donor solvents in biomass liquefaction. *Ind. Eng. Chem. Process Des. Dev.* 24:304–311, 1985.
52. Boocock, D. G. B., D. Mackay, M. McPherson, S. Nadeau, and R. Thurier. Direct hydrogenation of hybrid polar wood to liquid and gaseous fuels. *Can J. Chem. Eng.* 57:98–101, 1979.

53. Appell, H. R., Y. C. Fu, E. G. Illig, and F. W. Steffgen. Conversion of cellulosic wastes to oil. *U.S. Bureau of Mines, Report of Investigations 8013*, 1975.
54. Ergun, S. Biomass Liquefaction Efforts in the United States. Lawrence Berkeley Laboratory, Publ. No. LBL-10456, 1980, p. 20.
55. Figueroa, C., L. L. Schaleger, and H. G. Davis. LBL Continuous Bench-Scale Liquefaction Unit, Operation and Results. Lawrence Berkeley Laboratory, Publ. No. LBL-13709, 1982, p. 17.
56. Wender, I., F. W. Steffgen, and P. M. Yavorsky. Clean liquid and gaseous fuels from organic solid wastes. In *Recycling and disposal of solid wastes,* ed. T. P. Yen, Chapter 2. Ann Arbor, MI: Arbor Science Publishers, 1974.
57. Schultz, T. P., R. J. Preto, J. L. Pittman, and I. S. Goldstein. Hydrotreating of hydrochloric acid lignin in a hydrogen-donor solvent. *J. Wood Chem.* 2(1):17–31, 1982.
58. Stray, G., P. J. Cassidy, W. R. Jackson, F. P. Larkins, and J. F. Sutton. Studies related to the structure and reactivity of coals: The hydrogenation of lignin. *Fuel* 65:1524–1530, 1986.
59. Pott, A., and H. Broche. U.S. Patent 1,881,927, October 11, 1932.
60. Orchin, M., and H. H. Storch. Solvation and hydrogenation of coal. *Ind. Eng. Chem.* 40:1385, 1948.
61. Kamiya, Y., H. Sato, and T. Yao. Effect of phenolic compounds on liquefaction of coal in the presence of hydrogen donor solvent. *Fuel* 57:681–685, 1978.
62. Larsen, J. W., T. L. Sams, and B. R. Rodgers. Internal rearrangement of hydrogen during heating of coals with phenol. *Fuel* 60:335–341, 1981.
63. Sharma, R. K., K. P. Raman, and B. Miller. Catalysis of cleavage of di-β-naphthyl ether by electron-donating agents: An explanation for the effectiveness of phenols as promoters for coal conversion. *Fuel* 65:738–739, 1966.
64. Farcasiu, M., T. O. Mitchell, and D. D. Whitehurst. U.S. Patent 4,133,646, September 1, 1979.
65. Altieri, P., and R. W. Coughlin. Characterization of products formed during coliquefaction of lignin and bituminous coal at 400°C. *Energy & Fuels* 1:253–256, 1987.

CHAPTER
FIFTEEN

MATHEMATICAL ANALYSIS OF A MUNICIPAL SOLID WASTE ROTARY INCINERATOR

James T. Cobb, Jr., and Kunal Banerjee

INTRODUCTION

A number of the waste-to-energy facilities operating in the United States utilize a rotary kiln combustor as the principal treatment unit in their systems. The barrel of the kiln may be either refractory-lined or water-walled. Waste is fed to the kiln from a receiving system and the products of combustion are treated and released. The combustion gases are polished in a secondary combustor, cleaned of acidic components and particulates, and sent to a stack. Bottom ash is cooled, mixed with fly ash and scrubber effluent, stabilized, and landfilled [1].

Because of the concerns for full burnout of the combustibles in the feed, the destruction of toxic organics, the segregation and collection of toxic metals, and the efficient collection of energy, developers and operators of waste-to-energy processes must have a thorough understanding of the operation of each element of their facilities and of the integration of these elements. The mathematical model described in this chapter has been constructed and tested to assist in better understanding the operation of the rotary kiln combustor at the heart of many such facilities.

MODELING OF ROTARY KILN TREATERS

A number of theoretical, mathematical models for rotary kiln treaters have been developed. The two principal types of processes to which these treaters are applied are removal of volatiles from contaminated solids (frequently soils) and waste combustion for both volume reduction of solids and energy production from the heating value of the feed material.

The models themselves fall into three basic categories:

Elementary input/output analyses and expert systems.
Thermal transport analyses.
Chemical kinetic analyses coupled with heat and mass transport.

Three examples of the first category of models are present in the literature. An expert system for evaluating the incinerability of any of a wide variety of hazardous waste streams has been developed by Energy and Engineering Research Corporation of Irvine, California [2]. This methodology uses a combination of experimental and theoretical experiences to determine the compatibility of the waste and the destruction system and to predict the performance of a full-scale unit. A similar model has been developed at the University of California at Los Angeles [3]. This is a linear programming model, which includes only mass and energy balances, based on feed properties. Optimization is obtained using the SIMPLEX algorithm. The third example is a program developed by the Environmental Protection Agency [4]. In this program the EPA provides a complete spreadsheet of the elementary equations for hazardous waste incinerators. This code may be used by federal, state, and local regulators in evaluating projects seeking permits for commercial operation.

Predicting the heat transfer (thermal transport) from the hot combustion gases to the cooler waste solids is a critical element of six modeling efforts.

A model describing regenerative heat transfer was proposed by Gorog et al. [5]. It included a detailed formulation of the radiative and convective heat transfer coefficients and was employed to examine the effects of different kiln variables such as rotational speed, percent loading, gas and solid temperatures, and wall emissivity on the regenerative contribution to heat transfer. The model showed that this contribution was most significant at the cold end of a counter-current kiln.

More recently, Goshdastidar and co-workers [6] developed a heat transfer model for a solid waste incinerator. Radiation heat transfer between the refractory surface and the flame was considered. The refractory and flame were segmented into surface elements and radiation exchange was determined by the absorption factor technique. Temperature profiles for the solids in the bed, the refractory wall, and the gas phase were computed.

Oak Ridge National Laboratory has developed a comprehensive model with a number of state-of-the-art methods for predicting radiant transfer [7].

As part of its extensive modeling work, about which more will be said later, the University of Utah has developed a similar model to embed in its full simulation package [8, 9]. One improvement over earlier models was the inclusion of the demoisturizing step, which leads to a cooling effect in the heatup zone [8]. The model was used to predict gas, bed, and kiln wall temperatures under conditions typical of industrial operation. The authors state that

> at the conditions examined, the model predicts that coflowing gas and solid streams result in higher average burden temperatures than do counterflowing streams. The moisture level of the feed is predicted to be a key operating parameter [9]. Prediction based on this model for typical operating conditions indicate that 10 percent by weight moisture can lower peak soil temperatures by 200 to 300K. Peak soil temperatures for coflowing solids are 200K lower than those obtained under counterflowing conditions [8].

The Milan Polytechnic (Italy) focused on modeling the rate of evaporation of volatile contaminants from solids, as a result of heat transfer from combustion gas [10]. This group proposed a case-by-case approach to determining optimum kiln operating conditions during the incineration of hazardous wastes. Use of the model allowed the appropriate flow rates and feed conditions to be established, the temperature profile to be determined, and the kiln length to be set [11].

A similar, but much more comprehensive, modeling effort has been carried out by the chemical and mechanical engineering departments of the University of Utah in conjunction with Louisiana State University's mechanical and chemical engineering departments, the Louisiana Division of Dow Chemical USA, and the U.S. Environmental Protection Agency. The heat transfer portion of this program has already been mentioned [8, 9]. Two rotary kiln configurations were studied. One considers the destruction of carbon tetrachloride being volatilized from contaminated solid waste, using methane as the primary fuel [12]. An empirically specified release rate of volatile material from the bed is built into this version. A two-dimensional model of carbon tetrachloride incineration was formulated for a turbulent premixed methane–carbon tetrachloride flame. The destruction rate of the carbon tetrachloride was assumed to be governed by the reaction rate of the methane. The methane reaction was depicted by a phenomenological eddy breakup model. The other configurational study considers the additional volatilization of reactive hydrocarbon species from the solids in the kiln's bed [13]. Detailed calculations of two-dimensional reacting flow with three combustion models were presented for an industrial kiln. Inleakage of air was also considered. The authors state that

> the results [from the first study] are qualitatively compared with the measurements in a three dimensional kiln operating under similar conditions as the one in [the] study. In particular strong stratification of temperature and species concentration at the kiln exit is noted. This is validated by similar experimental observations [12]. Computational results [from the second study] show that hydrocarbons evolved from the bed may become caught in low temperature, high oxygen

streams resulting in limited hydrocarbon destruction in the kiln. Highly nonuniform concentration profiles may result across the exit [13].

The third category of mathematical models of rotary kiln combustors, chemical kinetic analysis, is represented in the open literature by a single model, which is central to the work presented in this chapter. This model was developed for the combustion of zirconium sponge. Lemieux et al. [14] adopted a shrinking-core model to describe the oxidation process. Their model computes the oxygen concentration profile in the bed and also the burning rate of the particles.

CHEMICAL RATE EXPRESSIONS

Incineration is a complex physical and chemical process [1]. As soon as waste enters the rotary kiln, it is demoisturized and begins to be heated. When it reaches its reaction onset temperature, pyrolysis, sublimation, and other solid volatilization reactions occur under starved-air conditions. After these are complete, the char remaining is oxidized under excess-air conditions. Thus, the reactor may be considered to contain two zones, a pyrolysis zone and a char combustion zone. A kinetic rate expression is required for each one.

For the pyrolysis zone a simple degradation model describing cellulose pyrolysis has been adopted [15]. This model is based on experimental work in the temperature range 300° to 1000°C. In its simplest version, the model assumes that cellulose is made up of a set of precursor components, each of which decomposes directly to each reaction product, i, by a single, independent reaction pathway,

$$\text{cellulose} \xrightarrow{k_i} \text{product } i \tag{1}$$

and that the kinetics of each reaction can be modeled by a unimolecular first-order reaction having a rate constant that may be written in the Arrhenius form,

$$(dV_i/dt) = k_i \exp(-E_i/RT)(V_i^* - V_i) \tag{2}$$

where V_i is the percent yield of gas i at any time t, V_i^* is the ultimate (maximum) yield of gas i, and k_i and E_i are kinetic parameters. The parameters shown in Table 1 were derived by correlation with gas-phase pyrolysis products, observed by experimentation, including light hydrocarbons, alcohols, aldehydes, acids, H_2, H_2O, CO, CO_2, and tar.

Two modifications of the kinetic model were required before it could be applied to MSW incineration. First, the rate expressions had to be rewritten to account for the decomposition of tar to smaller gaseous species. In order to achieve this, it was assumed that the tar decomposed in the same proportion as the ultimate yields of the various species. A modified ultimate yield was therefore defined according to

Table 1 Kinetic parameters for rapid cellulose pyrolysis[a]

Species	E (kcal/mol)	log k (s^{-1})	V^* (wt %)
CH_4	60.04	13.00	2.41
C_2H_4	49.82	10.82	2.07
C_2H_6	41.55	9.06	0.26
C_3H_6	60.67	14.93	0.67
H_2	27.29	6.17	1.16
CH_3OH	49.35	13.42	0.92
CH_3CHO	55.10	13.56	1.54
Butane/ethanol	42.54	9.90	0.32
Acetone/furan	43.04	11.07	0.81
CH_3COOH	58.18	12.80	1.19
H_2O	24.62	6.71	8.04
CO	52.74	11.75	21.64
CO_2	23.42	5.39	3.08
Tar	—	—	49.97
Cellulose	31.79	8.3	−94.08

[a]Data are for the temperature range 300°–1000°C and a pressure of 5 psig.

$$V^*_{m,i} = V^*_i + \frac{V^*_i}{\sum_{i=1}^n V^*_i} V^*_{tar} \qquad (3)$$

where n is the number of gaseous species evolved during pyrolysis. Second, the discrete rate equations were combined, using the modified ultimate yields, to achieve a mass balance.

For the combustion zone, oxygen must diffuse through the bed to the individual char particles and then through an inert ash layer to the surface of the carbon. The carbon core shrinks as the combustion reaction proceeds. However, the total particle radius remains constant because of the formation of the ash layer, with an accompanying increase in particle porosity. Although the model can account for a particle size distribution of char, for the purpose of this study a uniform particle size has been considered. A mass balance is performed on selected individual particles throughout the bed, using a shrinking-core modeling technique [16] to calculate the burning rate of the particles.

KILN REACTOR MODEL

As mentioned earlier, the kiln is divided into two active zones—a first zone for pyrolysis and a second zone for char combustion. The model for the pyrolysis zone relies on the following assumptions:

Bed height decreases linearly with distance along the kiln's centerline.
The bed is well mixed in the vertical direction.
Only convective heat transfer from the gas to the bed is considered.

The kiln wall is assumed to be refractory lined; thus no heat is transferred through the barrel.

The temperature in the product gas above the bed is constant throughout this zone of the kiln; its value is calculated by the method of Tillman et al. [1]. Both the solid and gas regions operate in the plug-flow regime.

The model for the combustion zone is based on the model for zirconium combustion developed by Lemieux et al. [14]. The bed depth and the uniform temperature throughout the bed are set at the values obtained at the exit of the pyrolysis zone. The model determines the burning rate of the char and calculates the char mass flow profile across the combustion zone. The bed is axially divided into slices of uniform surface oxygen concentration, and each axial slice is divided vertically into segments of uniform oxygen concentration and particle size. A mass balance is performed on the slices in the vertical direction, yielding an ordinary differential equation describing diffusion of oxygen through a porous solid, with boundary conditions at the wall-solid and gas-solid interfaces. The particle burning rate (calculated as defined earlier) is combined with equations used to described oxygen transport through the bed. Oxygen is assumed to be at a uniform concentration at each bed depth location, and the particles are assumed to be small enough that oxygen is at a uniform concentration in the gas surrounding each particle. The resulting finite-difference equations are solved using a tridiagonal matrix algorithm.

The residence time of char in the kiln was calculated according to [17]

$$\theta = 0.19L/NDS \qquad (4)$$

where θ is expressed in minutes, L is the kiln length, D the kiln diameter, N the number of revolutions of the kiln per minute, and S its inclination to the horizontal.

The complex three-dimensional nature of the kiln, with variations in the x, z, and ϕ directions, necessitates simplifications to enable a solution of the model equations. The bed height, t_b, is that corresponding to the one required for a bed with the same cross-sectional area and the same area exposed to the gas as in the actual kiln. It is reasonable to assume that the bed height remains constant in the burnout zone because char constitutes but a small fraction of the solids in this zone.

RESULTS FROM USING THE MODEL

Physical parameters required by the model were derived from the MSW incineration facility located at McKay Bay, a suburb of Tampa, Florida [18]. The sources of values of the various fundamental parameters required by the model are detailed in Reference 19. This reference also includes a discussion of the method used to solve the set of ordinary differential equations that constitute the pyrolysis model and the tridiagonal matrix, which is derived from the linear finite-difference equations that constitute the combustion model. The FORTRAN code is also provided.

The cellulose and char mass flow profiles for a base case where the McKay Bay facility is operating at 77% of its design capacity are shown in Figures 1 and 2. Three distinct regions exist in Figure 1, which shows the results of the pyrolysis model. A significant portion of the kiln length is taken up by the heatup zone, in which the MSW is raised to the reaction onset temperature. This is achieved by auxiliary fuel usage. Farther down the kiln is a short reaction zone in which pyrolysis of the waste occurs. This zone is characterized by a fall in the cellulose mass flow rate and a concomitant increase in that of the char. The third zone, where the pyrolysis model shows a constant char flow rate and is in reality the burnout zone, is more appropriately described in Figure 2 by the results of the combustion model. Note that the abscissa has been shifted between

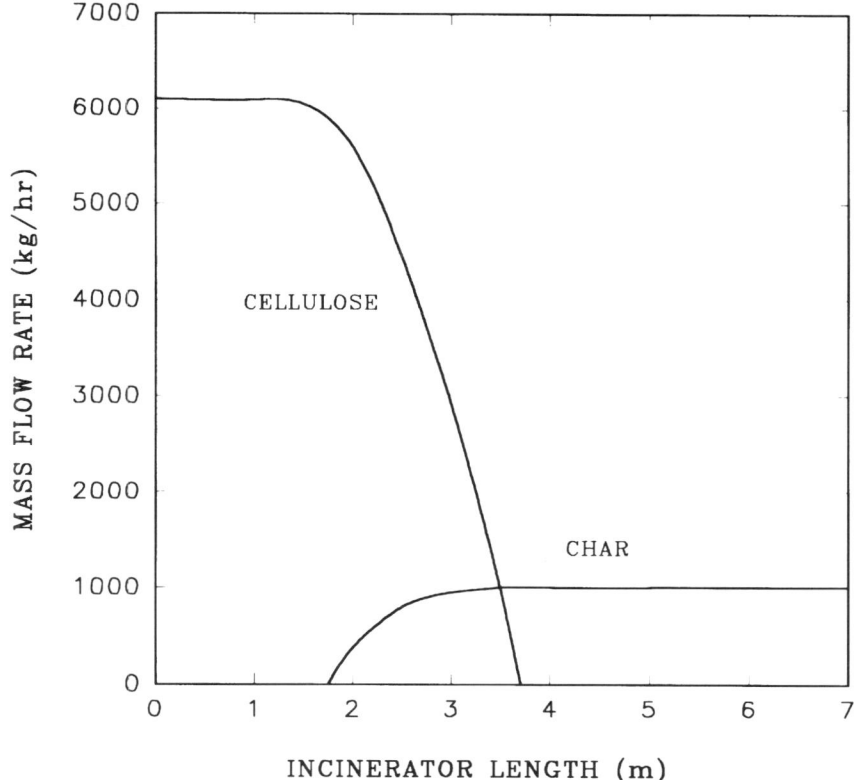

Figure 1 Mass flow profiles for the base case.

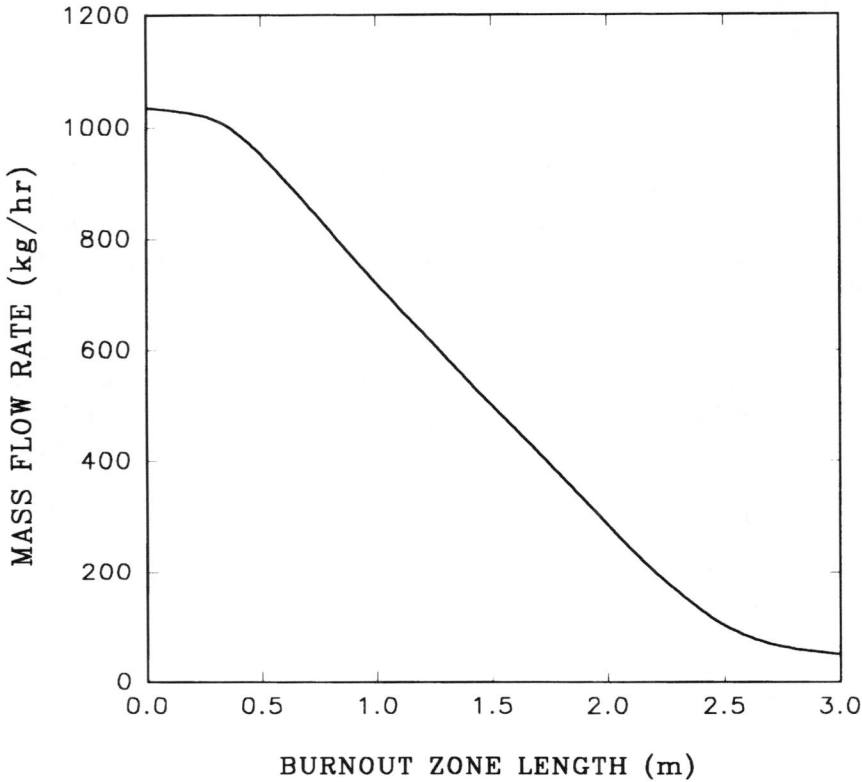

Figure 2 Char mass flow profile in the burnout zone.

Figures 1 and 2, such that the origin for Figure 2 is placed at the beginning of the burnout zone.

Bed temperature profiles through the heatup and reaction zones are shown in Figure 3 for the base case (9400 kg waste/h) and a case of further reduction in flow (4500 kg waste/h). This is an example of the use of the model to conduct parametric studies. Figure 3 indicates that the bed temperature is fairly insensitive to changes in the MSW feed rate. This could perhaps be justified by recalling that the bed is fairly well mixed and also that at a constant fuel-air ratio the flame temperature remains constant due to a balance between the reactive heat input and the convective heat removal. In operating incinerators the bed temperatures have been observed to be markedly different for different solids feed rates. This could result if the fuel-air ratio is not held constant during such experiments. Then again, incorporation of the radiative component of heat transfer could result in temperature profiles that match the actual trends better.

Another example of a parametric study is shown in Figure 4, which examines the relatively significant effect of bed porosity (0.54 void fraction, the base

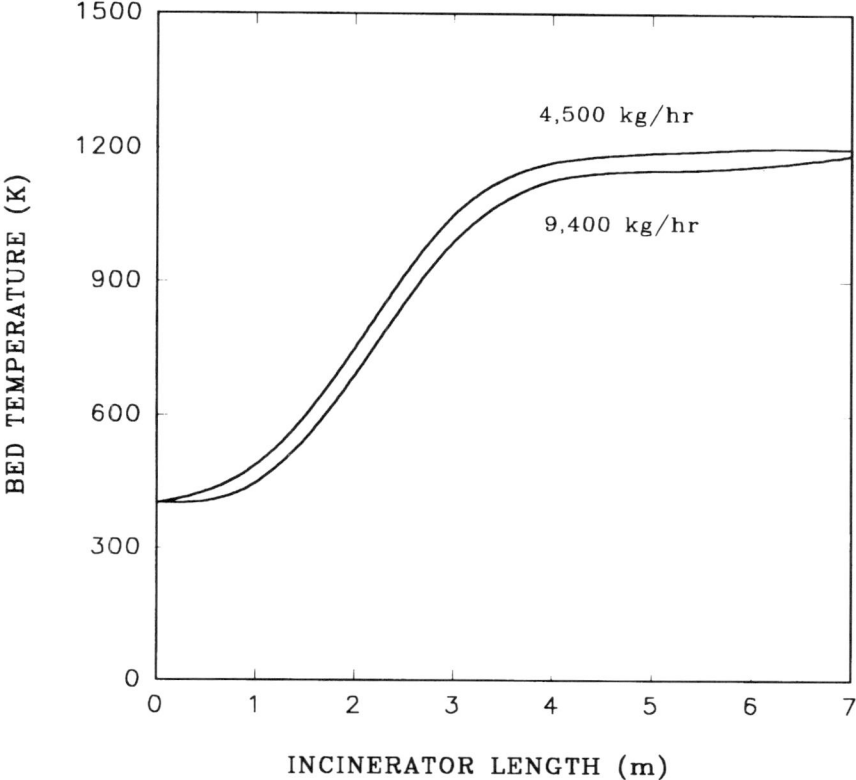

Figure 3 Bed temperature profiles.

case, versus 0.675 void fraction) on the rate of combustion in the burnout zone. Higher values of porosity reduce the length of this zone because the oxygen has better access to the char.

CONCLUSIONS

The mathematical model of a rotary kiln MSW combustor successfully predicts important aspects of its overall performance and several trends involving parametric variations in its operation. The size of the heatup zone, the size of the reaction zone, and the carbon burnout are quite accurate. However, temperature responses to changes in the MSW feed rate and the size of the burnout zone do not conform strictly to operational experience.

The current model, as well as improved future versions, should prove helpful in the analysis and design of waste-to-energy facilities employing this technology. The analysis of bed porosity has already shown that this parameter is a key

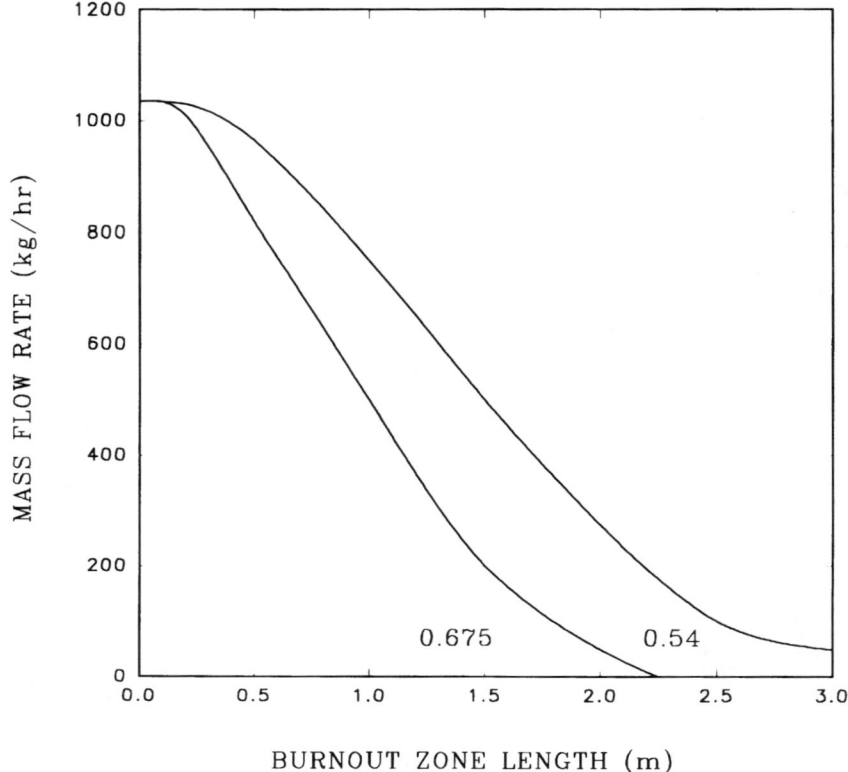

Figure 4 Effect of bed porosity.

factor in the performance of the burnout zone. Analysis of the heatup zone suggests that, because this region appears to require a major portion of the incinerator, preheat of the MSW feed by exiting flue gases might increase kiln capacity.

REFERENCES

1. Tillman, D. A., et al. *Incineration of municipal and solid hazardous wastes.* New York: Academic Press, 1989, p. 21.
2. Seeker, W. R., et al. Engineering analysis of hazardous waste incineration. *Proceedings of HAZMACON 86,* 1986, pp. 476–477.
3. Kirslis, S. J., and J. S. Watson. Heat transfer in a rotary kiln. Presented at the AIChE Spring National Meeting, Houston, TX, April 26, 1989.
4. Lee, C. C., and G. L. Huffman. Energy and mass balance calculations for incinerators. Preprints, 206th ACS National Meeting, Division of Fuel Chemistry, August 22–26, 1993, pp. 990–1000.

5. Gorog, J. P., T. N. Adams, and J. K. Brimacombe. Regenerative heat transfer in rotary kilns. *Metall. Trans.* 13B:153, 1982.
6. Ghoshdastidar, P. S., C. A. Rhodes, and D. I. Orloff. Heat transfer in a rotary kiln during the incineration of solid waste. Presented at the ASME National Heat Transfer Conference, Denver, CO, 1985.
7. Behmanesh, N., V. Manousiouthakis, and D. T. Allen. Optimizing the throughput of hazardous waste incinerators. *AIChE J.* 36(11):1707–1714, 1990.
8. Silcox, G. D., and D. W. Pershing. Heat transfer modeling in rotary kilns burning hazardous industrial wastes. Presented at the AIChE Annual Meeting, New Orleans, 1988.
9. Silcox, G. D., and D. W. Pershing. The effects of rotary kiln operating conditions and design on burden heating rates and determined by a mathematical model of rotary kiln heat transfer. *J. AWMA* 40(3):337–344, 1990.
10. Ghezzi, U., and L. D. A. Ferri. Theoretical optimization of rotary kiln for waste incineration. *Proceedings of the 8th Miami International Conference on Alternative Energy Sources,* Vol. 1–2, December 14–16, 1987, pp. 604–606.
11. Ghezzi, U., L. D. A. Ferry, and S. Pasini. *Proceedings of the 23rd Intersociety Energy Conversion Engineering Conference,* Piscataway, NJ, Vol. 4, July 31–August 5, 1988, pp. 433–437.
12. Jang, D. S., and S. Acharya. Two-dimensional modeling of waste combustion in a kiln. Presented at the AIChE Annual Meeting, New Orleans, 1988.
13. Wormeck, J. J., and D. W. Pershing. Computer modeling of combustion in a rotary kiln incinerator. *Proceedings of the ASME National Heat Transfer Conference,* Houston, 1988, pp. 97–101.
14. Lemieux, P. M., G. D. Silcox, and D. W. Pershing. Experimental and theoretical studies on the combustion of waste zirconium sponge in a rotary kiln simulator. *Waste Management* 9:125, 1989.
15. Hajalogol, M. R., J. B. Howard, J. P. Longwell, and W. A. Peters. Product compositions and kinetics for the rapid pyrolysis of cellulose. *Ind. Eng. Chem. Process Des. Dev.* 21:457, 1982.
16. Levenspiel, O. *Chemical reaction engineering.* New York, Wiley, 1977, p. 466.
17. Theodore, L., and J. Reynolds. *Introduction to hazardous waste incineration.* New York: Wiley, 1987, p. 248.
18. Waste Management, Inc. Fact Sheet and Project Description of the McKay Bay Refuse-to-Energy Facility, Tampa, FL, 1985.
19. Banerjee, K. M.S. thesis, University of Pittsburgh, 1990.

PART FOUR

COAL CONVERSION BY-PRODUCTS

CHAPTER
SIXTEEN

SELECTIVE LEACHING OF COAL AND COAL COMBUSTION SOLID RESIDUES

Catherine A. O'Keefe

INTRODUCTION

Billions of tons of coal are burned annually, producing millions of tons of fly ash. The fly ash must either be utilized, often as a cement replacement, or be disposed of, usually in a landfill. Thus the environmental impact of fly ash must be researched using a variety of techniques, one of which is selective leaching. Selective leaching procedures can be used in two ways: (1) as a predictive tool with chemical fractionation and (2) as a method of measuring the long-term leachability trends of elements. Chemical fractionation is used to determine the distribution of inorganic elements among the organic and mineral phases in coal, which is useful in predicting the composition of the fly ash produced by the combustion of coal. Millions of tons of fly ash are produced annually, and predicting the composition would be useful in determining how to utilize or dispose of the fly ash produced. If the fly ash is disposed of in a landfill, the leachability of the inorganic elements over an extended period of time needs to be determined to predict the environmental impact. Thus synthetic groundwater leaching and long-term leaching procedures were developed.

Leaching of materials to determine the mobility of major, minor, and trace constituents is a valuable laboratory tool for predicting the potential for environmental impact as well as other properties related to aqueous solubility of inorganic constituents. Leaching is generally used to determine bulk properties of materials, but with careful control of the leaching solution composition the technique of selective leaching can be used to determine more than simple solution mobility. Inorganic constituents in coal, coal ash, sediments, and soils can exist as crystalline minerals, soluble salts, amorphous phases, or organic complexes, and selective leaching under controlled conditions can be used to estimate the initial association of each component in the leachate. The information gained through selective leaching can be extremely useful in the development of predictive models. In addition, the determination of phase distributions (inorganic and organic) of materials is essential to the meaningful interpretation and understanding of leachate chemistry evolution and equilibrium concentration of inorganic constituents. This is particularly important with complex and variable materials such as coal conversion solid residues (CCSRs).

The Clean Air Act Amendments of 1990 cite 189 substances as potential hazardous air pollutants. Coal combustion is one anthropogenic source of several of the elements specified in the Clean Air Act Amendments. Therefore, it is important to understand the mobilization of the elements during coal combustion. According to Finkelman [1], there are three ways to reduce the amount of the element(s) mobilized during coal combustion: (1) to use coal containing a lower concentration of the element(s), (2) to extract the element(s) prior to combustion, and (3) to extract the element(s) from the effluent following combustion. In order to extract an element, several factors must be considered. These factors include the mode of occurrence, the leaching medium, and the formation as well as the leaching temperature and leaching time [2].

In the higher rank coals, the inorganic components are present primarily in crystalline phases. The inorganics in low-rank coals, however, can be associated as discrete mineral phases, coordinated metal ions, and cations bound to organic ion exchange sites or in clays [3]. The major mineral phases found in coal include aluminosilicates, carbonates, sulfides, and sulfates. Much of the Si, Al, and K as well as other alkali and alkaline earth elements are found in the aluminosilicate minerals such as clays, with Si also found in quartz. Clays may also adsorb some trace elements such as As, Cr, Li, Cd, and Rb [4–5]. Sulfide minerals such as pyrite can contain the trace elements As, Cd, Co, Cr, Cu, Hg, Mn, Mo, Ni, Pb, and Se [6, 7]. Carbonate minerals such as calcite, siderite, dolomite, and ankerite contain Mg, Ca, and Fe but may also have trace amounts of Mn, Sr, Cd, and Ba [4].

The organic portion of coal contains carboxylic acid and phenolic hydroxyl, mercapto, and imino groups that are able to bond with several trace elements. Cations of Na, Ca, Mg, K, Ba, Fe, Mn, Sr, and Zn can bond to the carboxyl group to produce carboxylate salts [7]. The trace elements are generally associated with both the organic and mineral matter. In the mineral portion of coal, the trace elements can often be found as discrete minor minerals, as replacement ions in minerals, or adsorbed on clays [6].

In order to predict the behavior of coal during combustion and the composition of the fly ash produced, the inorganic composition as well as the inorganic associations of major, minor, and trace elements must be determined. In the utilization of coal in combustion or gasification, several of the operational problems encountered can be attributed to the inorganic constituents in coal. Knowledge of the mineral concentrations and associations is useful in predicting the behavior of the coal in a given process under various conditions. During combustion, the inorganic materials are transformed into gases, liquids, and solids. Some of the inorganic components are vaporized to form gases that will condense upon cooling. Other low-melting-point inorganic components will melt to produce liquid particles in the combustion system, and some will remain as solids throughout the process. Studies of the final ash product (fly ash) indicate a bimodal size distribution [8–10]. The submicrometer-size particles form as a result of homogeneous condensation of flame-volatilized species. Flame-volatilized species may also condense heterogeneously on the surfaces of larger particles. The larger particles, sometimes referred to as residual ash, are largely derived from mineral grains. The composition and size distribution of the larger particles result from transformations or interactions between discrete mineral grains in higher rank coals. In lower rank coals, interactions occur between organically associated elements and minerals and between adjacent inorganic minerals. Processes such as mineral coalescence, partial coalescence, ash shedding, and char fragmentation during char combustion all play important roles in the size and composition of the final fly ash. Loehden and others [11] and Zygarlicke and others [12] indicate that three potential modes for fly ash generation can be used to describe fly ash particle size and composition evolution: (1) "fine limit" assumes that each mineral grain forms a fly ash particle and that the organically associated elements form fly ash particles of less than 2 μm, (2) "total coalescence" assumes that one fly ash particle forms per coal particle, and (3) "partial coalescence" suggests that the fly ash composition and particle size evolve as a result of partial coalescence.

After coal is burned to produce fly ash, a complete characterization, including phase assemblage, chemical components, and leaching properties, is necessary when fly ash is utilized as a cement replacement for concrete, road bed stabilization, low-density flowable material, and other engineering applications. Complete characterization defines not only the leachability of elements in the material but also the physical properties for engineering applications. In all utilization and disposal options, leachability of constituents is a key issue from the standpoint of environmental impact on groundwater and other resources.

For example, knowledge of fly ash particle size and composition distributions is necessary when fly ash is utilized as a cement replacement or in other engineering applications. Fly ashes can be cementitious as well as pozzolanic, and the phase associations and bulk chemistry define these properties. In addition, the leachability of the components of the fly ash used in a cement mixture is important not only because of the impact on the durability and strength of the cement but also because of the possible environmental impacts. Most of the fly ash produced in the United States is not utilized at this time, and the disposal of fly ash also poses a problem because of the leachability of possibly toxic materials under varying conditions of disposal. Fly ash presents unique problems in that the severe conditions of formation create a chemistry, particularly with low-rank ash, that leads to a highly heterogeneous material, which can be reactive. This must be taken into consideration in characterization protocols.

BACKGROUND

Leaching studies of soils and sediments to determine total mass of mobilizable elements have been conducted for years, typically using ammonium acetate, nitric acid, hydrochloric acid, and/or hydrofluoric acid. Soils and sediments are primarily composed of silicates, aluminosilicates (clay minerals), and carbonates, with minor amounts of organic matter and sulfide minerals. The carbonates are more readily leachable than the silicates. However, because of the variable degrees of weathering, the leachability of the silicates changes, with olivine the most susceptible to leaching and quartz the least susceptible [13, 14]. The inherent metals in the clay minerals and silicates are also more resistant to leaching than the absorbed minerals [15]. The complex sulfides are extremely difficult to leach, whereas the monosulfides are readily leached [16]. The cations bound to the minerals leach very differently from the organically bound cations [2]. In leaching procedures on soils, the determination of ion-exchangeable sodium, potassium, calcium, magnesium, and strontium is commonly done by using a 1 N ammonium acetate solution for extraction. Hydrochloric acid extraction, along with numerous other techniques, has been used as a reliable method for measuring the availability of micronutrients such as copper and zinc in soil [17–19].

Coal contains silicates, sulfides, carbonates, and organically bound cations, a composition similar to that of some soils; Huffman and Huggins [20], Miller and Given [21], and Renton [22] modified soil leaching techniques for use on coals. Huffman and Huggins [20] and Renton [22] concluded that carbonates and certain sulfides and oxides, which are minor constituents of coal, will dissociate in acid. Miller and Given [21] indicated that some of the cations in clays will be leached by hydrochloric acid, but the acid extraction is not entirely selective for complexed cations. Therefore, the acid extraction is not a reliable technique for leaching chelated ions from organic matter by itself. To supplement the acid extraction, Miller and Given [21] developed a fractionation method based on specific gravity using the assumption that if an element was concentrated in the fraction with the lowest specific gravity and was partially soluble in acid, then the element was organically bound in a phase of relatively low density. If the element was soluble in acid and found in a high-specific-gravity fraction, it was assumed to be associated with a mineral(s) having a higher density than organic molecules.

The ammonium acetate extraction was used to leach ion-exchangeable elements such as calcium, potassium, sodium, and magnesium attached to the carboxyl groups in coal. When the coal was separated using specific gravity, the heavier fraction contained some magnesium, which was insoluble, indicating that some of it is substituted into the lattice of clay minerals or other minerals [21]. Other studies indicated that metallic carboxylates in low-rank coals can promote liquefaction [23] and can act as catalysts in gasification of lignites [24]. Schafer [25] also found that the pyrolysis of lignite containing metallic carboxylates gave off more CO_2 than that of lignites containing the acid form.

A study by Wiltsee [26] was completed to determine the selectivity of ammonium acetate solution to leach cations. In this study, the findings indicated that within a chemical family, the susceptibility of removal from lignite coal was generally inversely proportional to the ionic radius. This parallels the findings of Miller and Given [21] and Stevenson [18]; even though Na^+ may be the dominant ion in the inherent water in lignite, the Ca^{2+} is apt to dominate in the absorbed phase. This is consistent with the selectivity series for cations in manufactured ion-exchange resins, which places the strength of divalent cation exchange above that of monovalent species.

Benson and Holm [3] performed a test to determine the fraction of the total mass of cations present in coal as water-soluble minerals or complexed with fulvic acids (a water-soluble fraction of the humic material). Coal was heated for 24 hours at 70°C. The filtrate from a water extraction contained all of the extractable potassium, about one quarter to one half of the extractable sodium, and a small portion of the extractable calcium and magnesium. The level of cations was much greater in the filtrate than the corresponding decrease in H^+ ions; thus the ratio of metal ions as soluble minerals or as fulvic acid salts could be determined. Because flocculation of the more soluble humic acid fraction occurred when the water extraction was heated, the water extraction had to be done at room temperature.

CHEMICAL FRACTIONATION PROCEDURES

A selective leaching process on coal known as chemical fractionation was used to determine the distribution of elements among the organic and mineral phases in coal based on solubility differences of the coal constituents in three separate, stirred solutions: (1) deionized water, (2) 1 M ammonium acetate (NH_4OAc), and (3) 1 M hydrochloric acid (HCl). The water-soluble minerals, such as halite and thenardite, are extracted by distilled deionized water. Elements bound to ion-exchange sites such as carboxyl groups are assumed to be removed by the ammonium acetate solution. Hydrochloric acid is used to leach elements associated with acid-soluble minerals, such as calcite, dolomite, and siderite, as well as organic coordination complexes. The remaining residue contains the insoluble minerals, such as clays, quartz, and pyrite. A representative 140-g sample of coal was ground to 80% −200 mesh and vacuum dried to constant weight. A 35-g portion of coal was removed and labeled CF1 (unleached coal).

The 105 g of remaining coal was subjected to successive extraction treatments. The first extraction treatment utilized a 4:1 deionized water-to-coal ratio and was stirred in a covered Nalgene beaker at room temperature for 24 hours. The slurry was vacuum filtered and rinsed with deionized water. About 35 g of the coal residue was removed, dried, and labeled CF2 after the water extraction. The filtrate was acidified to pH <1 for subsequent determination of elemental concentrations.

Ammonium acetate (1 M) was added to the remaining 70 g of coal residue in a 4:1 ratio. The slurry was covered, stirred, and heated to 70°C for 24 hours. The slurry was then vacuum filtered and rinsed with deionized water. The filtrate was acidified to pH <1 and stored. The 1 M ammonium acetate extraction procedure was performed three times. The filtrates from the three ammonium acetate extractions were combined for subsequent analyses. Approximately 35 g of residue coal was removed, dried, and labeled CF3 after the ammonium acetate extraction.

To the remaining 35 g of coal residue, 4 mL of 1 M hydrochloric acid per gram of coal was added. The slurry was covered, stirred, and heated to 70°C for 24 hours. The slurry was then vacuum filtered and rinsed with deionized water. The filtrate was stored. The 1 M hydrochloric acid extraction procedure was performed twice. The filtrates from the 2 days were combined and stored for subsequent analyses. The approximately 35 g of residue coal was dried and labeled CF4 after the hydrochloric acid extraction.

A portion of the CF1 residue (unleached) was analyzed using computer-controlled scanning electron microscopy (CCSEM) to determine the size and composition of the minerals. From each of the four residues, CF1–CF4, a portion of unashed coal was prepared for analyses by wavelength-dispersive x-ray fluorescence (WDXRF) for the trace elements of Pb, Cr, and Ni. Another portion from each fraction was used to determine the ash percentage. The remaining residue was ashed and analyzed by energy-dispersive x-ray fluorescence (EDXRF) for determination of the major and minor elements for each of the fractions. The three leachates from CF2, CF3, and CF4 were analyzed by inductively coupled argon plasma emission spectroscopy (ICAP) and/or atomic absorption (AA) for determination of the trace element concentrations of Hg, Se, Cd, and As.

After determination of the elemental composition of the four fractions, a mass balance was completed using Si as a tracer, because the solubility of Si is zero or close to zero. The results from the mass balance, which are normalized, indicate the percentage of the elements leached by the three extraction solutions as well as the percentage of the remaining elements that the extracts were unable to leach.

Table 1 Normalized bulk chemical composition of unleached coal ash using EDXRF

Oxides	Black Thunder	Rochelle
SiO_2	31	38
Al_2O_3	17	20
Fe_2O_3	4	5
TiO_2	1	1
P_2O_5	1	1
CaO	24	22
MgO	9	5
Na_2O	1	1
K_2O	<1	<1
SO_3	12	7

Chemical Fractionation Results

Chemical fractionation was completed on two subbituminous coal samples, Black Thunder and Rochelle. The residue from the four fractions was ashed at 750°C before being analyzed by EDXRF. The mineral compositions from CCSEM of the unleached coals are shown in Table 1, and Table 2 indicates the bulk chemical composition as analyzed by EDXRF. The data as shown in Table 3 for the two coals indicate that sodium is the only element that is consistently and appreciably extracted by the water treatment. The sodium is probably present in the coals as sodium chloride, sodium sulfate, or the salt of a water-soluble humic acid. From the CCSEM data of the coals studied, there were few water-

Table 2 Composition of 1- to 100-μm-diameter minerals using CCSEM, % of total

Minerals	Black Thunder	Rochelle
Quartz	47	49
Kaolinite	21	25
Montmorillonite	<1	2
Calcite	<1	em
Gypsum	—	—
Barite	2	<1
Pyrite	12	2
Dolomite	1	—
Iron oxide	<1	3
Pyrrhotite	—	—
Oxidized pyrrhotite	—	1

Table 3 Elements extracted by water, % of original content

Elements	Black Thunder	Rochelle
Aluminum	0	0
Iron	0	0
Titanium	0	0
Calcium	0	0
Magnesium	0	0
Sodium	27	11
Potassium	0	0

soluble minerals. Hence, the sodium removed by water is probably limited to halite (NaCl) in both coals.

Table 4 indicates that the ammonium acetate extraction primarily removed Ca, Mg, and Na in both coals; however, for Black Thunder, some K and Al were extracted. The ammonium acetate extraction primarily removes organically associated elements; thus Ca, Mg, and Na are associated with ion-exchange sites of the carboxylic acids in the organic portion of the coal.

The hydrochloric acid leaching, as shown in Table 5, extracted most of the Fe and Al removed during the chemical fractionation process. Much of the Ca and Mg left after the ammonium acetate leaching was removed by the hydrochloric acid extraction. The hydrochloric acid extraction typically removes elements that exist in organic coordination complexes, such as Fe and Al, and elements that are present as acid-soluble minerals, such as carbonates, hydroxides, and sulfates. Fe, Ca, and Mg are the most commonly removed elements in the acid leaching. The loss of these elements can be attributed to the dissolution of calcite ($CaCO_3$), dolomite [$CaMg(CO_3)_2$], and siderite ($FeCO_3$). Table 6 shows the elements that are not leached during chemical fractionation. These elements are associated with quartz, clays, and pyrite.

Chemical fractionation was also performed on SUFCo coal to determine the element associations for the major, minor, and trace elements. Twenty-five grams

Table 4 Elements extracted by ammonium acetate, % of original content

Elements	Black Thunder	Rochelle
Aluminum	26	0
Iron	0	0
Titanium	0	0
Calcium	72	64
Magnesium	64	57
Sodium	62	76
Potassium	48	0

Table 5 Elements extracted by hydrochloric acid, % of original content

Elements	Black Thunder	Rochelle
Aluminum	42	18
Iron	75	96
Titanium	0	0
Calcium	28	35
Magnesium	35	37
Sodium	8	5
Potassium	0	0

of residue from each of the four fractions was ashed at 500°C and prepared for analysis for major and minor elements using EDXRF. A lower ashing temperature was used, because Swaine [4] indicated that an ashing temperature between 450° and 500°C was preferable when analyzing for trace elements. Using the raw residue from each of the four fractions, 2 g was pressed into a pellet for analysis on a WDXRF system for trace elements and the other 2 g of residue was mounted in an epoxy plug for mineral composition analysis by CCSEM. The remaining residue was digested and analyzed for trace elements by ICAP and AA.

Table 7 indicates the results of the chemical fractionation on the major and minor elements of SUFCo coal using EDXRF and the mineral composition of the four fractions from the CCSEM with atomic number, absorption, and fluorescence (ZAF) correction analysis. The initial concentration of K, Ti, and P is so low that it increases the potential for error in the percent removed by the extractions; therefore, these elements are not included in the discussion of the chemical fractionation results. The three major elements leached by the water extract were Ca, Mg, and Na. The CCSEM data indicate that gypsum was leached by the water, which accounts for the removal of calcium. The sodium was probably present as sodium chloride, sodium sulfate, or the salt of a water-

Table 6 Elements remaining after extractions, % of original content

Elements	Black Thunder	Rochelle
Aluminum	32	82
Iron	25	4
Titanium	100	100
Calcium	0	1
Magnesium	1	6
Sodium	3	8
Potassium	52	100

Table 7 SUFCo coal chemical fractionation data: major elements and mineral composition

EDXRF

Element	Initial ppm	Removed by H_2O (%)	Removed by NH_4OAc (%)	Removed by HCl (%)	Remaining (%)
Aluminum	6,180	5	4	10	81
Iron	3,350	4	0	40	56
Titanium	505	0	4	0	96
Phosphorus	516	9	0	61	30
Calcium	12,326	11	80	4	5
Magnesium	1,147	13	40	20	27
Sodium	3,326	21	71	4	4
Potassium	68	29	24	5	42

CCSEM[a]

Mineral	Formula	Unleached (%)	After H_2O (%)	After NH_4OAc (%)	After HCl (%)
Gypsum	$CaSO_4 \cdot 2H_2O$	4	0	0	0
Kaolinite	$Al_2Si_2O_5(OH)_4$	1	9	25	34
Montmorillonite	$(0.5Ca,Na) \cdot 7(Al,Mg,Fe)_4(Si,Al_8O_{20})(OH)_4$	1	1	5	4
Aluminosilicate	$Al_4Si_4O_{10}(OH)_4$	1	1	14	4
Pyrrhotite	$Fe_{1-x}S$	1	0	0	0
Calcite	$CaCO_3$	25	2	0	0
Dolomite	$CaMg(CO_3)_2$	3	5	0	0
Ankerite	$(Ca,Fe)(CO_3)_2$	1	1	0	0
Apatite	$Ca_5F(PO_4)_3$	1	1	5	0
Iron oxide	Fe_2O_3			3	2
Quartz	SiO_2	32	32	32	32
Pyrite	FeS	7	8	14	20
Others		13	16	13	8

[a] Normalized to zero quartz loss.

soluble humic acid. However, most of the Ca, Mg, and Na was extracted by the ammonium acetate extraction. The Ca and Mg are associated with the carbonate minerals such as calcite, dolomite, and ankerite that were leached by ammonium acetate. However, looking at the initial concentration of Ca, Na, and Mg, the carbonate minerals cannot account for the total amount of these elements leached. Therefore, some of the Ca, Na, and Mg is also associated with the exchange sites of the ion carboxyl groups as well as the ion-exchange sites of the clays. The three appreciably leached elements in the hydrochloric acid extract are Fe, Mg, and Al, with lesser amounts of Ca and Na. Apatite accounts for the small amount of Ca and Na leached, and because of the low initial concentration of Mg, the montmorillonite leached probably accounts for most of the Mg removed.

Helble and others [27] have found that most of the Cr in coal is in the +3 oxidation state. The hydrochloric acid extract removed 44% of the Cr, as shown in Table 8, giving indirect evidence for the association of Cr with the coordinate complexes. The Cr is probably associated with the clay (aluminosilicate) minerals that were leached by the hydrochloric acid extract.

Arsenic and mercury appear to have similar distributions in the raw coal [1], with the percent of arsenic removed generally being higher than that of mercury [28]. Arsenic and mercury are associated with pyrite and other sulfides [1], which are part of the insoluble fraction of coal. Arsenic is also associated with clays [5] and carbonates [29]. The results in Tables 7 and 8 reflect that both As and Hg were associated with the insoluble fraction, because the hydrochloric extract removed 6% of the Hg and 30% of the As and the mineral pyrite was not leached. The fact that a greater amount of the As was leached along with the minerals leached provides some evidence that As is also associated with clays (aluminosilicates) that were extracted by hydrochloric acid.

Cadmium is commonly associated with sphalerite (ZnS) [1] and minor amounts with sulfides [1, 4], clays, carbonates [5], and organics [1, 30]. Because several carbonate minerals were leached by ammonium acetate, as shown in Table 7, Cd is probably associated with the carbonate minerals in SUFCo coal. The hydrochloric acid extraction provides evidence that Cd is associated with the clays found in the coal.

Lead is primarily associated with pyrite and galena (PbS) [4, 31], and according to Finkelman [32], association of Pb with other minerals is very rare. As the majority of pyrite from Table 7 remained after the leaching process and the Pb was found primarily in the insoluble fraction, the data support the previous finding that most of the PB is associated with pyrite and galena.

The mode of occurrence of Ni in coal is inconclusive at present. Some indirect evidence links Ni with organics, whereas other evidence indicates that the inorganically bound Ni is associated primarily with the sulfide minerals [1, 4]. The chemical fractionation data on SUFCo coal show that Ni was not leached, providing indirect evidence that the Ni is associated with the sulfide minerals.

Table 8 SUFCo coal chemical fractionation of trace elements

Element	Removed by H_2O (%)	Removed by NH_4OAc (%)	Removed by HCl (%)	Remaining (%)
As[a]	0	0	30	70
Cd[a]	0	12	37	51
Cr[b]	0	0	44	56
Pb[b]	0	0	25	75
Ni[b]	0	0	0	100
Se[a]	3	6	0	91
Hg[a]	0	0	6	97

[a]Determined by AA.
[b]Determined by WDXRF.

Most of the Se found in coal is organically associated, and a small but important portion of Se can easily substitute for S in the sulfide minerals [1]. The chemical fractionation data indicate the reverse of the published data.

ASH LEACHING PROCEDURES

In addition to selective leaching for the determination of associations of inorganic constituents, bulk leaching properties are important for gaining an understanding of the overall mobility and chemical evolution of leachate in an aqueous system. Current regulatory leaching is based on the application of a standard leaching test, usually the toxicity characteristic leaching procedure (TCLP). This test utilizes an acetic acid leaching solution, a 20:1 liquid-to-solid ratio, and an 18-hour equilibration time. This test may be inappropriate for two reasons. First, the TCLP utilizes an acetic acid or a dilute acetate buffer leaching solution at a pH near 5. This is designed to simulate leaching under codisposal conditions in a sanitary landfill. If disposal is in this scenario, the test is appropriate. However, if disposal is in a monofill, the use of a solution containing acetate or acetic acid is scientifically indefensible. Second, the duration of the test is extremely short with respect to geologic phenomena and duration of contact with water under most realistic disposal scenarios. It has been demonstrated that the application of several varying protocols is essential to a complete understanding of leaching characteristics. These include a leaching test developed at the Energy & Environmental Research Center (EERC) called the synthetic groundwater leaching procedure (SGLP) and the long-term leaching (LTL) procedure.

Synthetic Groundwater Leaching Procedure

The synthetic groundwater leaching procedure was developed at the EERC to simulate leaching of a material in contact with groundwater [33]. This test uses a protocol similar to EPA Method 1311, TCLP, with a 20:1 liquid-to-solid ratio, an 18-hour equilibration time, and end-over-end agitation. The only difference between the TCLP and the SGLP is the leaching solution. The TCLP utilizes an acidic solution containing acetic acid to simulate leaching in a sanitary landfill under codisposal conditions, while the SGLP specifies a solution approximating the chemistry of the solution most likely to be in contact with the disposed material under field conditions. The TCLP would be inappropriate in cases in which materials such as coal combustion solid residues (CCSRs) are being disposed of in a monofill or being used in construction materials, because in either case their contact with acetic acid is highly unlikely. Even in the event that acid rain infiltrates a landfill, the actual pH and chemistry of the leaching fluid likely to be in contact with placed material must be considered. Percolation of rainwater through the top layer of sediment and placed material would render the water alkaline from its initial weak acid rain composition. In addition, rainwater would probably acquire significant concentrations of dissolved constituents upon contact with ash. In the case of exposed material, the amount of material affected by acidic water would be minuscule compared with the amount that would be leached by a highly alkaline solution formed from the interaction of disposed material and water. If it is the case that an alkaline leaching solution will be doing the actual field leaching, the solution used in the laboratory must be alkaline with a similar chemistry. Even in the event that a neutral solution would be in contact with the ash, which is the case in ASTM Method D3987, the bulk/major element solution concentrations of leachate would be determined primarily by the composition of the fly ash being leached. The use of a leaching solution with the proper chemistry and pH cannot be overemphasized.

Long-Term Leaching Procedure

This protocol is a modification of the SGLP developed by researchers at the EERC to help determine long-term trends for the leachability of certain trace elements found in disposed materials. Most regulatory leaching tests are short term (18 or 24 hours) and provide inadequate information on the leachability of trace elements present in coal combustion by-products such as fly ash because of its mineralogical composition. When ash materials come in contact with water, hydration reactions occur with the different minerals and ash constituents. Many of these reactions are slow and can take up to 30 days or more to exert a measurable effect, as demonstrated in a previous research project [34] at the EERC. The effect of hydration reactions can be in excess of an order of magnitude with respect to solution concentration reductions of several potentially hazardous trace elements such as B, Cr, and Se. Notable among these materials is the mineral ettringite, which forms upon hydration in many low-rank coal ash types, products from advanced combustion systems such as fluidized-bed combustors, and residues from duct injection of lime for acid gas abatement. Ettringite is both an individual mineral, calcium aluminosulfate hydroxide hydrate $[Ca_6Al_2(SO_4)_3(OH)_{12} \cdot 26H_2O]$, and the group name for an analogous series of isostructural compounds. Ettringite formation has been shown to accompany the reduction of solution concentrations of trace elements such as boron, chromium, and selenium as well as other elements that exist as oxyanions in aqueous solution, many of which are often present in fly ash water systems [35].

Ash Leaching Results

Tables 9 and 10 show LTL procedures that present examples of the effect of formation of secondary hydrated phases. One problem in the study of ash for determination of potential environmental impact is that many ash specimens seen in the laboratory contain potentially problematic trace elements at concentrations

Table 9 Long-term leaching results: fly ash 1 with lime

Leaching time:	1 week	2 weeks	3 weeks	4 weeks	12 weeks
pH:	12.3	12.4	12.3	12.5	12.5
Conductivity:	12.4	13.0	12.7	13.2	15.5
RCRA elements		Concentration (mg/L)			
Arsenic	0.008	0.010	0.006	0.007	<0.005
Barium	0.18	0.14	0.30	0.32	1.51
Cadmium	<0.005	<0.005	<0.005	<0.005	<0.005
Chromium	0.100	0.092	0.084	0.074	<0.01
Lead	0.008	0.020	0.009	0.013	0.016
Mercury	<0.002	<0.002	<0.002	<0.002	<0.002
Selenium	0.048	0.043	0.039	0.037	0.029
Silver	<0.001	<0.001	<0.001	<0.001	<0.001
Aluminum	<0.2	<0.2	<0.2	<0.2	<0.2
Boron	3.76	1.71	0.85	0.48	<0.2
Bromide	1.51	1.63	1.54	1.55	5.14
Calcium	820	740	680	630	200
Chloride	410	440	430	410	400
Copper	<0.01	<0.01	<0.01	<0.01	<0.01
Iron	<0.2	<0.2	<0.2	<0.2	<0.2
Manganese	<0.1	<0.1	<0.1	<0.1	<0.1
Molybdenum	0.51	0.50	0.65	0.62	0.49
Nickel	<0.1	<0.1	<0.1	<0.1	<0.1
Strontium	1.51	2.00	2.35	2.60	3.35
Sulfate	1800	1600	1410	1190	30.9
Zinc	<0.025	<0.025	<0.025	<0.025	<0.025

so low that leachates contain elements below the detection limits of most analytical techniques. The examples shown are from a study of ash from processes that add lime for the control of acid gases.

It can be seen from these tables that several elements exhibit what is referred to as "anomalous" behavior. For the sake of interpretation, in normal elemental leaching behavior, when subjected to the batch LTL procedure, the elemental concentrations rise, usually rapidly at first, and then level off at some equilibrium concentration that remains constant. Anomalous behavior is usually seen as an initial rapid rise in concentration followed by a decrease, usually resulting in an equilibrium concentration well below the initial maximum solution concentration. Table 9 shows that there are several elements with anomalous behavior: Cr, Se, B, Ca, and sulfate. In these solids, x-ray diffraction indicated the formation of ettringite with time. With a composition of $Ca_6Al_2(SO_4)_3(OH)_{12} \cdot 26H_2O$, ettringite exerts an influence on Al, Ca, and sulfate solution concentrations. Because Al is sparingly soluble in these systems and not usually present in high concentrations in low-rank ash, the solution concentration remained at the detection limit of the method used and a change in its concentration was not

Table 10 Long-term leaching results: fly ash 2 with lime

Leaching time:	1 week	2 weeks	3 weeks	4 weeks	12 weeks
pH:	12.3	12.3	12.3	12.4	12.2
Conductivity:	8.00	8.25	8.06	7.80	7.65
RCRA elements		Concentration (mg/L)			
Arsenic	<0.005	<0.005	<0.005	<0.005	<0.005
Barium	0.42	0.48	0.43	0.50	0.82
Cadmium	<0.005	<0.005	<0.005	<0.005	<0.005
Chromium	0.047	0.058	0.053	0.043	<0.01
Lead	0.39	0.35	0.42	0.46	0.52
Mercury	<0.002	<0.002	<0.002	<0.002	<0.002
Selenium	0.020	0.025	0.021	0.017	<0.01
Silver	<0.001	<0.001	<0.001	<0.001	<0.001
Aluminum	<0.2	<0.2	<0.2	<0.2	<0.2
Boron	20.2	18.4	13.9	12.8	2.57
Bromide	<1.0	<1.0	<1.0	<1.0	<1.0
Calcium	1060	1170	1200	1150	800
Chloride	36.9	69.9	36.5	36.4	30.3
Copper	<0.01	<0.01	<0.01	<0.01	0.014
Iron	<0.2	<0.2	<0.2	<0.2	<0.2
Manganese	<0.1	<0.1	<0.1	<0.1	<0.1
Molybdenum	0.10	0.10	0.13	0.13	0.10
Nickel	<0.1	<0.1	<0.1	<0.1	<0.1
Strontium	2.80	3.20	2.73	3.06	4.32
Sulfate	450	480	410	380	53.7
Zinc	<0.025	<0.025	<0.025	<0.025	<0.025

noted. Cr, Se, B, Ca, and sulfate are also seen to exhibit anomalous behavior as shown in Table 10.

Numerous other examples of this unique ash leachate behavior have been noted. Virtually any ash that provides ettringite-forming conditions has the potential to reduce the leachate concentration of several potentially problematic trace elements. In addition, many ash specimens from advanced combustion processes tend to form ettringite. All five of these elements are either primary constituents of ettringite or are elements such as As, B, Cr, or Se that are not only known to substitute into the ettringite structure but also have been synthesized as fully substituted ettringites.

Ash materials, especially alkaline ash, resulting from the combustion of low-rank coals can be assumed to change mineralogically because of hydration reactions. Many of these reactions are slow, taking up to and in excess of 30 days. To assist in the scientifically valid interpretation of ash behavior, complete characterization must be performed. Physical, mineralogical, and chemical characterizations are essential components of this process. Key in this characterization is the determination of phase associations and leaching behavior, both of which have been discussed above.

SUMMARY

Organically associated elements tend to be liberated during the initial stages of coal combustion and thus react with other inorganic phases as combustion continues. This and other processes have an effect on fouling and slagging within the boiler, thus reducing efficiency. With an understanding of the fundamental mechanisms of mineral matter transformations by thorough characterization and careful selection of coals, coal-burning plants can increase efficiency and, in some cases, reduce pollution. In addition, understanding the behavior of inorganic constituents in coal and CCSRs may allow selective coal blending and tailored combustion processes to produce a residue with desirable engineering and leaching properties.

The trace elements are also influenced by associations in the coal matrix. Although trace constituents have little effect on combustion, ash formation, or efficiency of heat transfer, they are key in the determination of environmental impact. During coal conversion, trace elements are partitioned between gas and solid phases and reach equilibrium throughout the combustion process through mechanisms not yet clearly understood. Associations in the initial coal matrix along with combustion conditions and ash collection methods determine eventual partitioning and release of trace elements into the environment. It is theoretically possible, through a complete understanding of all aspects of coal characterization and with a complete understanding of combustion and ash collection processes, to model trace element transformations. Leaching characterization is a key element in this process.

Leaching techniques, as part of an overall characterization scheme, can greatly enhance the level of understanding of the inorganic components of coal, coal ash, sediments, soils, and other geologic materials. The enhanced and novel use of materials requires an understanding beyond that provided by traditional characterization and analysis schemes. The prediction of behavior is key in the rapid development of new applications and technologies. This is particularly important with respect to inorganic constituents of fuels and solid residue by-products such as CCSRs. Although this chapter has focused on the use of innovative leaching procedures for characterization, additional data are crucial for the correct and scientifically valid interpretation of results. Chemical fractionation is an indirect method that helps to identify the distribution of major, minor, and trace constituents and when used in conjunction with other direct techniques (electron microprobe to determine particle size and local microscale chemistry, x-ray diffraction to identify specific minerals, and optical microscopy using reflected visible, infrared, or ultraviolet light) gives conclusive evidence of the associations involved. The use of Fourier transform infrared spectroscopy has also aided in the identification and verification of the existence of specific oxyanions in hydrated coal residues. Numerous other evolving techniques such as microprobe laser Raman spectroscopy and imaging, as well as advanced laser and x-ray techniques, also hold the promise of adding to an even more complete understanding of the factors contributing to the leaching properties of solid materials.

REFERENCES

1. Finkelman, R. B. Mode of occurrence of hazardous elements in coal: Level of confidence. Paper presented at the Trace Element Transformations in Coal-Fired Power Systems Workshop, Scottsdale, AZ, 1993.
2. Kane, J. S., S. A. Wilson, J. Lipinski, and L. Butler. Leaching procedures: A brief review of their varied uses and their application to selected standard reference materials. *Am. Environ. Lab.* 5(4):14–15, 1993.
3. Benson, S. A., and P. Holm. Comparison of inorganic constituents in three low-rank coals. *Ind. Eng. Chem. Prod. Res. Dev.* 24:145, 1985.
4. Swaine, D. J. *Trace elements in coal.* London: Butterworth, 1990.
5. Kirsch, H., U. Schirmer, and G. Schwartz. The origin of the trace elements zinc, cadmium and vanadium in bituminous coals and their behavior during combustion. *VGB Kraftwerkstechniw* 60:734–744, 1980.
6. Martines-Tarazona, M., D. Spears, and J. Tascon. Organic affinity of trace elements in Austrian bituminous coals. *Fuels* 71:909–917, 1992.
7. Swaine, D. J. The organic association of elements in coals. *Org. Geochem.* 18(3):259, 1992.
8. Damie, A. S., D. S. Ensor, and M. B. Ranade. Coal combustion aerosol formation mechanisms: A review. *Aerosol Sci. Tech.* 1:119, 1982.
9. Flagan, R. C., and S. K. Friedlander. Particle formation in pulverized coal combustion—A review. In *Recent developments in aerosol science,* ed. D. T. Shaw, 25–59. New York: Wiley-Interscience, 1978.
10. Sarofim, A. F., J. B. Howard, and A. S. Padia. The physical transformation of the mineral matter in pulverized coal under simulated combustion conditions. *Combust. Sci. Technol.* 16:187, 1977.
11. Loehden, D., P. M. Walsh, A. N. Sayre, J. M. Beer, and A. F. Sarofim. Generation and deposition of fly ash in the combustion of pulverized coal. *J. Inst. Energy* June:119–127, 1989.
12. Zygarlicke, C. J., D. L. Toman, and S. A. Benson. Trends in the evolution of fly ash size during combustion. *Prepr. Pap. Am. Chem. Soc. Div. Fuel Chem.* 35(3):621, 1990.
13. Krauskopf, K. B. *Introduction to geochemistry.* New York: McGraw-Hill, 1976.
14. Mason, B. H. *Principles of geochemistry.* New York: Wiley, 1966.
15. Rendel, P. S., G. E. Batley, and J. A. Cameron. *Environ. Sci. Technol.* 14:314, 1980.
16. Sulcek, Z., and P. Povandra. *Methods of decomposition in inorganic analysis.* Boca Raton, FL: CRC Press, 1989.
17. Smith, R. T., and K. Atkinson. *Techniques in pedology.* London: Elek Science, 1975.
18. Stevenson, F. J. *Humus chemistry: Genesis, composition, reactions.* New York: Wiley, 1982.
19. Baker, D. E., and M. C. Amacher. Nickel, copper, zinc, and cadmium. In *Methods of soil analysis,* Part 2. *Chemical and microbiological properties,* 2d ed., ed. A. L. Page, R. H. Miller, and D. R. Keeney. Madison, WI: American Society of Agronomy and Soil Science, 1982.
20. Huffman, G. P., and F. E. Huggins. Analysis of the inorganic constituents of low rank coals. In *The chemistry of low-rank coals,* ed. H. H. Schobert. *Am. Chem. Soc. Symp. Ser.* 264:159–174, 1984.
21. Miller, R. N., and P. H. Given. The association of major, minor and trace inorganic elements with lignites. I. Experimental approach and study of a North Dakota lignite. *Geochim. Cosmochim. Acta* 50:2033, 1986.

22. Renton, J. J. Mineral matter in coal. In *Coal science,* ed. R. A. Meyers. New York: Academic Press, 1982.
23. Jackson, W. R., F. E. Larkins, M. Marshal, D. Rash, and N. White. Hydrogenation of brown coal: The effects of additional quantities of the inorganic constituents. *Fuel* 58:281-284, 1979.
24. Hippo, E. J., R. G. Jenkins, and P. L. Walker. Enhancement of lignite char reactivity to steam by cation addition. *Fuel* 58:338-344, 1979.
25. Schafer, H. Carboxyl groups and ion exchange in low-rank coals. *Fuel* 49:197-213, 1970.
26. Wiltsee, G. A. Quarterly Technical Progress Report DOE/FE/60181-1531, 15-1–15-6, 1984.
27. Helble, J. J., F. F. Huggins, C. L. Senior, S. Srinivasachar, N. Shah, and G. P. Huffman. The fate of chromium during pulverized coal combustion. *Proceedings of the Ninth Annual International Pittsburgh Coal Conference,* 1992, pp. 928-933.
28. Fonseca, A. G., P. R. Tumati, M. S. DeVito, M. S. Lancet, and G. F. Meenan. Trace element partitioning in coal utilization systems. Preprint, SME/AIME 1993 Annual Meeting.
29. Palmer, C. A., and R. H. Filby. Determination of modes of occurrence of trace elements in the Upper Freeport coal bed using size and density separation procedures. *Proceedings 1983 International Conference on Coal Science.* London: International Energy Agency, 1983, pp. 365-368.
30. Bogdanov, V. V. Zur genese der mikroelemente in den Kohlefuhrenden. Sammelwerk: *Materialy k9. sovest. Rabotn. geol.* Organ. Leningrad 7:90-94, 1965.
31. Finkelman, R. B. Mode of occurrence of accessory sulfide and selenide minerals in coal. In *Neuvième Congrès International de Stratigraphie et de Géologie du Carbonifère, Compte Rendu,* ed. A. T. Cross, Vol. 4, pp. 407-412, 1985.
32. Finkelman, R. B. Mode of occurrence of trace elements in coal. *U.S. Geological Survey Open-File Report* 81-99:312, 1981.
33. Hassett, D. J. Generic test of leachability: The synthetic groundwater leaching method. *Proceedings of the Waste Management for the Energy Industries Meeting,* University of North Dakota, Grand Forks, 1987, pp. 30-39.
34. Stevenson, R. J., D. J. Hasset, G. J. McCarthy, O. E. Manz, G. H. Groenewold, and F. W. Beaver. Solid Waste Codisposal Study. Final report to the Gas Research Institute, GRI 89/0280, 165, 1989.
35. Hassett, D. J., D. F. Pflughoeft-Hassett, and G. J. McCarthy. Ettringite formation in coal ash as a mechanism for stabilization of hazardous trace elements. Abstract in Proceedings of the 9th American Coal Ash International Coal Ash Utilization Symposium, Orlando, FL, 1991.

CHAPTER
SEVENTEEN

CHARACTERIZATION OF CONCRETES FORMULATED WITH BLENDS OF PORTLAND CEMENT AND OIL SHALE COMBUSTION ASH

James T. Cobb, Jr., and C. P. Mangelsdorf

INTRODUCTION

Through early 1991 Occidental Oil Shale, Inc. conducted a major development program for the production of shale oil in western Colorado. "In the Occidental [modified in situ (MIS) process], 20% or so of the shale to be processed is mined to form an underground cavity [or retort]" [1]. The shale around the retort is thermally treated to release oil and "a low-Btu MIS off-gas rich in hydrogen sulfide . . . . [The] raw shale mined to prepare the underground retorts [can be] burned together with the MIS off-gas to produce steam and power. . . . Coal may be added to increase capacity. The simultaneous combustion of these three fuels, in various combinations, [was] the subject of . . . tests in two circulating fluid-bed (CFB) pilot plants" [2]. Two major by-products from the CFB pilot plants were bottom ash and fly ash. The study reported in this chapter was undertaken to examine the replacement of portions of portland cement in standard concrete and of fine aggregate in zero-slump formulations by fly ash from one of the CFB pilot plants, located near Williamsport, Pennsylvania.

BACKGROUND

The beneficial use of various by-products from oil shale conversion projects has become an economically important aspect of the commercialization of this technology in several countries.

It should first be pointed out that three basic types of solid by-products can flow from oil shale conversion processes. They are (1) the overburden from surface mining of the oil shale, (2) residue from retorts in which shale oil is extracted from the oil shale, and (3) ash from the combustion of oil shale.

All of these are in commercial use in one or more countries. The Chinese, for example, produce lightweight aggregate, brick, and other products from "green shale," the cover layer of shale that is paragenetic with oil shale in the Fushun West Open Mine [3]. In addition, they use retort residue directly to produce bricks and, after "oxidative spontaneous combustion" (reburning), as an ingredient (10%) for the production of portland cement and as an additive (30%) for making portland-pozzolana cement [3, 4]. The Jordanians are also examining the use of their reburned retort residue both directly to produce lightweight, medium-strength masonry and building units or as an additive to portland cement [5].

But oil shale combustion ash is the most favored of the three for beneficial use, both as an admixture with portland cement for standard concrete and as a principal ingredient for zero-slump products. The Chinese again are active in both areas. For use as an admixture, they have found that ash from fluid-bed combustors operating at 1580°F (860°C) has optimal properties [6, 7]. Use of 30% oil shale combustor ash in the blend produces 525 grade portland-pozzolana cement, and reducing the proportion to 10% yields 625 grade portland cement [3]. They also produce two zero-slump materials. The first of these is ceramsite, a lightweight aggregate, made by grinding and granulating the ash, decarbonating the granules at low temperature, then rapidly sintering them at 2228°–2282°F (1220°–1250°C). The ceramsite has a density of 62.4 lb/ft^3 (1000 kg/m^3) and a compressive strength of 710–1070 psi (5–7.5 MPa) [3]. The other product is a brick that is 20% lighter than a normal clay brick; 100 million such bricks are manufactured each year at Fushun City [3].

Germany, Israel, and the former USSR are also active commercially in the beneficial use of oil shale combustor ash. The Germans produce an "oil shale cement," which is ground from a blend of 27% ash and 73% portland cement clinker and is used in the ready-mixed concrete industry of that country [8–14]. The Israelis are manufacturing concrete blocks, tiles, and mortars for rendering from this by-product and conducting extensive fundamental evaluations of the behavior of ash in the paste of these products [14]. The former Soviets have developed technology for use of shale ash as a binder and in concretes, but their reports on this work are available to the authors only in abstract form [15–17].

Table 1 Production parameters of ash samples

Sample period	Temperature		% Shale	% Coal
	°C	°F		
3	843	1550	54	46
4	868	1594	54	46
6	816	1500	65	35
7	843	1500	100	0

ASH PROPERTIES

Three samples of fly ash from the Occidental Oil Shale pilot combustor were used in this study. The combustion temperature and percentages of shale and coal being fed to the pilot unit during the period of collection of each sample are given in Table 1.

Samples of the fly ash from sample periods 3 and 4 were analyzed by the Coal By-Products Utilization Laboratory (CBUL) of the ND Mining and Mineral Resources Research Institute at the University of North Dakota. Elemental compositions and loss on ignition (LOI) of these two samples, which relate to those found for other ashes, are given in Table 2. The relatively high magnesia content of the fly ash from sample periods 3 and 4 does not appear to be a cause for concern. It is likely to be present in a bound form inside the ash particles. This is strongly suggested by CBUL's results, which show that the expansiveness of the fly ash was almost an order of magnitude lower than the ASTM C618 specification for both class C and class F fly ash.

CBUL also reported particle size distributions for which between 78% and 87% passed a No. 325 sieve. The fly ash thus meets the ASTM C618 specification that at least 66% must pass this sieve to be a class C or class F fly ash.

Table 2 Properties of six oil shale ashes

Comp.	Source of oil shale ash					
	Germany [14]	Israel [14]	China [6][a]	Jordan [5][b]	Sample 3	Sample 4
CaO	16–60	44.5	1.29	39.7	18.4	17.4
SiO_2	12–25	19.0	60.64	35.4	50.4	45.2
Al_2O_3	9–12	8.3	20.09	3.8	12.6	10.9
Fe_2O_3	6–7	4.3	11.89	2.0	4.6	4.5
MgO	1.4–2.0	0.7	0.83	4.0	6.6	6.6
SO_3	9–10	8.5	0.61	4.0	7.47	6.59
LOI	—	11.3	0.55	7.3	0.38	0.24

[a]1580°F (860°C) ash.
[b]Reburned spent oil shale.

CONCRETE FORMULATION AND COMPRESSIVE STRENGTHS

The basic concrete formulation used for 17 test batches of ash-containing concrete, made during this study, was a standard mix of

49% coarse aggregate
30% fine aggregate
13% type 1 portland cement
approximately 8% water

This and all other formulation percentages throughout this chapter are on a weight basis. The water content was adjusted to provide a slump of 2 inches (51 mm) [18]. Six batches also contained a workability (water reduction) additive. Two of these six batches also had type 3 portland cement in place of type 1 portland cement. Tables 3 and 4 record the formulations, characteristics, and compressive strengths of these 17 test batches. The parameters in Table 4 are

Percent substitution = percent of portland cement replaced by fly ash.
W/C = water/portland cement ratio; note that ash is *not* included in the denominator.
Slump = standard slump measurement in inches (1 inch = 25.4 mm); W/C was adjusted to obtain a slump of approximately 2 inches (51 mm), except for batches X-1, X-2, and X-3.

The compressive strength of cylinders [6 inches (15.2 cm) in length and 3 inches (7.6 cm) in diameter] are shown in Table 5.

Several trends may be noted from a comparison of the 3-day, 7-day, 14-day, and 28-day compressive strengths of the cylinders formed from these 17 batches (Table 5). At low levels of portland cement replacement (15%), oil shale fly ash

Table 3 Additives and special portland cements used in six batches

Batch	Additives
X-13	Polyheed (0.38 lb per 100 lb cement) added with cement in mixer
X-14	Polyheed (0.75 lb per 100 lb cement) mixed with the water
X-15	Polyheed (0.75 lb per 100 lb cement + ash) mixed with the cement separately
X-17	Type 3 Portland cement + Polyheed (0.75 lb per 100 lb cement + ash)
X-18	Polyheed (0.63 lb per 100 lb cement + ash)
X-19	Type 3 portland cement plus Polyheed (0.75 lb per 100 lb cement + ash)

Table 4 Production parameters of 17 batches

Batch	Sample period	Combustion temp (°F)[a]	Percent substitution	W/C[b]	Slump (inches)
X-1	—	—	—	0.50	1.25
X-2	3	1550	15	0.59	7.5
X-3	4	1594	15	0.59	9.0
X-4	3	1550	15	0.51	2.5
X-5	4	1594	15	0.44	1.5
X-6	3	1550	75	2.15	1.625
X-7	4	1594	75	2.00	1.875
X-8	3	1550	30	0.64	2.0
X-9	3	1550	45	0.82	2.0
X-10	3	1550	38	0.72	2.5
X-11	6	1500	15	0.57	2.875
X-13	3	1550	45	0.70	2.5
X-14	3	1550	45	0.95	2.0
X-15	3	1550	45	0.61	1.75
X-17	3	1550	45	0.76	1.75
X-18	3	1550	45	0.63	1.78
X-19	3	1550	60	1.04	2.0

[a] °C = (°F − 32)/(1.8).
[b] C = cement only, not including ash.

behaves as a workability (or water reduction) agent and leads to 9% higher compressive strength than that of standard concrete. This enhancement is lost by 30% replacement, which leads to 9% lower compressive strength than that of standard concrete. The addition of the workability agent restores the high strengths, even at 45% replacement, where it leads to a 9% higher compressive strength than that of standard concrete. Type 3 portland cement does not appear to be effective in building compressive strength.

It is difficult to make comparisons with the work of other investigators because of differences in methodologies and the lack of chemical component information in this preliminary study. The German "oil shale cement," for example, has compressive strengths up to 8800 psi (61.6 MPa) at 28 days [13]. However, the Israeli pastes of portland cement and oil shale ash reach only 2830 psi (19.8 MPa), when they have a water/cement ratio of 0.8 to 1.0 (which gives the same consistency as the portland cement paste), and portland cement pastes (water/cement ratios of 0.3 to 0.6) at 20 days reach 10,660 psi (74.6 MPa). However, in making cement blends (portland and oil shale ash together), the Israelis found that portland cement replacement by 15 to 25% oil shale ash yielded concrete with slightly increased strength [13]. The Chinese report 28-day strengths as high as 11,200 psi (78.4 MPa) with ash–portland cement blends. They also found an increase in strength at replacements up to 25% [6]. The Jordanians, on the other hand, saw a drop in compressive strength when reburned spent shale was added to portland cement. Their paste strengths were much

Table 5 Compressive strengths of 17 batches

Batch	Compressive strength (psi)[a]				
	3 days	7 days	14 days	28 days	90 days
X-1	3170	4140	4050	4870	3770[b]
X-2	2710	3530	3700	4490	5360
X-3	2760	3220	3500	4160	[c]
X-4	3650	4240	4510	5220	5730
X-5	3300	4480	4670	5270	5436
X-6	600	870	1160	1290	1770
X-7	690	1080	1810	1490	2020
X-8	3150	3760	3480	4280	5420
X-9	1360	2200	2820	3560	4010
X-10	1830	2710	[c]	2920[b]	4060
X-11	3380	4130	[c]	4300[b]	5710
X-12	1380	[c]	1970[b]	2880	3370
X-13	2850	[c]	3480[b]	4510	4480
X-14	1130[b]	1800[b]	3160	3430	3630
X-15	2850[b]	3840	5000	5440	5310
X-16	2970[b]	4460	4520	5520	6250
X-17	3230	3960	4450	4620	5430
X-18	2940	4000	3960	4210	5140
X-19	1860	2230	2930	3510	4190

[a]142.8 psi = 1 MPa.
[b]Alternate tester used.
[c]Equipment malfunction.

lower, peaking at 6530 psi (45 MPa) for pure portland cement paste at 28 days at a water/cement ratio of 0.4 [4].

MECHANICAL PROPERTIES OF CONCRETE

Two sets of concrete forms were prepared for physical testing. The first set, a control sample, was prepared with the basic concrete formulation without fly-ash replacement, and the second set with 45% fly-ash replacement (fly ash from sample period 3) was prepared with the workability agent (0.75 lb per 100 lb cement + ash) mixed with the cement separately.

The 17-day compressive strength of the standard formulation was 4010 psi (28.1 MPa), and that of the ash-containing formulation was 5050 psi (35.4 MPa). The length change test (ASTM C157-86) showed that from day 1 to day 7 the average shrinkage was 0.0023 inch (0.058 mm) for ash-containing specimens, compared with 0.0133 inch (0.34 mm) for specimens of standard concrete.

The abrasion test (ASTM C944-80) was performed on six 2-inch-high cylinders, cut from two 6-inch-diameter cylinders. One cylinder was standard concrete and the other was fly ash–containing concrete. The cut surfaces were abraded. The results are shown in Table 6.

Table 6 Abrasion tests for two batches

Specimen	Loss by abrasion (g)[a]			Average loss (g)[a]
X-1-1	0.7	0.5	0.6	0.60
X-1-2	0.5	0.5	0.6	0.53
X-1-3	0.8	0.9	1.0	0.90
X-15-1	0.7	0.8	0.8	0.77
X-15-2	1.0	1.1	1.3	1.13
X-15-3	0.9	0.8	0.8	0.83

[a] 1 lb = 454 g.

The freeze-thaw (ASTM C666-84) and deicing tests (ASTM C672-84) were performed by CBUL, using test specimens prepared at the University of Pittsburgh. The freeze-thaw tests were confounded by poor performance of the coarse aggregate. The samples containing oil shale ash had less resistance to scaling than those containing no ash.

The Chinese report that, for pastes in which 35 to 40% of the portland cement has been replaced by shale ashes, sulfate resistance increases by 166% and the bleeding ratio drops nearly to zero. Other physical properties of the oil shale–containing cements appear to the Chinese to be comparable to those of cements containing coal fly ash [5].

ZERO-SLUMP MATERIALS

Compressed and Autoclaved Ash/Lime Pastes

Three 2-sample sets of short, compressed cylinders [2.5 inches (6.4 cm) in diameter and 2 inches (5.1 cm) high], using fly ash, were prepared by R. I. Lampus Company, a member of the National Precast Concrete Association (NPCA) from Springdale, Pennsylvania (a suburb of Pittsburgh). Unfortunately, ash from sample period 7 was inadvertently selected for this work. Any comparisons with the results of the work reported in the previous two sections may not be fully relevant. The specimens had the formulations shown in Table 7. An optimum amount of water (10% based upon dry weight) was added for maximum com-

Table 7 Formulations for three batches

Specimen number	Percentage		
	Lime	Ash	Ash + sand
1	8	—	92
2	10	90	—
3	15	85	—

paction. The cylinders were autoclaved at the University of Pittsburgh for 8 hours at 170 psi (1.2 MPa) and 350°F (177°C) to obtain calcium silicate bonding. The compressive strengths achieved were 1500, 2130, and 2870 psi (10.5, 14.9, and 20.1 MPa), respectively. R. I. Lampus Company concluded that, based on this limited investigation, compressed and autoclaved oil shale ash/lime paste is a viable product. The company entered into negotiations with Occidental Oil Shale, Inc. to develop this product commercially.

Concrete Masonry Units

Two sets of zero-slump concretes, similar in formulation to concrete masonry units, were prepared at the University of Pittsburgh containing oil shale combustion fly ash. In both cases, cylinders 4 inches (10.2 cm) high and 2 inches (5.1 cm) in diameter were fabricated. Each set was prepared with 2.96 lb (1344 g) of sand, 0.33 lb (150 g) of water, and either 0.53 lb (240 g) type 1 portland cement (first set) or 0.37 lb (168 g) type 1 portland cement and 0.16 lb (72 g) oil shale fly ash (second set). The eight cylinders were rodded and allowed to set at room temperature in a humid environment for 5 hours. Next, they were heated to 150°F (66°C) at 1 atmosphere with steam over a 2-hour period and then cured at 150°F (66°C) and 1 atmosphere for 12 hours in a steamed environment before being cooled to room temperature. The average compressive strength of the vibrated, cured cylinders was 1590 psi (11.1 MPa) for both sets. This is somewhat lower than the 3-day strength of commercial concrete block, which is between 2000 psi (14.0 MPa) and 2400 psi (16.8 MPa).

Lightweight Concrete

The third, and final, zero-slump product line that was examined was lightweight concrete. This material generally has a very low density [down to 25 lb/ft^3 (400 kg/m^3)] and correspondingly low compressive strength [down to 60 psi (0.42 MPa)]. It is used for fireproofing and insulation in structures. The "foaming approach" to producing this material, applied in this project, uses a foaming agent in the concrete mix. The agent introduces and stabilizes air bubbles during mixing at high speed. A sample of oil shale ash was provided to Elastizell Corporation of America in Ann Arbor, Michigan. A test using 60% type 1 portland cement and 40% oil shale ash yielding a well-mixing, quick-setting formulation, which resulted in a hard, strippable product in 1 day. The cast density of the sample was 40.3 lb/ft^3 (646 kg/m^3). Elastizell concluded that, based on this limited investigation, oil shale combustion fly ash is a viable addition to lightweight concrete mixes.

CONCLUSIONS

The literature survey conducted within the framework of this project identified 13 articles on the use of oil shale combustion ash in cement. These come from Germany, Israel, China, and the former USSR.

The ash from the combustor tests with 54% shale and 46% coal is moderate in both CaO and MgO. The MgO is bound inside the particles and does not negatively affect the soundness of the ash.

Standard concrete with 15% substitution of portland cement with ash shows an improvement in compressive strength at the same consistency (slump) of the wet mix as standard portland cement concrete. Addition of a wettability agent allows an additional 30% replacement without a drop in compressive strength below that of standard portland cement concrete.

A first exploratory set of several physical tests were applied to the concrete that had 45% of the portland cement replaced by ash, along with the addition of a wettability agent. Generally, the ash-containing concrete exhibited somewhat poorer qualities than standard concrete.

Preliminary specimens of three zero-slump products were prepared: compressed and autoclaved ash/lime paste, concrete masonry units, and lightweight concrete. The properties of the first product were sufficiently interesting that one manufacturer entered into direct negotiations with Occidental Oil Shale, Inc. to test this product further. The third product met or exceeded one or more basic minimum standards, causing one manufacturer to desire continued testing.

It is regrettable that this promising activity ended when Occidental Oil Shale, Inc. was placed up for sale [19] upon Armand Hammer's passing [20] (the overall project losing $14 million the previous year [21]) and, no buyer being found, was disbanded.

ACKNOWLEDGMENTS

Occidental Oil Shale, Inc. provided the funds to carry out the majority of this preliminary study. Mr. Amin Tomeh prepared the 17 batches of ash-containing concrete and conducted the tests of compressive strengths at the University of Pittsburgh.

REFERENCES

1. Worthy, W. Synfuels: Uncertain and costly fuel option. *Chem. Eng. News,* August 27:20–26, 28, 1979.
2. Moore, R. E., et al. Simultaneous combustion of oil shale, low-Btu gas, and coal in a circulating fluid-bed combustor. *Clean Energy for the World: Proceedings of the International Conference*

on Fluidized Bed Combustion, Vol. 1, 553–557. New York: American Society of Mechanical Engineers, 1991.
3. Guangshu, T., Z. Xingshan, H. Zhongpu, and X. Zhihua. Multipurpose use of shale ash and its paragenetic mineral in China. *Proceedings of the International Conference on Oil Shale and Shale Oil,* Colorado School of Mines 21st Oil Shale Symposium, Beijing, China, May 16–19, 1988, pp. 547–551.
4. Guangshu, T., Z. Xingshan, H. Zhongpu, and L. Xianglin. Investigation on the Portland Pozzolana Cement with More Shale Ash Content. *Proceedings of the International Conference on Oil Shale and Shale Oil,* Colorado School of Mines 21st Oil Shale Symposium, Beijing, China, May 16–19, 1988, pp. 559–564.
5. Smadi, M., A. Yeginobal, and T. Khedaywi. Potential uses of Jordanian spent oil shale ash as a cementive material. *Magazine of Concrete Research* 41(148):183–190, 1989.
6. Zhongpu, H., and X. Qihai. Properties of Shale Ash (Taken from the Boiling Furnace) Portland Cement, *Proceedings of the International Conference on Oil Shale and Shale Oil,* Colorado School of Mines 21st Oil Shale Symposium, Beijing, China, May 16–19, 1988, pp. 552–558.
7. Zhimin, Wang, et al. Studies on Strength and Workability of High-Strength and Flowing Concrete Utilizing Low Calcium Oil Shale Ash. *Proceedings of the International Conference on Oil Shale and Shale Oil,* Colorado School of Mines 21st Oil Shale Symposium, Beijing, China, May 16–19, 1988, pp. 565–572.
8. Wuhrer, J. The composition of oil shale ash and the reasons for its hydraulic characteristics. *Zement-Kalk-Gips,* April:61–65, 89–93, 1949.
9. Wuhrer, J. Oil shale hydraulic cements. *Zement-Kalk-Gips,* March:45–48, 1950.
10. Rohrbach, R. Manufacture of oil shale cement and electric power generation with oil shale according to the Rohrbach-Lurgi process. *Zement-Kalk-Gips* 22(7):293–296, 1969.
11. Rechmeier, H. The five stage preheater kiln for the burning of clinker from limestone and oil shale. *Zement-Kalk-Gips* 23(6):249–253, 1970.
12. Rapp, G. Architectural concrete made with oil shale cement. *Betonwerk Fertigteil-Tech.* 50(7):468–473, 1984.
13. Mathauer, W. Oil shale cement in ready-mix concrete. *Betonwerk Fertigteil-Tech.* 50(9):613–619, 1984.
14. Bentur, A., M. Ish-Shalom, M. Ben-Bassat, and T. Grinberg. Properties and application of oil shale ash. *Proceedings of the Second International Conference on Fly Ash, Silica Fume, Slag and Natural Pozzolans in Concrete,* Madrid, Spain, April 21–25, 1986, SP 91-37, American Concrete Institute, pp. 779–802.
15. Puzyrev, Yu. A. Ways for using electrostatic precipitator shale ash as a binder. *Tekhnol. Izgot. i Svoistvanov. Kompozits. Stroit. Mater., L.,* 63–68, 1986. Referenced in *Ref. Zh. Khim. 1987,* Abstract No. 7M415.
16. Remnev, V. A. Effect of plasticizing additives on the water-solid ratio and gas evolution in a shale-ash binder. *Issled. po Str-vu Stroit. Teplofiz. Dolgovech. Konstruktsii, Tallin,* 75–85, 1987. Referenced in *Ref. Zh. Khim. 1988,* Abstract No. 8P95.
17. Puzyrev, Yu. A., A. B. Klyuev, and I. I. Zinchenko. Development of technology for use of shale ash in concretes. *Intensif. Tekhnol. Protsessov v Pr-ve Sbor. Zhelezobetona, L.,* 92–97, 1988. Referenced in *Ref. Zh. Khim. 1989,* Abstract No. 11M340.
18. Mindess, S., and J. F. Young. Proportioning concrete mixes. Chapter 9 of *Concrete.* Englewood Cliffs, NJ: Prentice-Hall, 1981.
19. Mindess, S., and J. F. Young. Oxy maps plan to trim debt, non-core assets. *Oil Gas J.,* January 21:18, 1991.
20. Cook, J. The high cost of Hammer. *Forbes,* May 27:104, 1991.
21. Mindess, S., and J. F. Young. Unocal plan closing spells end to shale oil development era. *J. Commerce,* April 17:12B, 1991.

CHAPTER EIGHTEEN
UTILIZATION OF COAL-TAR PITCH IN INSULATING-SEAL MATERIALS

Janusz Zieliński and George Górecki

BACKGROUND

Substances obtained from coal, in light of their physiochemical properties, can find a number of interesting applications. Among these substances are tars and coal-tar pitch. Collin [1] has described a variety of applications using coal-tar pitch (Figure 1). The majority of these applications, however, are realized only in a narrow scope.

With a high carbon content (93%) and having only trace levels of inorganic components (in contrast to coal), pitch is predestined for the production of technically pure carbon materials. Specific examples include carbon anodes for aluminum and magnesium production by fusion electrolysis; graphite electrodes for electric steel manufacture and chlorine-alkali electrolysis; carbon materials for heat exchangers, for refractories, and for acid-resistant cladding for chemical reactors; and aggregates for flue gas desulfurization. Nuclear graphite for graphite-moderated nuclear reactors and high-temperature reactors using the pebble-bed system is produced from special coal-tar pitch cokes of high purity. Fine cokes are excellently suited as carburizers for carbonizing steel and cast iron. High-melting, anisotropic mesophase pitches with softening points around 300°C are intermediates for the production of fine-grain graphites and for spinning and

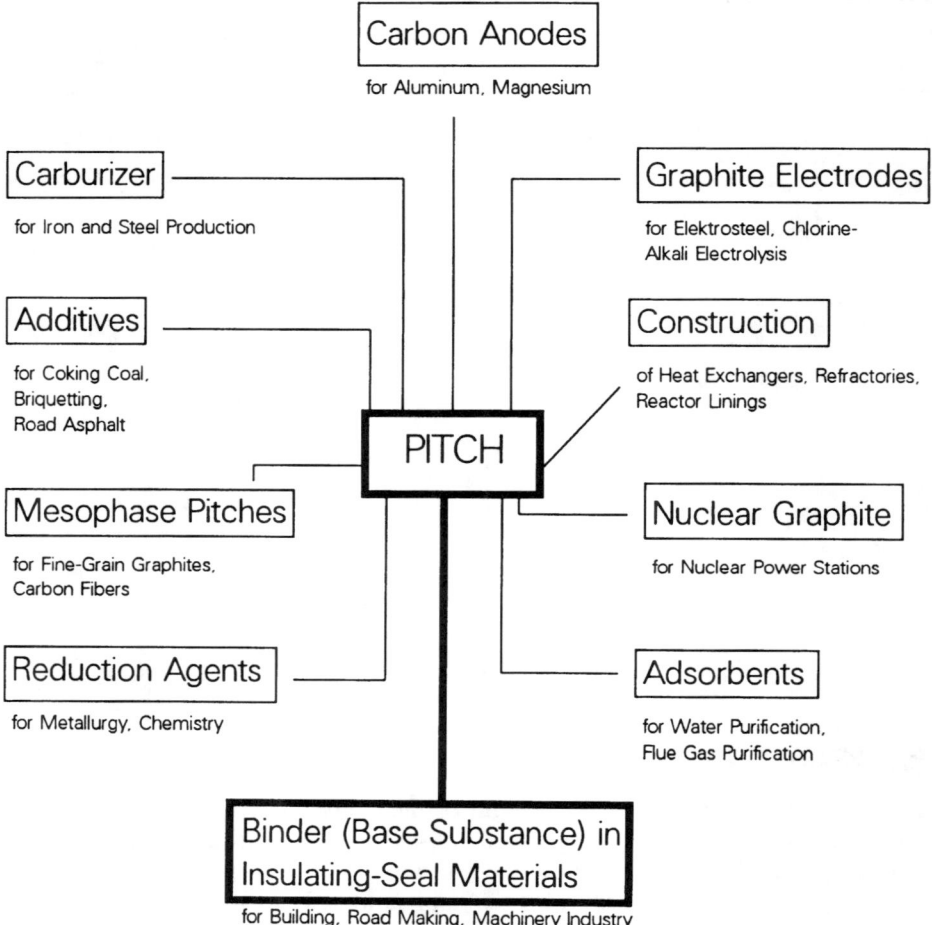

Figure 1 Applications of coal-tar pitch. (From Reference 11.)

subsequent carbonization to high-modular carbon fibers for high-performance composite materials. Coal-tar pitch is further used in briquetting coal, ores, and quartz sand (to produce silicon); as an additive for road asphalt; and as an additive to improve coking coal blends.

At the industrial level, pitch is most often treated as a by-product of the process of continuous distillation of coal-based tar. The process of coal-tar transformation by the method of continuous distillation is shown in simplified form in Figure 2. One path that leads to the improvement of thermal and rheological properties of bitumens derived from coal is their modification by means of the

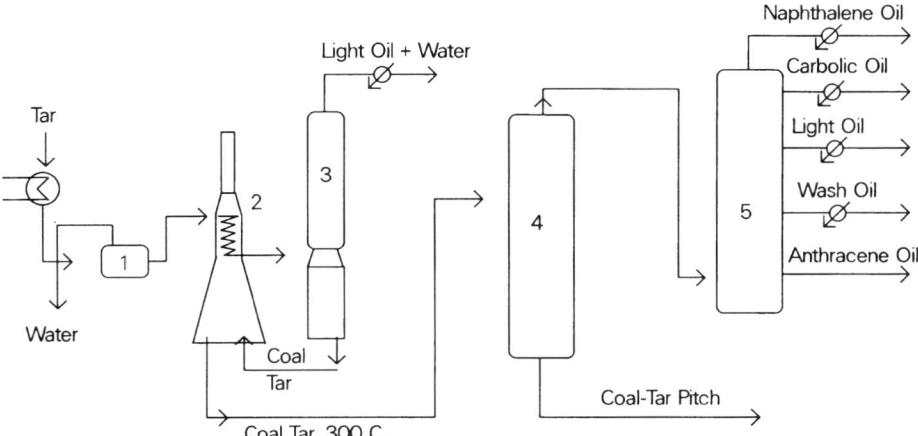

Figure 2 The flow of continuous distillation of high-temperature coal tar. 1, Pressure tank; 2, pipe furnace; 3, dehydration column; 4, coal-tar pitch column; 5, oil column.

addition of polymers. This creates the possibility of attaining greater utility from coal-tar pitch. One example is the improvement of binder quality in insulating-seal materials. This capability permits, in a relatively straightforward manner, the application of coal-tar pitch on a large scale. We feel that, in addition to the applications of pitch described by Collin, it is important to understand that pitch can be used as a binder or as a base substance in insulating-seal materials for the building, road construction, and machinery industries.

Compositions based on petroleum asphalts are examples confirming the usefulness of this approach. Petroleum asphalts result from the distillation of naphthalene oil. The heaviest fractions are separated at temperatures above 460°C. As a rule, these fractions yield oxides as well. These oxides are used to ascertain the corresponding physicochemical properties.

The literature pertaining to the modification of petroleum bitumens by polymers is very extensive, and the majority of it has been published in patents. For the most part, these works deal with the physical modification of asphalts by means of plastics, with the intention of obtaining homogeneous mixtures [2–6]. Polymers produce diverse changes in the properties of petroleum bitumen. This is related mostly to an increase in stability and durability of the bitumen's physicocolloidal structure [7].

On the other hand, combining coal-tar pitch with polymers has generated a significantly smaller level of interest. Nevertheless, various types of polymers have been utilized to modify the properties of tars and coal-tar pitch. These polymers provided the means of attaining homogeneous and stable compositions. In particular, the goal of these works was to obtain binders used in building roads [8] and in tar-based lacquers [9]. The goal of the investigations was to determine the pitch-polymer composition that would fill the role of the matrix, having as its basis insulating-seal materials of high molecular weight. The investigations were carried out in two stages.

EXPERIMENTAL STUDIES

In the first stage, problems related to the attainment of homogeneous and stable compositions made from the chosen types of polymers, plasticizers, and coal-tar pitch were investigated. The first stage is outlined in Figure 3. In the second stage, the formulations of the compositions were established. These mixtures defined beneficial characteristics for bituminous insulating-seal materials.

The following materials were studied:

1. Normal coal-tar pitch at its softening point of 87°C as well as the contents of components insoluble in toluene (21.45%).

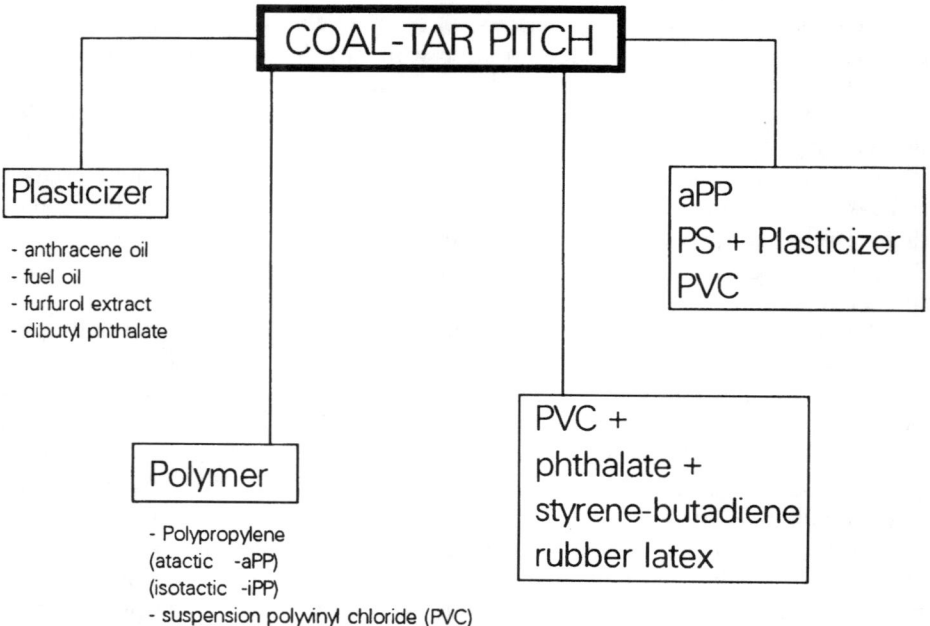

Figure 3 Essential directions of investigation concerning the modification of coal-tar pitch properties.

2. Polymers: atactic polypropylene [molecular weight (MW) 32,600], isotactic polypropylene (MW 220,000), polyvinyl chloride (MW 139,000), polystyrene (MW 304,000), styrene-butadiene rubber latex.
3. Plasticizers: dibutyl phthalate, anthracene oil, furfurol extract.
4. Fillers: milled chalk, talc, colloidal silica.

The following determinations were performed on the obtained compositions:

1. Fragility point (Fraass).
2. Softening point (Pierscien and Kula).
3. Penetration and stability at elevated temperatures according to the tube test [10]. The stability index was determined from the mutually relative densities of upper and lower layers of samples heated to 150°C for 4 days.
4. Adhesion. Concrete, glass, wood, and aluminum were used as substrates.
5. Flow along vertical surfaces.
6. Resistance to freezing.
7. Elasticity at negative temperatures, as low as −20°C.
8. Aging tests. Twenty cycles, in which one cycle lasted 48 hours and involved exposure to ultraviolet radiation, artificial rain, and heating to 70°C and freezing to −25°C.

In the case of compositions made up of coal-tar pitch and polymers, certain intrinsic factors play a particular role with regard to obtaining mixtures that are homogeneous and that will not delaminate. These factors include temperature, the time needed to reach homogeneity, and the order of addition of components. Another important factor is the elimination of circumstances that may destroy the polymers.

Beneficial properties were attained only in the case in which pitch was combined with polar polyvinyl chloride (PVC). Mixtures containing 10% (w/w) PVC in pitch were homogeneous and stable. The addition of PVC increased the softening point to 114°C, with a concurrent lack of improvement in elasticity.

Information in the literature and our own results show that the mixing of the components of pitch-polymer compositions are determined by the following factors:

1. Character of the bituminous raw product. Bitumen should contain large oil fractions for the purpose of dissolving and expanding the polymer. At the same time, bitumen should also contain a relatively large amount of condensed components in order to ensure the durability of the composition's structure. Coal-tar pitches typically contain large amounts of aromatic hydrocarbons, which are often condensed. It is especially advantageous to mix pitches with polar polymers containing an aromatic chain structure.

2. Structure and amount of polymer. These limit, among other things, the polymer's ability to dissolve and to swell in aliphatic and aromatic fractions of bitumen. The structure and amount of polymer also limit the formation of new colloidal structures, resulting in mutual separation between fragments of the polymer chain and the active components of the bitumen.
3. Mixing parameters. Especially important are time and temperature, so as to prevent polymer destruction.

Plastification of pitch by oil fractions that are coal based (anthracene oil) and petroleum based (fuel oil, furfurol extract), as well as by artificially produced dibutyl phthalate obtained from industry, was studied. It was demonstrated that the most beneficial changes in properties were obtained using dibutyl phthalate at 25–30% (w/w). The dibutyl phthalate–modified pitch provided new characteristics in pitch-plasticizer binders. Mixtures of pitch and furfurol extract or anthracene oil at 20–25% (w/w) provided similar improvements.

For eventual application as base substances for insulating-seal materials, two-component pitch-polymer and pitch-plasticizer mixtures required further modification. This was made possible by improving their mechanicorheological properties. The three-component systems comprised coal-tar pitch, polymers, and plasticizers. The best properties were obtained with compositions having 25% (w/w) dibutyl phthalate relative to pitch and 5–8% (w/w) PVC relative to pitch-plasticizer binder.

The compositions were also modified with styrene-butadiene rubber in order to impart resistance to elevated temperatures and to provide high elasticity at negative temperatures. The following composition provides advantageous properties: coal-tar pitch (58.6% w/w), phthalate (22%), PVC (7.4%), styrene-butadiene rubber latex (7.4%). The fragility point of this mixture was $-36°C$ and the softening point was 58°C. The penetration at 25°C was about 100 × 10^{-4} m and the stability index was 0.4%. These properties permitted the mixture to behave like an insulating-seal material.

The second part of the investigation involved determining the formulation of the final mixture, which had coal-tar pitch as a fundamental component. The composition was defined by standard specifications used for dilatant putties and hydroinsulation linings used in the building industry. As a rule, fillers and adhesion agents are materials of this type. The influence of the types and amounts of fillers (milled chalk, talc, colloidal silica), as well as that of adhesive media (balmy rosin, coumarone-indene resin), on the properties of butadiene-styrene pitch-phthalate-PVC-latex mixtures was studied. It was found that, regardless of the choice of filler, the most beneficial changes in applied properties were attained with filler mixtures containing chalk or talc with colloidal silica in a mass ratio of 4:1. Testing was performed by examining leveling at higher temperature, vertical flow at 60°C, and resistance to freezing at $-10°$ and $-20°C$. At the same time, the amount of filler in the entire substance, depending on the application,

Table 1 Optimal quantities of components

Component	Sealing compound (% w/w)	Sealing putty (% w/w)	Pitch-polymer lining (% w/w)
Coal-tar pitch	56.10	48.90	48.30
Dibutyl phthalate	18.70	16.30	16.10
PVC	6.50	5.70	7.10
Styrene-butadiene rubber latex	6.50	5.70	7.10
Talc	9.76	—	17.10
Milling chalk	—	17.00	—
Colloidal silica	2.44	4.30	4.30
Balmy rosin	—	2.10	—

should be 15–30% (w/w). On the other hand, balmy rosin, in quantities of 2–3% (w/w), was used as the adhesive agent.

RESULTS AND APPLICATIONS

The results of the investigation provided optimal formulations, which are listed in Tables 1 and 2. These compositions have many practical applications, such as filling horizontal slits in concrete as well as strengthening the bonds of metal-glass and wood-glass interfaces. Other applications include bituminous dilatant putties, sealing vertical surfaces in building elements, and hydroinsulating linings (1–3 mm thickness) formed by the hot rolling method, for use in the building industry.

The thermomechanical, rheological, and practical properties of these compositions are comparable to the properties of analogous mixtures based on petroleum asphalts. In some instances, the properties of the pitch-based mixtures exceed those of the petroleum-based ones (e.g., adhesion to concrete, metallic, or glass substrates). Modified pitch-polymer compositions can be used in general

Table 2 Properties of selected coal-tar pitch polymer compositions

Composition	Softening point (°C)	Fragility point (°C)	Penetration (25°C), 10^{-4} m	Stability index (%)
Pitch + 10% aPP	88.0	−9.5	8.6	10.25
Pitch + 90% aPP	113.6	−11.3	10.2	0.44
Pitch + 5% iPP	87.5	—	2.6	2.02
Pitch + 5% PVC	88.5	—	2.5	1.00
Pitch + 10% PVC	114.0	—	0.0	0.05
Pitch + 5% PS	133.5	—	0.0	0.95
Pitch + 10% PS	108.0	—	0.0	13.27

and specialized building applications. Examples include the construction of roads, airports, and underground objects. Other examples can be found in the machine industry, where these mixtures are used as materials to prevent water penetration and to deaden vibrations.

REFERENCES

1. Collin, G. Production and application of coal-tar pitch. *International Conference on Structure and Properties of Coals.* 1991, p. 154.
2. U.S. Patent 4,430,464, 1984.
3. U.K. Patent 1,457,999, 1976.
4. Zielinski, J. Investigations on thermal properties of asphalt-polymer compositions. *Erdol Kohle Erdgas Petrochem.* 42:456, 1989.
5. Zielinski, J., and A. Bukowski. Bestimmung der Eigenschaften von Bitumen-Polymer-Gemischen. *Erdol Kohle Erdgas Petrochem.* 39:513, 1986.
6. Zielinski, J., and A. Bukowski. Bitumen-polymer compositions, characteristics and structure. *Ropa Uhlie* 29:404, 1987.
7. Zielinski, J. The studying of structure and the properties of bitumen-polymer compositions. Wyd. Politechniki Warszawskiej, Seria Chemia, 1991.
8. Stompel, Z., G. Collin, A. Szen, and G. Herion. Pitch-asphalts; a new binder for road building. *Koks Smola Gaz* 33:24, 1988.
9. Kozakiewicz, J., A. Orzechowski, I. Krowicka, and E. Woczynska. Modification of urethane prepolymers with coal-tar pitch leading to moisture-curing coatings and sealants for the building industry. *Polimery* 37:534, 1992.
10. Zenke, G. Tubon-Verfahren. Ein einfaches Hilfsmittel zur Bestimmung der Heisslagerstabilitat bituminoser Stoffe. *Bitumen* 1:11, 1973.
11. Collin, G. Production and application of coal-tar pitch. *Proceedings of the International Conference on Structure and Properties of Coals,* Institute of Chemistry and Technology of Petroleum and Coal, Technical University of Wrocław, Wrocław, Poland, 1991, Figure 2, p. 155.

CHAPTER
NINETEEN

BRIQUETTING ANTHRACITE FINES FOR RECYCLE

Salustio Guzman and John T. Price

INTRODUCTION

QIT-Fer et Titane Inc.'s metallurgical operations in Quebec consist of an open pit mine located in the Allard Lake region, 500 miles north of Montreal, where an ilmenite ore is extracted, and the operations in Sorel, where the ore is beneficiated before smelting in electric arc furnaces. Smelting produces a titanium dioxide–rich slag that is used as feedstock by pigment-producing companies and liquid iron metal, which is processed further for production of different grades of pig iron, production of steel billets at QIT's steel plant, and production of iron and steel powder products at QIT's metal powders plant, Quebec Metal Powders (QMP).

Smelting at QIT in electric furnaces consists of a carbothermic reduction of iron oxides present in the ilmenite in a molten bath. The reducing agent employed is Pennsylvania anthracite, which is delivered to Sorel by boat.

Generation of fines occurs in any metallurgical operation in which large volumes of diverse types of raw materials are processed. Preparation of raw materials in the unit stages (drying, calcining, roasting) preceding smelting and during handling (feeding and transporting) and smelting of ores or mineral concentrates constitute the main stages associated with the generation of fines.

At QIT, Pennsylvania anthracite reductant is dried prior to feeding to the electric arc furnaces. The drying operation on anthracite generates fines, which cannot be directly recycled to the electric furnaces; because of their fine size, during feeding they would be lost to the wet-gas scrubbers. Briquetting of anthracite fines would allow recycling of this material as reductant for smelting ilmenite at QIT.

This chapter presents results of a laboratory study conducted at the Energy Research Laboratories of CANMET and also discusses the proposed flowsheet for briquetting.

FINES GENERATION

Anthracites by nature are considered friable materials; thus they have an intrinsic tendency to generate fines in varying quantities during handling, transportation, and storage, all of which are practiced in processing anthracite for any metallurgical operation or application.

The anthracite received at Sorel contains moisture that must be removed prior to feeding the anthracite to the arc furnaces. QIT employs three rotary louver–type driers similar to the schematics shown in Figure 1. Each drier has a capacity to produce dried anthracite product at up to 17 MT per hour; the product is subsequently fed to the electric arc furnaces as a reducing agent.

The driers remove 8 to 8.5% of the moisture from the anthracite shipped by the suppliers by boat to Sorel. Heat is supplied to the driers by burning fuel gas produced from the reduction of iron oxides at the arc furnaces, which contains 85% CO and 15% H_2. The fuel gas, after cleaning as it exits the arc furnaces, is stored for use as fuel throughout the plant. The anthracite fines entrained in the gas exiting the driers is removed by dry cyclones and wet scrubbers. Fines represent approximately 3.4% by weight of the total annual amount of anthracite used at QIT.

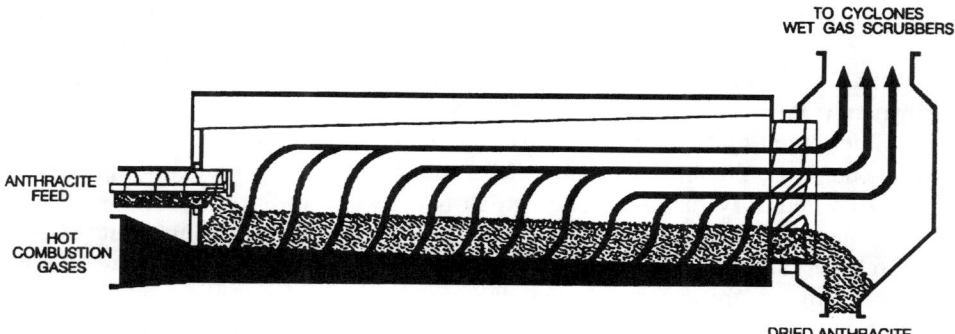

Figure 1 Anthracite dryer schematics.

Typical anthracite fines contain by weight 75.38% fixed carbon, 5.7% volatile matter, 0.68% sulfur, and 18.92% ash, and 79% of the particles are bigger than 38 μm. Tables 1 and 2 summarize the chemical composition and particle size distribution for anthracite fines. The high ash content in the fines compared with the typical 9.6–10 wt % ash in dry anthracite is consistent with the fact that coal fines in general are associated with a higher ash content.

BRIQUETTING OF ANTHRACITE FINES

In general, briquetting is a viable method for consolidation of fine powdery materials, particularly fine coal powders. Briquetting of anthracite fines in particular allows the use of low-melting-point carbonaceous binders, which have the potential to produce green briquettes with acceptable mechanical strength and physical integrity, to withstand transportation and handling practiced in metallurgical applications.

The laboratory experimental work was conducted at the Energy Research Laboratories of CANMET. A batch laboratory size twin-roll briquetting machine

Table 1 Chemical analysis of anthracite fines

Compound	Percent by weight
Proximate analysis	
Volatile matter	5.7
Fixed carbon	75.38
Sulfur	0.68
Ash	18.92
Ash analysis	
TiO_2	7.29
Fe (total)	16.92
FeO	Nil
Fe_2O_3	24.2
SiO_2	43.31
Al_2O_3	19.13
CaO	1.45
MgO	1.70
MnO	0.11
Cr_2O_3	0.07
V_2O_5	0.27
P_2O_5	0.19
Na_2O	0.53
K_2O	1.59
Sulfur	0.15
TOTAL (dry)	99.99

Table 2 Particle size distribution of anthracite fines

Mesh size	% by weight retained	Cumulative wt % retained
+48M	13.55	13.55
−48M + 65M	8.08	21.63
−65M + 100M	11.18	32.81
−100M + 150M	13.59	46.40
−150M + 200M	15.57	61.97
−200M + 270M	10.65	72.62
−270M + 325M	6.46	79.08
−325M + 400M	2.04	81.12
−400M	18.89	100.01
Total	100.01	

manufactured by K. R. Komarek, model B-100, was employed for the tests. Rolls 130 mm in diameter and 50 mm in width were employed to produce pillow-shaped briquettes with dimensions 12 × 12 × 40 mm.

The only binder tested was type II roofing asphalt, because this carbonaceous binder is compatible with the nature of anthracite reductant and is commercially available. The carbonaceous nature of roofing asphalt does not contaminate or significantly downgrade the quality of the anthracite, as other inorganic binders would.

A typical briquetting procedure consisted of adjusting the moisture in anthracite fines to about 3 wt % by adding water; this was necessary to minimize dusting during handling. The required amount of pitch binder preheated to about 140°C was added to 7 kg of moist fines and mixed by stirring for 5 minutes in a stainless steel bowl. The mixture was quickly dumped into the feeder bin, which was equipped with a feeder intake tube wrapped with heating tape, and briquettes were produced with the corresponding experimental conditions chosen for each experiment, as summarized in Table 3.

RESULTS

Briquettes produced within 10–12 wt % pitch and 6880 to 13,760 pounds pressure had a dull-shine to gloss-shine physical appearance, compared with the opaque pitted appearance and poor physical properties associated with briquettes produced outside the 10–12 wt % pitch range, as summarized in Table 4.

The highest weight, 90.1–92.6%, of briquettes produced meeting the 12 × 12 × 40 mm size specification, relative to the initial feed weight, corresponds once again to the use of 10–12 wt % binder as illustrated in Table 4. The

Table 3 Experimental conditions for briquetting anthracite fines

Rolls speed (rpm)	Auger rate (rpm)	Briquet. force (lb)	Moisture content (wt %)	Binder content (wt %)	Fines weight (g)
1.818	32.25	10,320	3	5	7077
1.818	32.25	10,320	3	7	7077
1.818	32.25	10,320	3	9	7077
1.818	32.25	10,320	7	5	7000
1.818	32.25	10,320	7	7	7000
1.818	32.25	10,320	7	9	7000
1.818	32.25	6,880	3	12	7077
1.818	32.25	10,320	3	12	7077
1.818	32.25	13,760	3	12	7077
1.818	32.25	6,880	3	10	7077
1.818	32.25	10,320	3	10	7077
1.818	32.25	13,760	3	10	7077
1.818	32.25	6,880	3	12	7077
1.818	32.25	10,320	3	12	7077
1.818	32.25	13,760	3	12	7077
1.818	32.25	6,880	3	14	7077
1.818	32.25	10,320	3	14	7077
1.818	32.25	13,760	3	14	7077

difference from 100% represent mostly fines and/or briquettes smaller than the target size and material squeezed at the edges of the briquettes.

Binder below 10 wt % is not enough for efficient production of briquettes with the physical properties needed for recycling, and above 12 wt % is not practical because with the high binder content it is difficult to operate the briquetting machine under normal conditions, as indicated in Table 4.

Green briquettes with the highest drop and abrasion resistance properties were produced with 12 wt % binder addition and 10,320 pounds pressure. These briquettes withstood 9-foot drop tests without breakage, whereas the rest of the briquettes withstood drop tests of only 6 feet or less. Thus the briquettes with 12 wt % binder are approximately 30% stronger than the rest, as shown in Table 4.

The green briquettes with 10 and 12 wt % binder additions were crushed to produce briquetted anthracite material in a size smaller than the original 12 × 12 × 40 mm. A roll crusher with a 6.3-mm gap was employed to crush the briquettes, and the resulting crushed material had a particle size ranging from 1.13 cm (3/8 inch) to 130 μm (100 mesh), as shown in Table 5 and Figure 2. The +1.651 mm (10 mesh) fraction was subject to tumbling tests. A modified ASTM tumbler test for coal was used to accommodate the amount, size of material, and number of screens employed for this particular case.

Once again, the crushed green briquettes containing 12 wt % binder and briquetted at 10,320 pounds pressure exhibited the highest abrasion resistance,

Table 4 Briquetting conditions and results

	Briquetting conditions									
Moisture (%)	3	3	3	3	3	3	3	3	3	3
Binder (%)	5	7	9	10	10	10	12	12	12	14	14	14
Pressure (lb)	10,320	10,320	10,320	6,880	10,320	13,760	6,880	10,320	13,760	13,760	10,320	6,880
					Results					
Briquettes within size spec. (wt %)	1.7	10.1	69.3	91	92	91.7	92.6	90.8	90.1	73.4	—	—
Drop test Drop height without breakage (ft)	2.5	6	6	6	6	6	9	9	9	6	—	—
					Observations					
Appearance	Looked poor	Dull and pitted	Semishine, slightly pitted	Dull shine	Dull shine	Semishine	Dull shine	Semishine	Gloss shine	Pitted finish	Extreme difficult operation due to bridging in press	

Table 5 Particle size distribution for crushed briquettes

		10% wt Binder								12% Binder			
		3% moisture 6,880 (lb)		3% moisture 10,320 (lb)		3% Moisture 13,760 (lb)		3% Moisture 6,880 (lb)		3% Moisture 10,320 (lb)		3% Moisture 13,760 (lb)	
No. of passes through rolls:		1		1		1		2		2		2	
Tyler mesh size	Average diameter (µm)	% on	% cum.	% on	% cum.	% on	% cum.	% on	% cum.	% on	% cum.	% on	% cum.
3/8 inch	11,377	27.01	27.01	14.73	14.73	14.31	14.31	8.55	8.55	7.12	7.12	8.32	8.32
3M	8,052	22.76	49.77	25.94	40.67	24.88	39.19	31.68	40.23	27.22	34.34	28.23	36.55
4M	5,690	12.07	61.84	14.79	55.46	14.82	54.01	16.99	57.22	19.08	53.42	18.44	54.99
6M	4,013	7.92	69.76	8.47	63.93	9.62	63.63	10.46	67.68	10.84	64.26	11.19	66.18
8M	2,845	5.05	74.81	6.23	70.16	6.28	69.91	7.03	74.71	7.69	71.95	7.50	73.68
9M	2,190	2.04	76.85	2.55	72.71	2.66	72.57	2.86	77.57	2.93	74.88	2.90	76.58
14M	1,595	4.66	81.51	5.60	78.31	5.91	78.48	6.05	83.62	6.61	81.49	6.58	83.16
20M	1,001	2.14	83.65	2.69	81.00	2.92	81.40	2.86	86.48	3.18	84.67	3.05	86.21
28M	711	1.97	85.62	2.37	83.37	2.49	83.89	2.29	88.77	2.63	87.30	2.43	88.64
35M	499	2.07	87.69	2.43	85.80	2.60	86.49	2.30	91.07	2.60	89.90	2.47	91.11
48M	352	1.44	89.13	1.79	87.59	1.79	88.28	1.44	92.51	1.61	91.51	1.53	92.64
60M	274	0.72	89.85	0.78	88.37	0.86	89.14	0.74	93.25	0.85	92.36	0.75	93.39
100M	200	2.34	92.19	3.52	91.89	3.84	92.98	2.38	95.63	2.40	94.76	2.46	95.85
−100M	130	7.81	100.00	8.11	100.00	7.02	100.00	4.37	100.00	5.24	100.00	4.15	100.00
Ave. Part. Size (µm)		6328		5431		5358		5425		5075		5262	

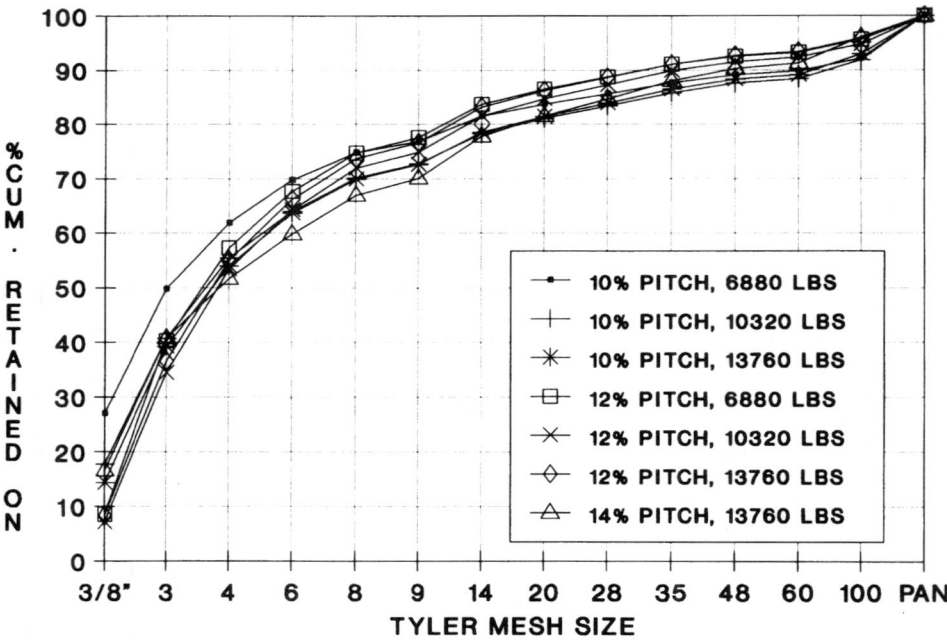

Figure 2 Particle size distribution of crushed briquettes.

with only a 16.6% decrease in the average particle size, from 6420 μm before to 5355 μm after tumbling. Larger average particle size reductions ranging from 29.4 to 37.2% were experienced, after tumbling the +10 mesh fraction of crushed briquettes containing 10 wt % binder, as shown in Table 6.

The final chemical composition of the briquettes produced with 12 wt % binder additions, which corresponds to the briquettes with the highest drop and abrasion properties, have a proximate analysis of 16.5% ash, 13.3% volatile matter, and 70.2% fixed carbon, and an ultimate analysis of 77.0% carbon, 2.4% hydrogen, 0.7% nitrogen, 1.04% sulfur, 2.4% oxygen, and 16.5% ash, as shown in Table 7. The respective proximate and ultimate analysis for briquettes with 10 and 14% binder are also shown in Table 7.

As an illustration of the laboratory briquetting practice, Figure 3 shows a photograph of the initial anthracite fines from which briquettes were produced, and Figure 4 shows the crushed briquettes along with the fines and briquettes.

In summary, use of 12 wt % binder and 10,320 pounds pressure produced strong green briquettes, which have the drop and abrasion properties required to withstand the handling and transportation needed for recycle. In addition, the 12 × 12 × 40 mm pillow-shaped briquettes can be crushed to reduce their size and tailor the size distribution to meet any particular potential need.

Table 6 Particle size distribution of +10 mesh fraction crushed briquettes

Tyler mesh size	Average diameter (μm)	Before tumbling						After tumbling					
		3% Moisture 6,880 (lb)		3% Moisture 10,320 (lb)		3% Moisture 13,760 (lb)		3% Moisture 6,880 (lb)		3% Moisture 10,320 (lb)		3% Moisture 13,760 (lb)	
		% on	% cum.	% on	% cum.	% on	% cum.	% on	% cum.	% on	% cum.	% on	% cum.
						10% Binder							
3/8 inch	11,377	35.15	35.15	24.38	24.38	19.72	19.72	7.69	7.69	5.16	5.16	4.98	4.98
3M	8,052	29.62	64.76	31.55	55.93	34.28	54.00	27.45	35.14	25.19	30.35	25.88	30.86
4M	5,690	15.71	80.47	20.34	76.28	20.42	74.42	18.10	53.24	21.65	52.00	22.25	53.11
6M	4,013	10.31	90.77	11.65	87.92	13.26	87.68	10.76	64.00	13.63	65.63	14.26	67.37
8M	2,845	6.57	97.35	8.57	94.69	8.65	96.33	7.81	71.81	9.42	75.05	9.37	76.74
9M	2,190	2.65	100.00	3.51	100.00	3.67	100.00	2.93	74.74	3.64	78.69	3.72	80.46
14M	1,595							5.42	80.16	5.15	83.84	4.95	85.41
20M	1,001							1.84	82.00	1.64	85.48	1.37	86.78
28M	711							1.66	83.66	1.38	86.86	1.17	87.95
35M	499							1.84	85.50	1.49	88.35	1.32	89.27
48M	352							1.40	86.90	1.15	89.50	0.92	90.19
60M	274							0.64	87.54	0.51	90.01	0.49	90.68
100M	200							2.48	90.02	2.19	92.20	2.10	92.78
−100M	130							9.98	100.00	7.80	100.00	7.22	100.00
Ave. Part. Size (μm)		7936		7260		7024		5066		4958		5045	
						12% Binder							
3/8 inch	11,377	11.02	11.02	9.51	9.51	10.86	10.86	3.39	3.39	3.72	3.72	4.19	4.19
3M	8,052	40.84	51.86	36.47	45.98	36.86	47.73	29.38	32.77	29.58	33.30	29.26	33.45
4M	5,690	21.90	73.77	25.36	71.34	24.08	71.81	26.24	59.01	25.94	59.24	25.45	58.90
6M	4,013	13.48	87.25	14.48	85.82	14.61	86.42	14.91	73.92	15.21	74.45	15.71	74.61
8M	2,845	9.06	96.31	10.27	96.09	9.79	96.21	8.69	82.61	10.26	84.71	10.13	84.74
9M	2,190	3.69	100.00	3.91	100.00	3.79	100.00	4.73	87.34	3.96	88.67	3.89	88.63
14M	1,595							3.94	91.28	3.66	92.33	3.75	92.38
20M	1,001							0.95	92.23	0.81	93.14	0.82	93.20
28M	711							0.83	93.06	0.70	93.84	0.71	93.91
35M	499							0.90	93.96	0.76	94.60	0.78	94.69
48M	352							0.63	94.59	0.52	95.12	0.51	95.20
60M	274							0.37	94.96	0.30	95.42	0.31	95.51
100M	200							1.25	96.21	1.44	96.86	1.00	96.51
−100M	130							3.79	100.00	3.14	100.00	3.49	100.00
Ave. Part. Size (μm)		6668		6420		6522		5366		5433		5450	

Table 7 Chemical analysis of briquettes

Proximate Analysis				Ultimate Analysis					
Briquettes Description	Ash %	Volatile Matter %	Fixed Carbon %	Carbon %	Hydrogen %	Nitrogen %	Sulfur %	Ash %	Oxygen (by diff.) %
10% binder	17.1	12.9	70.0	76.4	2.4	0.7	0.97	17.1	2.4
12% binder	16.5	13.3	70.2	77.0	2.4	0.7	1.04	16.5	2.4
14% binder	16.4	15.6	68	76.8	2.6	0.7	1.09	16.4	2.4

At QIT, the crushed or uncrushed briquettes can be recycled to the electric arc furnaces and used as a reductant for the smelting of ilmenite.

PROPOSED PROCESS FLOWSHEET

The successful completion of the laboratory testing has led to the proposal of a flowsheet for briquetting anthracite fines. The schematic of the proposed flowsheet, shown in Figure 5, has bins for storing premoistened anthracite fines (3% H_2O) to minimize dust during handling and feeding. The fines are mixed with the preheated asphalt binder (about 12% type II roofing asphalt) and the mixture fed to the roll briquetting machine. The briquette product is sized for collection (12 × 12 × 40 mm briquette size), and the undersize fraction is treated for size reduction, if needed, or could be recycled directly to the fine storage bins for briquetting, as shown in Figure 5.

It is estimated that a briquetting plant producing 5 metric tonnes per hour would be more than sufficient for processing the anthracite fines generated at QIT. The produced briquettes can be stockpiled for recycle, or they can be reduced in size by crushing to obtain a smaller size than the original briquettes.

Figure 3 Anthracite fines and briquettes.

Figure 4 Anthracite fines, briquettes, and crushed briquettes.

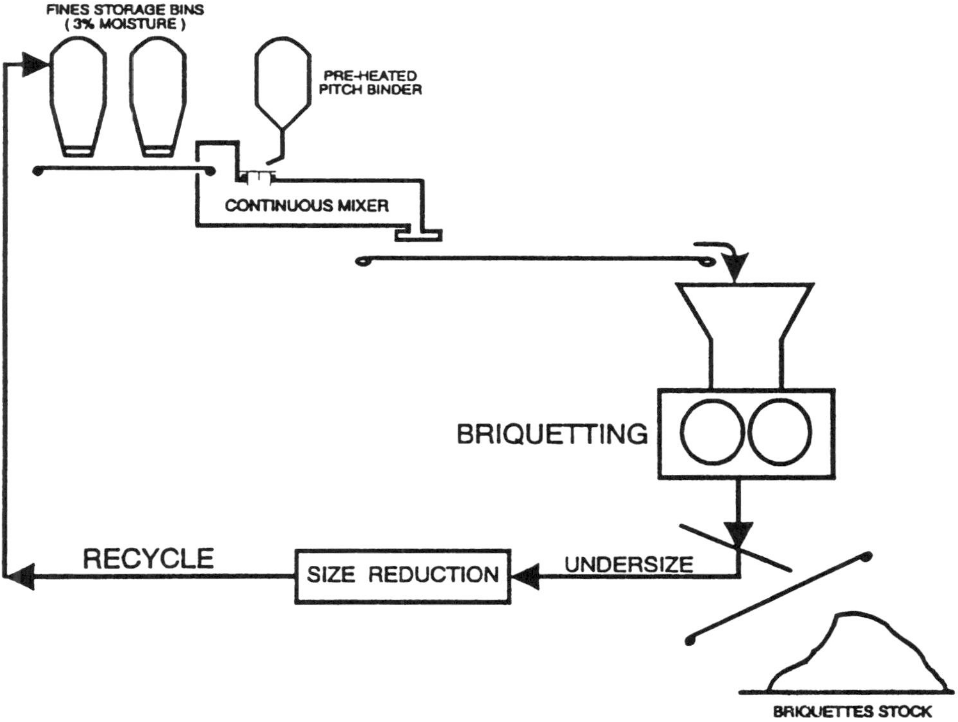

Figure 5 Schematic of proposed flowsheet.

REFERENCES

1. Koerner, R. M., and J. A. McDougall. Elements II, Briquetting and Aglomeration. Institute for Briquetting and Agglomeration, June 1983.
2. Harry, L. S., and M. K. Ferguson. Briquetting bulk solids. *Chemical Engineering,* April:114–122, 1992.

INDEX

Acid hydrolysis, 170–171
Acrylonitrile butadiene styrene (ABS), 98
Activated carbon, 106–118
Advanced recycling, 54
Advanced Recycling Technology, Inc., 4, 54
Agricultural waste, 41
Alcoholysis, 56
American Plastic Council, 61
Anthracite fines, 271–280
Apatite, 244
Asphaltene, 42
Auburn University, 39
Automotive shredder residue, 77–100
Automotive shredder residue (ASR), 78, 81

BASF, 59
Battelle, 63
Bureau of Mines, 85
Best demonstrated technology (BDT), 182
Biocatalysts, 158–164
Biochemical oxygen demand (BOD), 32
Biomass, 157–166
Biosolids, 147–155
Bishydroxyethylterephethalate (bis-HET), 57
Boudouard reactor, 111
Boudouard, carbon, 112, 116, 117

Briquetting, 217–280
Business, small waste recovery, 3–13

Calcite, 242, 244
Carbon black, 40, 107
Carbon residue, 107
Cation exchange capacity, 151–152
Cellulose, 10, 223–225
Cellulosic waste, 41
Char activation, 109–115
Chemical kinetics, 222
Chemical oxygen demand (COD), 32
Chloride capture, 72
Chlorofloroform carbon (CFC), 22
Clean Air Act, 236
Coal, 4
Coal tar, 263–269
Coal tar, pitch, 263–269
Coke bottle, 45
Coliquefaction, 17
Colloidal silica, 269
Combustion efficiency, 194–196
Commingled, 35
Community Environmental Services (CES), 5
Composites, 79
Compost bin, 9

Composting, 8–9
Computer monitors, 11
Computer recycling, 4
Conrad Industries, process, 63, 66
Cost, for recycling, 26
Costeam, 201–226
Curbside, collected, 6

Department of Energy (DOE), 61
Depolymerized scrap rubbr (DSR), 119
Design, for recycling, 25
Dibutyl phthalate, 269
Diesel, 40
Disposal fees, 24, 25
Disposal of waste, 3
Dolomite, 242, 244

Electrochemical gasification, 91
Electron spin resonance (ESR), 123–131
Energy & Environmental Research Center (EERC), 61
Environmental Protection Agency (EPA) (*see* U.S. Environmental Protection Agency)
Enzymatic process, 11
Enzymes, 6
Ethanol, 158–178
Extruder technology, 4

Ferridydrite, 41
Fiberglass, 11
Fiscal incentives, 12
Fischer-Tropsch process, 200
Fluff, 78, 93
Fly ash, 253–261
Fly ash, 235–250
Food and Drug Administration (FDA), 57
Food waste, 41
Forestry, 158
Formate, 203–215
Fuji, 62

Gas turbines, 185
Gasification, waste, 5, 135–144
Gasoline, 40
Glass, 79

Green Dot, 59
Green PET, 20

Halogens, 62
High density polyethylene (HDPE), 4, 20, 35, 36, 67
Home furnishings, 5
Hydrolysis, 89
Hydrothermal treatment effects on
    sludge gel structure, 142
    functional groups, 139–140
    sludge viscosity, 139
Hydrothermal treatment of sludge, 137, 139–143
Hydrothermal treatment temperature effects, 139, 143

In-situ electron spin, 123–131
Incineration, 84–85, 222–230
Infrared analysis, 107
Insulating materials, 263–269
Intrinsic value, 21
Ion exchange, 46, 47, 48

Keep America Beautiful (KAB), 16

Land application, 147–155
Landfill, 15, 199
Landfill, gas, 182–187
Leaching, 239–249
Lead, 245–250
Lignin, 204–208
Lignocellulosic waste, 199–216
Liquefaction, 46
Long-term leaching, 246–250
Low-rank coal, 46

Materials recovery facility (MRFs), 22, 30, 31, 32, 36
Medium density polyethylene (MDPE), 41, 43
Methanolysis, 56
Methyl methacrylate, 59
Methyl tert-buty ether (MTBE), 63
Milk jug, 45
Mixed waste plastic (MWP), 41
Montmorillonite

Municipal solids waste (MSW), 5, 86–87, 181–196, 199, 199–216, 244

New Source Performance Standards (NSPS), 182
Not-in-my-backyard (NIMBY), 15

Oil shale, 253–261
Olefins, 63
Oxford Energy Corporation, 105
Oxygenated fuels, 159–166

Packaging materials, 22
Pallets, 5
Paper, 41
Paper, recycling, 6, 53
Penn State Coal Bank, 126
PET, 35, 36
Pittsburgh Energy Technology Center (PETC), 202
Plastic, 5
Plastic lumber, 87
Plastic, recovery, 4
Polyamides, 55
Polychlorinated di-benzo-P-dioxins (PCDD), 92
Polychlorinated biphenyl (PCB), 80
Polyester, 55
Polyethylene, 4, 40, 41, 43, 52–72, 98
Polyethylene terephthalate (PET), 4, 20, 40, 68, 69
Polymer, 265–269
Polypropylene, 35, 40, 41, 43, 63, 67, 71, 96, 266
Polystyrene, 40, 63, 67, 266
Polystyrene foam, 21
Polyurethane, 55–59
Polyurethane foam (PUF), 96, 98
Polyvinyl chloride, 55, 63, 64, 96, 98, 266
Preasphaltene, 42
Product derived from wastes
    compost, 8–9
    furniture, lumber, building materials, 4–5, 8, 10
    irrigation systems, 6–7
    fuels, 5, 7
    shoes, 6
Propylene, 52–72
Proximate analysis, 42

PS, 35
Public Utility & Regulatory Policy Act (PURPA), 92
PVC, 35, 95, 96, 98
Pyrrhotite, 202–205

Quarternary gasification, 91

Reclaimed products, 106
Reclamation, 28
Recycling, 3, 85
Recycling of
    computers, 5–6, 11
    construction materials, 9–11
    fiberglass, 11
    film, 10–11
    glass, 11–12
    paper, 8, 10
    plastics, 4–6, 10
    rubber, 6
    sewage, 5, 7–8
    tires, 5–7
Refineries, 66–68
Refuse-derived fuel, 200
Resource Conservation and Recovery Act (RCRA), 182
Reuse, 17
Road asphalt, 264
Roofing materials, 10
Rotary kiln, 223–230

Sewage sludge, 4, 5, 7–8, 137–138
Siderite, 242, 244
Sintering, 253–261
Solvent dissolution, 91
Steam activation, 108–117
Steam cracking, 61
Styrene, 106
Styrene butadiene rubber, 106, 266
Surface area, char, 108
Syntal, 4
Synthetic groundwater leaching procedure (SGLP), 246–248

Tertiary recycling, 88
Tetrahydrofuran (THF), 42
Texaco gasification process, 4

Timberex, 10
Tipping fees, 83
Tires, 4, 105–120, 124–132
Tires, scrap, 4, 105
Total organic halogens (TOX), 93
Toxicity characteristic leaching procedure (TLCP), 246

U.S. Environmental Protection Agency (EPA), 92
Ultimate analysis, 42
University of West Virginia, 39
University of Kentucky, 39
University of Utah, 39
Utilization, 3

Veba Oel, 60
Viscosity, of sewage sludge, 138, 140

Volatile organic carbon (VOC), 185

Waste recovery, 3
Waste-to-energy, 15, 221
White goods, 77
Wood, 5–9
Wood, chips, 9
Wood, waste, 9

Xylose isomerase, 173–174

Yard waste, with sludge, 154
Yard, waste, 41
Yard, clippings, 5

Zeolite, 40, 62